THE ULTIMATE
BMAT COLLECTION

UniAdmissions

Published by *RAR Medical Services Limited*
www.uniadmissions.co.uk
info@uniadmissions.co.uk
Tel: +44 (0) 208 068 0438

THE ULTIMATE
BMAT COLLECTION

DR. ROHAN AGARWAL

MATTHEW WILLIAMS

EDITED BY CHANTAL DUCHENNE

ABOUT THE AUTHORS

Rohan is the **Director of Operations** at *UniAdmissions* and is responsible for its technical and commercial arms. He graduated from Gonville and Caius College, Cambridge and is a fully qualified doctor. Over the last five years, he has tutored hundreds of successful Oxbridge and Medical applicants. He has also authored ten books on admissions tests and interviews.

Rohan has taught physiology to undergraduates and interviewed medical school applicants for Cambridge. He has published research on bone physiology and writes education articles for the Independent and Huffington Post. In his spare time, Rohan enjoys playing the piano and table tennis.

Matthew is **Resources Editor** at *UniAdmissions* and a 5th year medical student at St Catherine's College, Oxford. As the first student from Barry Comprehensive School in South Wales to receive a place on the Oxford medicine course he embraced all aspects of university life, both social and academic. Matt Scored in the **top 5% for his UCAT and BMAT** to secure his offer at the University of Oxford.

Matt has worked with UniAdmissions since 2014 – tutoring several applicants successfully into Oxbridge and Russell group universities. His work has been published in international scientific journals and he has presented his research at conferences across the globe. In his spare time, Matt enjoys playing rugby and golf.

TABLE OF CONTENTS

PREFACE

First of all, I'd like to thank you for trusting UniAdmissions to help you with your university application.

It's a tough and confusing process to approach in many ways. As time has passed, the application has become more competitive and convoluted with the introduction of Admissions Tests and an extremely high calibre of students applying, making differentiating yourself increasingly challenging.

When I was applying to Cambridge, there weren't any good resources that I felt I could trust my future with. Unsurprisingly, many of my peers felt the same way and UniAdmissions was born! We started in 2013 and have had an amazing response from the thousands of students we support with their applications each year.

Since UniAdmissions' inception in 2013, we have made it our mission to improve our students' Oxbridge acceptance rate year on year. We've been quite successful in doing that; while the average Oxbridge acceptance rate stands at around 18%, UniAdmissions' success rate is consistently at 61% - triple the national average. Everyone here at UniAdmissions is always looking at how we can push this success rate even higher.

Overwhelmingly, our focus is on applications to highly competitive universities and courses. These are world-renowned and draw the brightest and best pupils from around to globe to their gates. To set yourself apart from this exceptional group of applicants, you must show yourself to be truly remarkable. Over the years, we have made great headway to cracking this formula – as evidenced with our success rates, which we're extremely proud of.

The purpose of this book is to help you prepare for your upcoming admissions test and, ultimately, help you gain your place to study at great universities like Oxford or Cambridge, or on your dream course. I sincerely believe this book will help you prepare, but words on a page can only take you so far.

In the past we have offered our students books, physical courses, tuition, mock interviews, online resources and much more. All of these forms of support make their mark on your application, although what we have discovered is that a holistic approach works much better. Which is why, nowadays, we have shifted our focus at UniAdmissions completely towards our Oxbridge and Medical Programmes. The Programmes each represent a structured syllabus that covers everything you need to know and practice in order to get your dream offer.

The support we provide splits broadly into four key categories: one-to-one teaching, intensive courses, materials, and enrichment seminars. Each represents a different style of learning which has its own positives and negatives.

We believe in offering the very best support we can provide to each student who entrusts a portion of their application to us, as you have with this book

All that's left is for you to read on, I hope you enjoy this book and I wish you the very best of luck!

– Dr Rohan Agarwal

THE ULTIMATE BMAT GUIDE

THE BASICS

What is the BMAT?

The BioMedical Admissions Test (BMAT) is a 2-hour written exam for medical and veterinary students who are applying for competitive universities.

What does the BMAT consist of?

Section	SKILLS TESTED	Questions	Timing
ONE	Problem-solving skills, including numerical and spatial reasoning. Critical thinking skills, including understanding argument and reasoning using everyday language.	32 MCQs	60 minutes
TWO	Ability to recall, understand and apply GCSE level principles of biology, chemistry, physics and maths. Usually the section that students find the hardest.	27 MCQs	30 minutes
THREE	Ability to organise ideas in a clear and concise manner, and communicate them effectively in writing. Questions are usually but not necessarily medical.	One essay from three	30 minutes

Why is the BMAT used?

Medical and veterinary applicants tend to be a bright bunch and therefore usually have excellent grades. For example, in 2013 over 65% of students who applied to Cambridge for Medicine had UMS greater than 90% in all of their A level subjects. This means that competition is fierce, so universities must use the BMAT to help differentiate between applicants.

When can I sit BMAT?

There are two sittings for the BMAT – the second week of September and the first week of November (normally Wednesday mornings). You can generally sit the BMAT on either date. Be aware, however that some universities will ask that you sit the test on a specific date e.g. Oxford, Lee Kong Chian and Chulalongkorn University will only accept results from the November BMAT sitting. You're highly advised to check which date you should sit the BMAT depending on your university choices. This information is available online, on the official BMAT page of the Cambridge Assessment admissions testing site, as well as on individual university websites.

When should I sit the BMAT?

The difficulty and content of both sittings is the same so the answer will depend on how much time you have over the summer and how important it is for you to know your BMAT result before you submit your UCAS application. In general, if you're applying for two or more BMAT universities, it's a good idea to sit the BMAT in September if circumstances allow.

	SEPTEMBER	NOVEMBER
Positive	Get BMAT Results before UCAS Deadline Can use the summer to prepare thoroughly	More time to prepare overall Will have covered more science topics in school
Negative	Less time to prepare overall May conflict with UCAT, Personal Statement etc	Hard to balance school-work with BMAT revision Won't get BMAT results until after UCAS Deadline

Who has to sit the BMAT?

Applicants to the following universities must sit the BMAT:

University	Course
University of Cambridge	Medicine and Veterinary Medicine
University of Oxford	Medicine, Graduate Medicine, Biomedical Sciences
University College London	Medicine
Imperial College London	Medicine, Graduate Medicine, Biomedical Science
Brighton and Sussex	Medicine
University of Leeds	Medicine, Gateway Year to Medicine, Dentistry
Lancaster University	Medicine
Keele University	Medicine (International Applicants)
Royal Veterinary College	Veterinary Medicine
Lee Kong Chian (Singapore)	Medicine
Thammasat University	642901(Medicine) and 642902 (Dental Surgery)
Universidad de Navarra	Medicine
Mahidol University	Medicine

How is the BMAT Scored?

Section 1 and Section 2 are marked on a scale of 0 to 9. Generally, 5 is an average score, 6 is good, and 7 is excellent. Very few people (<5%) get more than 8.

The marks for sections 1 + 2 show a normal distribution with a large range. The important thing to note is that the difference between a score of 5.0 and 6.5+ is often only 3-4 questions. Thus, you can see that even small improvements in your raw score will lead to massive improvements when they are scaled.

Section 3 is marked on 2 scales: A-E for Quality of English and 0-5 for Strength of Argument

The marks for section 3 show a normal distribution for the strength of argument; the average mark for the strength of argument is between 3 – 3.5. The distribution of quality of English marks is negatively skewed. In other words, the vast majority of students will score A or B for quality of English. The ones that don't tend to be students who are not fluent in English.

This effectively means that the letter score (A-E) is used to flag students who have a comparatively weaker grasp of English- i.e. it is a test of competence rather than excellence like the rest of the BMAT. In practice this means that essays scoring a C or below are therefore likely to be more heavily scrutinised by admissions tutors than essays graded A or B.

Finally, section 3 is marked by two different examiners. If there is a large discrepancy between their marks, it is marked by a third examiner.

Can I resit the BMAT if I'm unhappy with my score?

No, you can only sit the BMAT once per admissions cycle. You can resit the BMAT if you apply for medicine again in the future.

Do I have to resit the BMAT if I reapply?

You must resit the BMAT with each admissions cycle, if you applying to a university that requires it. You cannot use your score from any previous attempts.

Where do I sit the BMAT?

For the September sitting, you will need to register yourself and sit the test at one of 20 authorised centres. In November, your school will normally register you and you can usually sit the BMAT at your school or college (ask your school exams officer for more information). Alternatively, if your school isn't a registered test centre or you're not attending a school or college, you can sit the BMAT at an authorised test centre.

When do I get my results?

For the September sitting, you will get your results by the end of September online. You are then responsible for informing the University of your BMAT score. For the November sitting, the BMAT results are usually released to universities in mid-late November and then to students in late November.

How is the BMAT used?

Cambridge: Cambridge interviews more than 90% of students who apply so the BMAT score isn't vital for making the interview shortlist. However, it can play a huge role in the final decision – for example, 50% of overall marks for your application may be allocated to the BMAT. Thus, it's essential you find out as much information about the college you're applying to, as each college places a slightly different weighting on each part of your application.

Oxford: Oxford typically receives thousands of applications each year and they use the BMAT to shortlist students for interview. Typically, 450 students are invited for interview for 150 places. Thus, if you get offered an interview- you are doing very well! Oxford centralise their short listing process and use an algorithm that uses your % A*s at GCSE along with your BMAT score to rank all their applicants of which the top are invited to interview. BMAT sections 1 + 2 count for 40% each of your BMAT score whilst section 3 counts for 20% [the strength of argument (number) contributes to 13.3% and the quality of English (letter) makes up the remaining 6.7%].

UCL: UCL make offers based on all components of the application and whilst the BMAT is important there is no magic threshold that you need to meet in order to guarantee an interview. Applicants with higher BMAT scores tend to be interviewed earlier in the year.

Imperial: Imperial employs a BMAT threshold to shortlist for interview. This exact threshold changes every year but in the past has been approximately 4.5-5.0 for sections 1 + 2 and 2.5 B for section 3.

Leeds: The BMAT contributes to 15% of your academic score at Leeds. You will be allocated marks based on your rank in the BMAT. Thus, applicants in the top 20% of the BMAT will get the full quota of marks for their application and the bottom 20% will get the lowest possible mark for their application. Thus, you can still get an interview if you perform poorly in the BMAT (it's just much harder!). Leeds will calculate your BMAT score by attributing 40% to section 1, 40% to section 2 and 20% to section 3 (lower weighting as it can come up during the interview).

Brighton & Sussex: BSMS recently started using the BMAT as part of the shortlisting process to decide which students to interview. They assign a total score of 28 to the BMAT (9 marks for Section 1, 9 marks for Section 2 and 5 marks for each component of Section 3 – letter and number grade). An example score at BSMS for a BMAT score of 5.0 in Section 1, 4.0 in Section 2 and 3.5A in Section 3 would be 17.5/28. The medical school rank all applicants based on their total score out of 28 and work down the rankings until all interview spots are filled. There is no specific cut off score, so it will vary each year depending on applicants' BMAT scores. In addition, they state on their website that it "may also be used as a final discriminator if needed after interview."

Royal Veterinary College: It is unclear how the RVC use the BMAT- it has influenced applications both before and after interview and it's likely that they use it on a case-by-case basis rather than as an arbitrary cut-off.

GENERAL ADVICE

Start Early

It is much easier to prepare and do well if you practice little and often. Start your preparation well in advance; ideally by mid September but at the latest by early October for the November sitting, or 8-12 weeks prior to the earlier sitting. This way you will have plenty of time to complete as many papers as you wish to feel comfortable with the exam content and style and won't have to panic and cram just before the test, which is a much less effective and more stressful way to learn. In general, an early start will give you the opportunity to identify the complex issues and work at your own pace to better your understanding of the material.

Prioritise

Some questions in sections 1 + 2 can be long and complex – and given the intense time pressure in the BMAT you need to know your limits. It is essential that you don't get stuck and waste valuable time on very difficult questions. If a question looks particularly long or complex, mark it for review and move on. You don't want to be caught 5 questions short at the end just because you took more than 3 minutes in answering a challenging multi-step physics question. If a question is taking too long, choose a sensible answer, mark for review if you have time later, and move on. Remember that each question carries equal weighting and therefore, you should adjust your timing accordingly. With some disciplined practice under timed conditions, you can get very good at this and learn to maximise your efficiency. In short, exam technique is a crucial part of the BMAT, and matters almost as much as learning the content, so it is wise to spend your preparation time developing a consistent approach to the exam.

Positive Marking

There are no penalties for incorrect answers in the BMAT; you will gain one mark for each right answer and will not lose a mark for a wrong or unanswered question. This affords you the luxury of being able to guess if you are completely unable to figure out the right answer to a question or find yourself running out of time. Since each question provides you with 4 to 6 possible answers, you have a 16-25% chance of guessing correctly. Therefore, if you aren't sure (and are running short of time), you can make an educated guess and move on without risking being penalised for an incorrect answer. Before 'guessing' you should try to eliminate a couple of answers to increase your chances of getting the question correct. For example, if a question has 5 options and you manage to eliminate 2 options, your chances of getting the question correct increase from 20% to 33%!

It is important also to try to avoid losing easy marks on other questions because of poor exam technique. The BMAT is very time pressured, so you must attempt to get through as many questions as you can by doing plenty of practice under timed conditions well in advance of the exam. If you have failed to finish the exam on the day, take the last 10 seconds to guess the remaining questions to at least give yourself a chance of getting them right.

Practice

This is the best way to familiarise yourself with the style of questions and the timing for BMAT. Although the BMAT tests only GCSE level knowledge, you are unlikely to be familiar with the style of questions in all 3 sections when you first encounter them. Although you will have previously encountered all of the scientific ideas in the BMAT specification, often the questions use a different approach to questions on the school exams syllabus. This means you will be required to apply principles rather than knowledge acquired from rote-learning, so it is important to ensure you fully understand the material on the test specification.

Practising questions will put you at ease and make you more comfortable with the exam. The more comfortable you are, the less you will panic on the test day when confronted with tricky questions, and the more likely you are to score highly. Initially, work through the questions at your own pace, and spend time carefully reading the questions and looking at any additional data provided on the test paper. As you get closer to the test, **make sure you practice the questions under exam conditions**.

Past Papers

Official past papers and answers from 2003 onwards are freely available online and once you've worked your way through the questions in this book, it's a good idea to attempt as many of them as you can. You should aim to complete at least 5 full BMAT papers from the most recent specification under timed conditions prior to taking your test. Keep in mind that the specification was changed in 2009 so some things asked in earlier papers may not be representative of the content that is currently examinable in the BMAT. If a topic has been removed from the BMAT specification, this is usually made clear on the older past papers, so they are still a useful resource for practice. In general, **it is worth doing at least all the papers from 2009 onwards**. If time permits, you could work backwards from 2009 for additional Section 1 & 2 practice, although there is little point doing the section 3 essays pre-2009 as they are significantly different to the current style of essay questions.

Scoring Tables

Use these to keep a record of your scores – you can then easily see which paper you should attempt next to build on your syllabus knowledge and exam technique (always the one with the lowest score).

SECTION I	1st Attempt	2nd Attempt
2003		
2004		
2005		
2006		
2007		
2008		
2009		
2010		
2011		
2012		
2013		
2014		
2015		
2016		
2017		
2018		
2019		
2020		

SECTION 2	1st Attempt	2nd Attempt
2003		
2004		
2005		
2006		
2007		
2008		
2009		
2010		
2011		
2012		
2013		
2014		
2015		
2016		
2017		
2018		
2019		
2020		

Repeat Questions

When checking through your answers to practice papers, pay particular attention to questions you have got wrong. If there is a worked solution, look through that carefully until you feel confident that you understand the reasoning, and then repeat the question without help to check that you can do it. If only the answer is given, have another look at the question and try to work out why that answer is correct. You can refer to worked solutions for similar types of questions for some pointers if the answer doesn't seem obvious. This is the best way to learn from your mistakes, and means you are less likely to make similar mistakes when it comes to the real BMAT. The same applies for questions which you were unsure of and made an educated guess which was correct, even if you got it right. When working through this book, **make sure you highlight any questions you are unsure of**, to ensure you know to spend more time looking over them once marked.

No Calculators

You aren't permitted to use a calculator in the BMAT – thus, it is essential that you have strong numerical skills and feel confident manipulating figures with pen and paper. For instance, you should be able to rapidly convert between percentages, decimals and fractions. You will seldom get questions that would require calculators but you would be expected to be able to arrive at a sensible estimate. Consider for example:

Estimate 3.962×2.322;

3.962 is approximately 4 and 2.323 is approximately 2.33 = 7/3.

Thus, $3.962 \times 2.322 \approx 4 \times \frac{7}{3} = \frac{28}{3} = 9.33$

It is an important part of exam technique to know that in the BMAT you will rarely be asked to perform difficult calculations. In fact, you can use this as a marker of whether you are tackling a question correctly. For example, when solving a physics question, if you end up having to divide values such as 8,079 by 357- this should raise alarm bells as calculations in the BMAT are rarely this difficult.

A word on timing...

"If you had all day to do your BMAT, you would get 100%. But you don't."

Whilst this isn't completely true, it illustrates a very important point. Once you've practiced and know how to answer the questions, the clock is your biggest enemy. This seemingly obvious statement has one very important consequence. The way to improve your BMAT score is to improve your speed. There is no magic bullet. But there are a great number of techniques that, with practice, will give you significant time gains, allowing you to answer more questions and score more marks.

Timing is tight throughout the BMAT – mastering the timing of each section is the first key to success.

> ***Top tip!*** In general, students tend to improve the fastest in Section 2 and slowest in Section 1; Section 3 usually falls somewhere in the middle. Thus, if you have very little time left to prepare for the BMAT, it's best to prioritise Section 2 in order to maximise your overall score.

Some candidates choose to work as quickly as possible through every question to save up time at the end to check their answers, but this is generally not the best approach.

BMAT questions can have a lot of information in them – each time you start answering a question it takes time to get familiar with the instructions and information given. By splitting the question into two sessions (the first run-through and the return-to-check) you double the amount of time you spend familiarising yourself with the data, as you have to do it twice instead of only once. This costs valuable time. In addition, candidates who do check back may spend 2–3 minutes doing so and yet not make any actual changes.

While this can feel reassuring in a high-stakes exam, it is actuallly false reassurance as this approach is unlikely to have a significant impact on your actual score. Therefore, it is usually best to pace yourself very steadily, aiming to spend the same amount of time on each question and finish the final question in a section just as time runs out. This reduces the time spent on re-familiarising yourself with the content of the questions and maximises the time spent arriving at the answer. This approach is the most efficient way to gain the maximum marks you are capable of.

It is essential that you don't get stuck with the hardest questions – no doubt there will be some tough ones scattered through each section of the BMAT. In the time you might spend attempting to answer only one of these questions you could miss out on answering three easier questions, worth three times as many marks! If a question is taking too long, choose a sensible answer (through educated guesswork) and move on. Never see this as giving up or in any way failing – rather, it is the smart way to approach a high stakes test with a tight time limit. With practice and discipline, you can get very good at this and learn to maximise your efficiency. It is not about being a hero and aiming for full marks – this is almost impossible and very much unnecessary (even Oxbridge will regard any score higher than 7 as exceptional). It is about maximising your efficiency and gaining the maximum possible number of marks within the time you have. In summary, perfecting your exam timing and technique is an important way for you to give yourself the best chance of achieving a score that will secure you an interview at medical school.

> *Top tip!* Ensure that you take a watch that can show you the time in seconds into the exam. This will allow you have a much more accurate idea of the time you're spending on a question. In general, if you've spent >150 seconds on a section 1 question or >90 seconds on a section 2 questions – move on regardless of how close you think you are to solving it, to ensure you have enough time to answer all the other questions too.

Use the Options:

Some questions may try to overload you with information. You could see this as a distraction technique designed to intimidate candidates, so knowing what to do in this situation is critical.

When presented with large tables and data, it's essential that you look at the answer options so you can focus your mind. This can allow you to reach the correct answer a lot more quickly. Consider the example below:

The table below shows the results of a study investigating antibiotic resistance in staphylococcus populations. A single staphylococcus bacterium is chosen at random from a similar population. Resistance to any one antibiotic is independent of resistance to others.

Calculate the probability that the bacterium selected will be resistant to all four drugs.

A. 1 in 10^6

B. 1 in 10^{12}

C. 1 in 10^{20}

D. 1 in 10^{25}

E. 1 in 10^{30}

F. 1 in 10^{35}

Antibiotic	Number of Bacteria tested	Number of Resistant Bacteria
Benzyl-penicillin	10^{11}	98
Chloramphenicol	10^9	1200
Metronidazole	10^8	256
Erythromycin	10^5	2

Looking at the answer options first makes it obvious that there is **no need to calculate exact values**- only in powers of 10. This makes your life a lot easier. If you hadn't noticed this, you might have spent well over 90 seconds trying to calculate the exact value when it wasn't even being asked for.

In other cases, you may actually be able to use the answer options to arrive at the solution quicker than if you had tried to solve the question from scratch as you normally would. Consider the example below: A region is defined by the two inequalities: $x - y^2 > 1$ and $xy > 1$. Which of the following points is in the defined region?

A. (10,3)   B. (10,2)   C. (-10,3)   D. (-10,2)   E. (-10,-3)

Whilst it's possible to solve this question both algebraically or graphically by manipulating the identities, by far **the quickest way is to simply use the options given to you in the question**.

Note that options C, D and E violate the second inequality, narrowing down to answer to either A or B. For A: $10 - 3^2 = 1$ and thus this point is on the boundary of the defined region and not actually in the region. Thus, the answer is B (as $10 - 4 = 6 > 1$)

In general, it pays dividends to look at the answer options briefly to see if they can help you arrive at the solution more quickly. Get into this habit early on in your BMAT preparation – it may feel unnatural at first, but it's guaranteed to save you time in the long run.

Key Words

If you're stuck on a question; pay particular attention to the options that contain key modifiers like **"always"**, **"only"**, **"all"** as examiners like using them to test if there are any gaps in your knowledge. For example, the statement "arteries carry oxygenated blood" would normally be true. However, the statement "<u>all</u> arteries carry oxygenated blood" would be false because there is an exception: the pulmonary artery carries deoxygenated blood.

SECTION 1

This is the first section of the BMAT and as you walk in, it is inevitable that you will feel nervous. Make sure that you have been to the toilet prior to the exam because once it starts you cannot simply pause and go due to the time pressure. Take a few deep breaths and calm yourself down. Remember that panicking will not help and may negatively affect your marks - so try and avoid this as much as possible.

You have one hour to answer 32 questions in section 1. The questions fall into two categories:
- Problem solving (16 questions)
- Critical thinking (16 questions)

Whilst this section of the BMAT is renowned for being difficult to prepare for, there are a number of powerful shortcuts and techniques that you can use. Learning specific exam techniques and practising plenty of example questions will train you to work efficiently and make the best of the time available for Section 1.

You have approximately 100 seconds per question. Although this may initially sound like plenty of time the questions in Section 1 often require you to read and analyse passages or graphs to find the correct answer, meaning it is important to work as quickly and efficiently as possible.

Nevertheless, this part of the BMAT is not as time pressured as Section 2 so most students usually finish the majority of questions in time. However, some questions in Section 1 are very tricky and can be a big drain on your limited time. **The people who fail to complete section 1 are those who get bogged down on a particular question**.

Therefore, it is vital that you start to get a feel for which questions are going to be easy and quick to do and which ones should be left until the end. The best way to do this is through practice and the questions in this book will offer extensive opportunities for you to do so.

SECTION 1: CRITICAL THINKING

BMAT critical thinking questions require you to understand the components of a good argument and be able to pick them apart. The majority of BMAT critical thinking questions tend to fall into 3 major categories:

1. Identifying Conclusions
2. Identifying Assumptions + Flaws
3. Strengthening and Weakening arguments

> **Top tip!** Though it might initially sound counter-intuitive, it is often best to read the question **before** reading the passage. Then you'll have a much better idea of what you're looking for in the text and are therefore likely to pick out the relevant information quickly.

Having a good grasp of language and being able to filter unnecessary information quickly and efficiently is a vital skill in medical school – you simply do not have the time to sit and read vast numbers of textbooks cover to cover. Instead, you will need to be able to filter the information to quickly identify the key points. Ultimately this skill will contribute to your success in your studies, which is why it forms part of the BMAT.

Selecting the most relevant pieces of information from a larger passage is also a key skill for qualified doctors, who simply do not have the time to read pages and pages of notes on the wards to make important healthcare decisions. As such, it is important not to underestimate the importance of getting to grips with verbal reasoning skills at this stage and throughout university.

Key Tips & Tricks for Section 1

1. Only Use the Passage
Your answer must only be based on the information available in the passage provided by the question. Do not try and guess the answer based on your general knowledge as this can be a trap. For example, if the passage says that spring is followed by winter, then take this as true even though you know that spring is followed by summer.

2. Take your time
Unlike the problem solving questions, critical thinking questions are less time pressured. Most of the passages are well below 300 words and therefore don't take long to read and process (unlike the UCAT in which you should skim read passages). Thus, your aim should be to understand the intricacies of the passage and identify key information so that you don't lose easy marks.

Section 1 Question Types

Identifying Conclusions

Students often struggle with these types of questions because they confuse a premise for a conclusion. Let's start by defining the differences between the two:

- A **conclusion** is a summary of the arguments being made and is usually explicitly stated or heavily implied.
- A **premise** is a statement from which another statement can be inferred or a statement that leads the reader to a conclusion. A premise would always be explicitly stated within a passage of text.

Hence a conclusion is shown/implied/proven by a premise. Similarly, a premise shows/indicates/establishes a conclusion.

Consider for example: *My mom, being a woman, is clever as all women are clever.*

Premise 1: My mom is a woman + **Premise 2:** Women are clever = **Conclusion:** My mom is clever.

This is fairly straightforward as it's a very short passage and the conclusion is explicitly stated. Sometimes the latter may not happen.

Consider: My mom is a woman and all women are clever.

Here the conclusion is not explicitly stated, yet both premises still stand and can be used to reach the same conclusion as in the first statement.

You may sometimes be asked to identify if any of the options cannot be "reliably concluded". This is effectively asking you to identify why an option **cannot** be the conclusion.

There are many reasons why a statement cannot be reliably concluded from a passage of text but the most common ones are:

1. Over-generalising

    My mom is clever therefore all women are clever.

2. Being too specific or narrow:

    All kids like candy thus my son also likes candy.

3. Confusing correlation and causation:

    Example: Lung cancer is much more likely in patients who drink water. Hence, water causes lung cancer.

4. Confusing cause and effect:

    Example: Lung cancer patients tend to smoke so it follows that having lung cancer must make people want to smoke.

Note how conjunctives like hence, thus, therefore, and it follows give you a clue as to when a conclusion is being stated in a passage. More examples of these phrases that indicate a conclusion include: it follows that, implies that, whence, entails that.

Similarly, words and phrases like "because, as indicated by, in that, given that, due to the fact that" usually identify premises from which a conclusion or linking statement could be drawn.

Assumptions + Flaws:

Other types of critical thinking questions may require you to identify assumptions and flaws in the reasoning given to reach a conclusion in a passage of text. Before proceeding with examples, let's clarify the definitions:

- An assumption is a reasonable assertion that can be made based on the available evidence. It is an unstated piece of information that the rest of the argument relies upon.

- A flaw is an element of an argument that is inconsistent with the rest of the available evidence. A flaw undermines the crucial components of the overall argument being made.

Consider for example: My mom is clever because all doctors are clever.
Premise 1: Doctors are clever. **Assumption:** My mom is a doctor. **Conclusion:** My mom is clever.
Note that the assumption will **never** be explicitly stated within a passage of text.

In this short passage, the conclusion follows naturally even though there is only one premise because of the assumption. The argument relies on the assumption to work.
If you are unsure if a given answer option is an assumption, just ask yourself:

1) Is it in the passage? If the answer is **no,** then proceed to ask:
2) Does the conclusion rely on this piece of information in order to work? – If the answer is **yes** – then you've identified an assumption.

You may sometimes be asked to identify flaws in an argument – it is important to be aware of the types of flaws to look out for. In general, these are broadly similar to the flaws discussed earlier in the conclusion section (over-generalising, being too specific, confusing cause and effect, confusing correlation and causation). Remember that an assumption may also be a flaw, generating an inappropriate or inconsistent conclusion.

Note that an argument may be sound (no flawed reasoning or assumptions) and still reach a conclusion that you know is false. For instance, a sound argument may arrive at the conclusion that the earth is flat. The important thing to remember is that Section 1 of the BMAT is testing your ability to analyse the components of an argument and **not** testing your knowledge - Section 2 is a test of knowledge.

For example, consider this short extract again: *my mom is clever because all doctors are clever.*
What if the mother was not actually a doctor? The argument would then break down as the assumption would be incorrect or **flawed**.

> ***Top tip!*** Don't get confused between premises and assumptions. A **premise** is a statement that is explicitly stated in the passage. An **assumption** is an inference that is made from the passage and will never be stated.

Strengthening and Weakening Arguments:

You may be asked to identify an answer option that would <u>most</u> strengthen or weaken the argument being made in the passage. Normally, you'd also be told to assume that each answer option is true – even if you know them to be false (flat earth, pink elephants etc). Before we can discuss how to strengthen and weaken arguments, it is important to understand what constitutes a good argument:

1. **Evidence:** arguments which are heavily based on value judgements and subjective statements tend to be weaker than those based on facts, statistics and the available evidence.
2. **Logic**: a good argument should flow, and the constituent parts should fit well into an overriding view or belief. There should not be any questionable jumps to conclusions from flawed assumptions or premises.
3. **Balance:** a good argument must concede that there are other views or beliefs (counter-argument). The key is to carefully dismantle these ideas and explain why they are wrong. Look out for words like 'the majority, often, usually' to signal an argument that is more balanced than one using 'always, every, never' etc.

Thus, when asked to strengthen an argument, look for options that would: increase the evidence basis for the argument, support or add a premise, address the counter-arguments.

Similarly, when asked to weaken an argument, look for options that would: decrease the evidence basis for the argument or create doubt over existing evidence, undermine a premise, strengthen the counter-arguments.

In order to be able to strengthen or weaken arguments, you must completely understand the conclusion of a passage. This means you can quickly test the impact of each answer option on the conclusion to see which one strengthens or weakens it the most i.e. is the conclusion stronger/weaker if I assume this information to be true and included in the passage.

Often, you'll have to decide which option strengthens/weakens the passage most. To do this, you can determine the most powerful part of a given argument and compare the impact of the additional statements. Is it the evidence provided in the original argument? Is this additional information fully addressing any counter-arguments, to make the argument watertight? Perhaps the additional statement proves that a counter argument is more powerful than the conclusion in the passage, or reveals a major flaw in the logic. The best way to learn how to tackle these questions is through practice, as eventually you will learn to recognise the patterns in the questions. Thankfully, you have plenty of time for these questions so can consider the options carefully in the exam.

CRITICAL THINKING QUESTIONS

Question 1-6 are based on the passage below:

People have tried to elucidate the differences between the genders for many years. Are they societal pressures or genetic differences? In the past it has always been assumed that it was programmed into our DNA to act in a certain more masculine or feminine way but now evidence has emerged that may show it is not our genetics that determine the way we act, but that society pre-programmes us into gender identification. Although it is generally acknowledged that not all boys and girls are the same, why is it that most young boys like to play with trucks and diggers while young girls prefer dolls and pink?

The society we live in has always been an important factor in our identity, from cultural differences, the languages we speak, the food we eat, to the clothes we wear. All of these factors influence our identity. New research shows that the people around us may prove to be the biggest influence on our gendered behaviour. It shows our parents buying gendered toys may have a much bigger influence than the genes they gave us. Girls are being programmed to like the same things as their mothers and this has lasting effects on their personality. Young girls and boys are forced into their gender stereotypes through the clothes they are bought, the hairstyle they wear and the toys they are given to play with.

The power of society to influence gendered behaviour explains the cases where children have been born with different external sex organs to those with the organs matching their sex determining chromosomes. Despite the influence of their DNA they identify as the gender they have always been told they are. Once the difference has been detected, how then are they ever to feel comfortable in their own skin? The only way to prevent society having such a large influence on gender identity is to allow children to express themselves, wear what they want and play with what they want without fear of not fitting in.

Question 1:
What is the main conclusion from the first paragraph?
A. Society controls gendered behaviour.
B. People are different based on their gender.
C. DNA programmes how we act.
D. Boys do not like the same things as girls because of their genes.

Question 2:
Which of the following, if true, points out the flaw in the first paragraph's argument?
A. Not all boys like trucks.
B. Genes control the production of hormones.
C. Differences in gender may be due to an equal combination of society and genes.
D. Some girls like trucks.

Question 3:

According to the passage, how can culture affect identity?

A. Culture can influence what we wear and how we speak.

B. Our parents act the way they do because of culture.

C. Culture affects our genetics.

D. Culture usually relates to where we live.

Question 4:

Which of these is most implied by the passage?

A. Children usually identify with the gender they appear to be.

B. Children are programmed to like the things they do by their DNA.

C. Girls like dollies and pink because their mothers do.

D. It is wrong for boys to have long hair like girls.

Question 5:

What does the passage say is the best way to prevent gender stereotyping?

A. Mothers spending more time with their sons.

B. Parents buying gender-neutral clothes for their children.

C. Allowing children to act how they want.

D. Not telling children if they have different sex organs to their chromosomal sex.

Question 6:

What, according to the passage, is the biggest problem for children born with different external sex organs to those born with sex organs matching their sex chromosomes?

A. They may have other problems with their DNA.

B. Society may not accept them for who they are.

C. They may wish to be another gender.

D. They are not the gender they are treated as which can be distressing.

Questions 7-11 are based on the passage below:

New evidence has emerged to say that the most important factor in a child's development could be their napping routine. It has come to light that regular napping could be the key factor in determining toddlers' memory and learning abilities. This new countrywide survey of 1000 toddlers, all born in the same year, showed around 75% had regular 30-minute naps. Parents cited the benefits of their child having a regular routine (including mealtimes) such as decreased irritability, and stated the only downfall of regular naps was occasional problems with sleeping at night. Research indicating that toddlers were 10% more likely to suffer regular nighttime sleep disturbances when they regularly napped supported the parents' view.

Those who regularly took 30-minute naps were more than twice as likely to remember simple words such as those of new toys than their non-napping counterparts, who also had higher incidences of memory impairment, behavioural problems and learning difficulties. Toddlers who regularly had 30-minute naps were tested on whether they were able to recall the names of new objects the following day, and then compared to a control group who did not regularly nap. These potential links between napping and memory, behaviour and learning ability provides exciting new evidence in the field of child development.

Question 7:

If 5% of 100 toddlers who did not nap were able to remember a new teddy's name, how many out of 100 regularly napping toddlers would be expected to remember?

A. 8   B. 9   C. 10   D. 12

Question 8:

Assuming that the incidence of nighttime sleep disturbances is the same for all toddlers independent of all characteristics other than napping, what is the percentage of toddlers who suffer regular nighttime sleep disturbances as a result of napping?

A. 10%   B. 14%   C. 20%   D. 50%

Question 9:

Using the information from the passage above, which of the following is the most plausible alternative reason for the link between memory and napping?

A. Children who have bad memory abilities are also likely to have trouble sleeping.
B. Children who regularly nap are born with better memories.
C. Children who do not nap were unable to concentrate on the memory testing exercises for the study.
D. Parents who enforce a routine of napping are more likely to conduct memory exercises with their children.

Question 10:

Which of the following is most strongly indicated?

A. Families have more enjoyable mealtimes when their toddlers regularly nap.
B. Toddlers have better routines when they nap.
C. Parents enforce napping to improve their toddlers' memory ability.
D. Napping is important for parents' routines.

Question 11:

Which of the following, if true, would strengthen the conclusion that there is a causal link between regular napping and improved memory in toddlers?

A. Improved memory is also associated with regular mealtimes.

B. Parents who enforce regular napping are more inclined to include their children in studies.

C. Toddlers' memory development is so rapid that even a few weeks can make a difference to performance on tests.

D. There is a significant improvement and more consistency in memory test performance amongst toddler playgroups with a higher incidence of napping, compared to playgroups where napping is discouraged.

Question 12:

Tom's father says to him: 'You must work for your A-levels. That is the best way to do well in your A-level exams. If you work especially hard for geography, you will definitely succeed in your geography A-level exam'.

Which of the following is the best statement Tom could say to prove a flaw in his father's argument?

A. 'It takes me longer to study for my history exam, so I should prioritise that.'

B. 'I do not have to work hard to do well in my geography A-level.'

C. 'Just because I work hard, does not mean I will do well in my A-levels.'

D. 'You are putting too much importance on studying for A-levels.'

E. 'You haven't accounted for the fact that geography is harder than my other subjects.'

Question 13:

Today the NHS is increasingly struggling to be financially viable. In the future, the NHS may have to reduce the services it cannot afford. The NHS is supported by government funds, which come from those who pay tax in the UK. Recently the NHS has been criticised for allowing fertility treatments to be free, as many people believe these are not important and should not be paid for when there is not enough money to pay the doctors and nurses.

Which of the following is the most accurate conclusion of the statement above?

A. Only taxpayers should decide where the NHS spends its money.

B. Doctors and nurses should be better paid.

C. The NHS should stop free fertility treatments.

D. Fertility treatments may have to be cut if NHS finances do not improve.

Question 14:

'We should allow people to drive as fast as they want. By allowing drivers to drive at fast speeds, through natural selection the most dangerous drivers will kill only themselves in car accidents. These people will not have children, hence only safe people will reproduce and eventually the population will only consist of safe drivers.'

Which one of the following, if true, most weakens the above argument?

A. Dangerous drivers harm others more often than themselves by driving too fast.

B. Dangerous drivers may produce children who are safe drivers.

C. The process of natural selection takes a long time.

D. Some drivers break speed limits anyway.

Question 15:

In the winter of 2014, the UK suffered record levels of rainfall, which led to catastrophic damage across the country. Thousands of homes were damaged and even destroyed, leaving many homeless in the chaos that followed. The government faced harsh criticism that they had failed to adequately prepare the country for the extreme weather.

The government regularly assesses the likelihood of adverse events such as extreme weather happening in the future and balance the risk against the cost of advanced measures to reduce the impact should they occur. This is then compared to the cost of the event with no preparative defences in place. The risk of flooding is usually low, so it could be argued that the costs associated with anti-flooding measures would have been pre-emptively unreasonable. Should the government be expected to prepare for every conceivable threat that could come to pass? Are we to put in place expensive measures against a seismic event as well as a possible extra-terrestrial invasion?

Which of the following best expresses the main conclusion of the statement above?

A. The government has an obligation to assess risks and costs of possible future events.

B. The government should spend money to protect against potential extra-terrestrial invasions and seismic events.

C. The government should have spent money to protect against potential floods.

D. The government was justified in not spending heavily to protect against flooding.

E. The government should assist people who lost their homes in the floods.

Question 16:

Sadly, the way in which children interact with each other has changed over the years. While children once used to play sports and games together in the street, they now sit alone in their rooms on the computer playing games on the internet. In the past, young children learned human interaction from active games with their friends, yet this is no longer the case. How then, when these children are grown up, will they be able to socially interact with their colleagues?

Which one of the following is the conclusion of the above statement?

A. Children who play computer games now interact less outside of them.

B. The internet can be a tool for teaching social skills.

C. Computer games are for social development.

D. Children should be made to play outside with their friends to develop their social skills for later in life.

E. Adults will in the future play computer games as a means of interaction.

Question 17:

Between 2006 and 2013 the British government spent £473 million on Tamiflu antiviral drugs in preparation for a flu pandemic, despite there being little evidence to support the effectiveness of the drug. The antivirals were stockpiled for a flu pandemic that never fully materialised. Only 150,000 packs of Tamiflu were used during the swine flu episode in 2009, and it is unclear if this improved outcomes. Therefore, this money could have been much better spent on drugs that would actually benefit patients.

Which option best summarises the author's view in the passage?

A. Drugs should never be stockpiled, as they may not be used.

B. Spending millions of pounds on drugs should be justified by strong evidence showing positive effects.

C. We should not prepare for flu pandemics in the future.

D. The recipients of Tamiflu in the swine flu pandemic had no difference in symptoms or outcomes to patients who did not receive the antivirals.

Question 18:

High BMI and central obesity are risk factors associated with increased morbidity and mortality. Many believe the development of cheap, easily accessible fast-food outlets is partly responsible for the increase in rates of obesity. Unhealthy weight is commonly associated with a generally unhealthy lifestyle, such as a lack of exercise. The best way to tackle the growing problem of obesity is for the government to tax unhealthy foods so they are no longer a cheap alternative.

Why is the solution given, to tax unhealthy foods, not a logical conclusion from the passage?

A. Unhealthy eating is not exclusively confined to low-income families.

B. A more general approach to unhealthy lifestyles would be optimal.

C. People do not only choose to eat unhealthy food because it is cheaper.

D. People have personal responsibility for their own health.

E. E. None of the above

Question 19:

As people are living longer, care in old age is becoming a larger burden. Many elderly people require carers to come into their home numerous times a day or need full time residential care. It is not right that the NHS should be spending vast funds on the care of people who are sufficiently wealthy to fund their own care. Some argue that they want their savings kept aside to give to their children; however this is not a right, simply a luxury. It is not right that people should be saving and depriving themselves of necessary care, or worse, making the NHS pay the bill, so they have money to pass on to their offspring. People need to realise that there is a financial cost to living longer.

Which of the following statements is the main conclusion of the above passage?

A. We need to take personal financial responsibility for our care in old age.

B. Caring for the elderly is a significant burden on the NHS.

C. The reason people are reluctant to pay for their own care is that they want to pass money onto their offspring.

D. The NHS should limit care to the elderly to reduce their costs.

E. People shouldn't save their money for old age.

Question 20:

There is much interest in research surrounding production of human stem cells from non-embryo sources for potential regenerative medicine, and a huge financial and personal gain at stake for researchers. In January 2014, a team from Japan published two papers in *Nature* that claimed to have developed totipotent stem cells from adult mouse cells by exposure to an acidic environment. However, there has since been much controversy surrounding these papers. Problems included: inability by other teams to replicate the results of the experiment, an insufficient protocol described in the paper and issues with images in one of the papers. It was dishonest of the researchers to publish the papers with such problems. A requirement of a scientific paper is a sufficiently detailed protocol, so that another group could replicate the experiment.

Which statement is most implied in this passage?

A. Research is fuelled mainly by financial and personal gains.

B. The researchers should take responsibility for publishing the paper with such flaws.

C. Rivalry between different research groups makes premature publishing more likely.

D. The discrepancies were in only one of the papers published in January 2014.

Question 21:

The placebo effect is a well-documented medical phenomenon in which a patient's condition improves after being given an ineffectual treatment that they believe to be a genuine treatment. It is frequently used as a control during trials of new drugs/procedures, with the effect of the drug being compared to the effect of a placebo. If the drug on trial does not have a greater effect than the placebo, then it is classed as ineffective. However, this analysis discounts the fact that the drug treatment still has more of a positive effect than no action, and so we are clearly missing out on the potential to improve certain health conditions. It follows that where there is a demonstrated placebo effect, but treatments are ineffective, we should still give treatments, as there will still be some benefit to the patient.

Which of the following best expresses the main conclusion of this passage?

A. In situations where drugs are no more effective than a placebo, we should still give drugs, as they will be more effective at improving a patient's condition than not taking action.

B. Our current analysis discounts the fact that even if drug treatments have no more effect than a placebo, they may still be more effective than no action.

C. The placebo effect is a well-recognised medical phenomenon.

D. Drug treatments may have negative side effects that outweigh their benefit to patients.

E. Placebos are better than modern drugs.

Question 22:

The speed limit on motorways and dual carriageways has been 70mph since 1965, but this is an out-dated policy and needs to change. Since 1965, car brakes have become much more effective, and many safety features have been introduced into cars, such as seatbelts (which are now compulsory to wear), crumple zones and airbags. Therefore, it is clear that cars no longer need to be restricted to 70mph, and the speed limit can be safely increased to 80mph without causing more road fatalities.

Which of the following best illustrates an assumption in this passage?

A. The government should increase the speed limit to 80mph.

B. If the speed limit were increased to 80mph, drivers would not begin to drive at 90mph.

C. The safety systems introduced reduce the chances of fatal road accidents for cars travelling at higher speeds.

D. The roads have not become busier since the 70mph speed limit was introduced.

E. The public wants the speed limit to increase.

Question 23:

Despite the overwhelming scientific evidence for the theory of evolution, and even acceptance of the theory by many high-ranking religious ministers, there are still sections of many major religions that do not accept evolution as true. One of the most prominent of these in western society is the Intelligent Design movement, which promotes a religious-based idea as if it were a theory based on strong scientific evidence. Intelligent Design proponents often point to complex issues of biology as proof that God is behind the design of human beings, much like a watchmaker is inherent in the design of a watch.

One part of anatomy that has been identified as supposedly supporting Intelligent Design is fingerprints, with some proponents arguing that they are a mark of individualism created by God, with no apparent function except to identify each human being as unique. This is incorrect, as fingerprints do have a well documented function – namely channelling away water to improve grip in wet conditions. In wet conditions, hairless, smooth skinned hands would struggle to grip smooth objects. The individualism of fingerprints is accounted for by the complexity of thousands of small grooves on each digit. Development is inherently affected by stochastic or random processes, meaning that the body is unable to uniformly control its development to ensure that fingerprints are the same in each human being. Clearly, the presence of individual fingerprints does nothing to support the so-called-theory of Intelligent Design.

Which of the following best illustrates the main conclusion of this passage?

A. Fingerprints have a well-established function.

B. Evolution is supported by overwhelming scientific proof.

C. Fingerprints do not offer any support to the notion of Intelligent Design.

D. The individual nature of fingerprints is explained by stochastic processes inherent in development that the body cannot uniformly control.

E. Intelligent design is a credible and scientifically rigorous theory.

Question 24:

High levels of alcohol consumption are known to increase the risk of many non-infectious diseases, such as cancer, atherosclerosis and liver failure. James is a PhD student, and is analysing the data from a large-scale study of over 500,000 people to further investigate the link between heavy alcohol consumption and health problems. In the study, participants were asked about their alcohol consumption, and then their medical history was recorded. His analysis displays surprising results, concluding that those with high alcohol consumption have a *decreased* risk of cancer. James decides that those carrying out the study must have incorrectly recorded the data.

Which of the following is **NOT** a potential reason why the study has produced these surprising results?

A. Previous studies were incorrect, and high alcohol consumption does lower the risk of cancer.

B. The studies analysed didn't take account of other cancer risk factors in comparing those with high and low alcohol consumption.

C. James has made some errors in his analysis, and thus his conclusions are erroneous.

D. The participants involved in the study did not truthfully report their alcohol consumption, leading to false conclusions being drawn.

E. The study's control group data was mixed up with the test group data.

Question 25:

A train is scheduled to depart from Newcastle at 3:30pm. It stops at Durham, Darlington, York, Sheffield, Peterborough and Stevenage before arriving at Kings Cross station in London, where the train completes its journey. The total length of the journey between Newcastle and Kings Cross was 230 miles, and the average speed of the train during the journey (including time spent stood still at calling stations) is 115mph. Therefore, the train will complete its journey at 5:30pm.

Which of the following is an assumption made in this passage?

A. The various stopping points did not increase the time taken to complete the journey.

B. The train left Newcastle on time.

C. The train travelled by the most direct route available.

D. The train was due to end its journey at Kings Cross.

E. There were no signalling problems encountered on the journey.

Question 26:

There have been many arguments over the last couple of decades about government expenditure on healthcare in the devolved regions of the UK. It is often argued that, since spending on healthcare per person is higher in Scotland than in England, people in Scotland will be healthier. However, this view fails to take account of the different needs of these two populations within the UK. For example, one major factor is that Scotland gets significantly colder than England. The cold weakens the immune system, leaving people in Scotland at much higher risk of infectious disease. Thus, Scotland requires higher levels of healthcare spending per person simply to maintain the health of the population at a similar level to that of England.

Which of the following is a conclusion that can be drawn from this passage?

A. The higher healthcare spending per person in Scotland does not necessarily mean people living in Scotland are healthier.

B. Healthcare spending should be increased across the UK.

C. Wales requires more healthcare spending per person simply to maintain population health at a similar level to England.

D. It is unfair on England that there is more spending on healthcare per person in Scotland.

E. Scotland's healthcare budget is a controversial topic.

Question 27:

Vaccinations have been hugely successful in reducing the incidence of several diseases throughout the 20th century. One of the most spectacular achievements was arguably the global eradication of smallpox, once a deadly worldwide killer, during the 1970s. Fortunately, there was a highly effective vaccine available for smallpox, and a major factor in its eradication was an aggressive vaccination campaign. Another disease that is potentially eradicable is polio. However, although there is a highly effective vaccine for polio available, attempts to eradicate it have so far been unsuccessful. It follows that we should plan and execute an aggressive vaccination campaign for polio, in order to ensure that this disease too is eradicated.

Which of the following is the main conclusion of this passage?

A. Polio is a potentially eradicable disease.

B. An aggressive vaccination campaign was a major factor in the eradication of smallpox.

C. Both polio and smallpox have been eradicated by effective vaccination campaigns.

D. We should execute an aggressive vaccination campaign for polio.

E. The eradication of smallpox remains one of the most spectacular achievements of medical science.

Question 28:

The Y chromosome is one of two sex chromosomes found in the human genome, the other being the X chromosome. As the Y chromosome is only found in males, it can only be passed from father to son. Additionally, the Y chromosome does not exchange sections with other chromosomes during cell division (as happens with most chromosomes), meaning it is passed on virtually unchanged through the generations. All of this makes the Y chromosome a fantastic tool for genetic analysis, both to identify individual lineages and to investigate historic population movement. The identification of Genghis Khan as a descendant of up to 8% of males in 16 populations across Asia is just one famous achievement of genetic research using the Y chromosome, and provides further evidence of its utility.

Which of the following best illustrates the main conclusion of this passage?

A. The Y chromosome is a useful tool for genetic analysis.

B. Mutations in the Y chromosome will be passed onto sons but not daughters.

C. X chromosome linked genetic diseases are more common than Y chromosome linked diseases.

D. Y chromosomal analysis is a recent achievement that will help scientists investigate the movements of populations over time.

E. Analysis of the Y chromosome will enable scientists to determine the family tree of Genghis Khan.

Question 29:

In order for a bacterial infection to be cleared, a patient must be treated with antibiotics. Rachel has a minor lung infection, which is thought by her doctor to be a bacterial infection. She is treated with antibiotics, but her condition does not improve. Therefore, it must not be a bacterial infection.

Which of the following best illustrates a flaw in this reasoning?

A. It assumes that a bacterial infection would definitely improve after treatment with antibiotics.

B. It ignores the other conditions that could potentially be treated by antibiotics.

C. It assumes that antibiotics are necessary to treat bacterial infections.

D. It ignores the actions of the immune system, which may be sufficient to clear the infection regardless of what has caused it.

E. It assumes that antibiotics are the only option to treat a bacterial infection.

Question 30:

The link between smoking and lung cancer has been well established for many decades by overwhelming numbers of studies and conclusive research. The answer is clear and simple; the single best measure that can be taken to avoid lung cancer is to not smoke, or to stop smoking if one has already started. However, despite the evidence and clearly communicated health risks, many smokers continue to smoke, and seek to minimise their risk of lung cancer by focusing on other, less important risk factors, such as exercise and healthy eating. This approach is obviously severely flawed, and the fact that some smokers feel this is a good way to reduce their risk of lung cancer shows that they are delusional.

Which of the following best illustrates the main conclusion of this passage?

A. Eating healthily and exercising can also help to reduce the risk of lung cancer.

B. Some smokers consider exercise and healthy eating an easier option than cessation of smoking.

C. The best way to minimise risk of lung cancer is avoidance of smoking.

D. The link between lung cancer and smoking is undeniable.

E. Banning smoking would reduce the incidence of lung cancer.

Question 31:

The government should invest more money into outreach schemes in order to encourage more people to go to university. These schemes allow students to meet other people who went to university, which they may not always be able to do otherwise, even on open days.

Which of the following is the best conclusion of the above argument?

A. Outreach schemes are an effective way for encouraging people to go to university.

B. People will not go to university without seeing it first.

C. The government wants more people to go to university.

D. Meeting people who went to a university is a more effective way to encourage students than university open days.

E. It is easier to meet people at university on outreach schemes than on open days.

Question 32:

The illegal drug cannabis was recently upgraded from a class C drug to class B, which means it will be taken less in the UK, because people will know it is more dangerous. It also means if people are caught in possession of the drug they will face a longer prison sentence, which will also discourage its use.

Which **TWO** statements if true, most weaken the above argument?

A. Class C drugs are cheaper than class B drugs.

B. Upgrading drugs in other countries has not reduced their use.

C. People who take illegal drugs do not know what class they are in.

D. Cannabis was not the only class C drug before it was upgraded.

E. Even if they are caught possessing class B drugs, people do not think they will go to prison.

Question 33:

Schools with better sports programmes such as high-performing football and netball teams tend to have better academic results, less bullying and have overall happier students. Thus, if we want schools to have the best results, reduce bullying and increase student happiness, teachers should start more sports clubs.

Which one of the following best demonstrates a flaw in the above argument?
A. Teachers may be too busy to start sports clubs.
B. Better academic results may be a precondition of better sports teams.
C. Better sports programmes may prevent students from spending time with their family.
D. Some sports teams may be seen to encourage internal bullying.
E. Sport teams that do not perform well lead to increased bullying.

Question 34:

The legal age for purchasing alcohol in the UK is 18. This should be lowered to 16 because the majority of 16-year-olds drink alcohol anyway without any fear of repercussions. Even if the police catch a 16-year-old buying alcohol, they are unable to enforce any consequences. If the drinking limit was lowered the police could spend less time trying to catch underage drinkers and deal with other more important crimes. There is no evidence to suggest that drinking alcohol at 16 is any more dangerous than at 18.

Which one of the following, if true, most weakens the above argument?
A. Most 16-year-olds do not drink alcohol.
B. If the legal drinking age were lowered to 16, more 15-year-olds would start purchasing alcohol.
C. Most 16-year-olds do not have enough money to buy alcohol.
D. Most 16-year-olds are able to purchase alcohol currently.

Question 35:

There has been a recent change in the way the government helps small businesses. Previously small businesses were given non-repayable grants to help them grow their profits. Now, small businesses only receive government loans that must be repaid with interest when the business turns a certain amount of profit. The government wants to support small businesses, but studies have shown they are less likely to prosper under the new scheme as they have been deterred from taking government money for fear of loan repayments.

Which one of the following can be concluded from the passage above?
A. Small businesses do not want government money.
B. The government cannot afford to give out grants to small businesses any longer.
C. All businesses avoid accumulating debt.
D. The government's change in policy is likely to do more harm than good to small businesses.
E. Big businesses do not need government money.

Questions 36-41 are based on the passage below:

Despite the innumerable safety measures in place within medical practice, there can still be disastrous results or 'never events'. Safety measures can fail when the weaknesses in the layers of defence align to create a clear path for failure. This is known as the 'Swiss cheese model of accident causation'. One such occurrence occurred where the wrong kidney was removed from a patient due to a failure in the line of defences designed to prevent such an incident occurring.

When a kidney is diseased it is removed to prevent further complications. This operation, known as a 'nephrectomy', is regularly performed by experienced surgeons. Where normally the consultant who knew the patient would have conducted the procedure, in this case he passed the responsibility to his registrar, who was also well experienced but had not met the patient previously. The person who had copied out the patient's notes had poor handwriting had accidentally written the 'R' for 'right' in such a way that it was read as an 'L' and subsequently copied, and not noticed by anyone who further reviewed the notes.

The patient had been put under anaesthetic before the registrar had arrived and so he proceeded without checking the procedure with the patient, as he normally would have done. The nurses present noticed this error but said nothing, fearing repercussions for questioning a senior medical professional. A medical student was present who had met the patient on the ward previously. The student tried to alert the registrar to the mistake he was about to make, however the registrar shouted at the student that she should not interrupt surgery; she did not know what she was talking about and asked her to leave. Consequently, the surgery proceeded with the end result being that the patient's healthy left kidney was removed, leaving them with only their diseased right kidney, which would eventually lead to the patient's premature death. Frightening as these cases appear what is perhaps scarier is the thought of how those reported may be just the 'tip of the iceberg'.

When questioned about his action to allow his registrar to perform the surgery alone, the consultant had said that it was normal to allow capable registrars to do this. 'While the public perception is that medical knowledge steadily increases over time, this is not the case with many doctors reaching their peak in the middle of their careers.' He had found that his initial increasing interest in surgery had enhanced his abilities, but with time and practice the similar surgeries had become less exciting and so his lack of interest had correlated with worsening outcomes, thus justifying his decision to devolve responsibility in this case.

Question 36:
Which of the following, if true, most weakens the argument above?
A. Only the most severe adverse events in clinical practice will be reported.
B. Doctors undergo extensive training to reduce risks.
C. Thousand of operations happen every year with no problems.
D. Some errors are unavoidable.
E. The patient could have passed away even if the operation had been a complete success.

Question 37:

Which one of the following is the overall conclusion of the passage above?

A. The error that occurred was a result of the failure of safety precautions in place.
B. Surgery should only be performed by surgeons who know their patients well.
C. The human element to medicine means errors will always occur.
D. The safety procedures surrounding surgical procedures need to be reviewed.
E. Some doctors are overconfident.

Question 38:

Which of the following is attributed as the original cause of the error?

A. The medical student not having asserted herself.
B. The poor handwriting in the chart.
C. The hierarchical system of medicine.
D. The registrar not having met the patient previously.
E. The lack of surgical skill possessed by the registrar.

Question 39:

What does the 'tip of the iceberg' refer to in the passage?

A. Problems we face every day.
B. The probable large numbers of medical errors that go unreported.
C. The difficulties of surgery.
D. Reported medical errors.
E. Problems within the NHS.

You may use the graphs below once, more than once, or not at all.

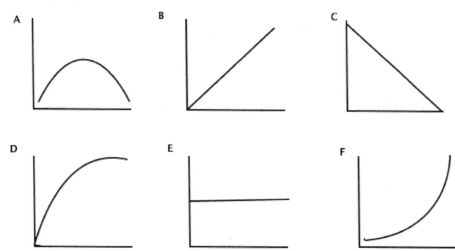

Question 40:
Which graph best describes the consultants' performance versus emotional investment in cases over his career?

A. A   B. B   C. C   D. D   E. E   F. F

Question 41:
Which graphs best describe the medical knowledge acquired over time?

Option	Public Perception	Consultant's Perception
A	B	B
B	B	D
C	D	B
D	D	D
E	F	B

Question 42:
Sadly, in recent times, the number of people taking less exercise and having an associated sedentary lifestyle has increased in the developed world. The lack of opportunity for exercise is endemic and these countries have also seen a rise of diseases such as diabetes even in young people. In these developed countries, conditions usually associated with old age, like high blood pressure, are rapidly increasing in younger patients. These are however still uncommon in undeveloped countries, where most people are physically active throughout the entirety of their lives.

Which one of the following can be concluded from the passage above?
A. Exercise has a greater effect on old people than young people.
B. Maintenance of good health is associated with lifelong exercise.
C. Changes in lifestyle will be necessary to cause increased life expectancies in developed countries.
D. Exercise is only beneficial when continued into old age.
E. Obesity and diabetes are the result of lack of exercise.

Questions 43 - 45 are based on the passage below:

'Midwives should now encourage women to give birth at home as often as possible. Not only is there evidence to suggest that normal births at home are as safe as in hospital, but it removes the medicalisation of childbirth that has emerged over the years. With the increase in availability of health resources we now use services such as a full medical team for a natural process that women have been completing single-handedly for thousands of years. Midwives are extensively trained to assist women during labour at home and are capable of assessing when there is a problem that requires a hospital environment. Expensive hospital births must and should move away from being standard practice, especially in an era where the NHS has far more demands on its services that it can currently afford.'

Question 43:

Which one of the following is the most appropriate conclusion from the statement?

A. People are over dependent on healthcare.
B. Some women prefer to have their babies in hospital.
C. Having a baby in hospital can actually be more risky than at home.
D. Childbirth has been over-medicalised.
E. Encouraging women to have their babies at home may relieve some of the financial pressures on the NHS.

Question 44:

Which one of the following, if true, most weakens the argument presented in the passage above?

A. Some women are scared of home births.
B. Home births are associated with poorer outcomes.
C. Midwives do not like performing home visits.
D. Some home births result in hospital births anyway.
E. We should have more midwives than doctors.

Question 45:

Which one of the following describes what the statement cites as the cause for the 'medicalisation of childbirth'?

A. Women fear giving birth without a full medical team present.
B. Midwives are incapable of aiding childbirth without help.
C. Giving birth at home is not as safe as it used to be.
D. Easy access to and availability of health services.
E. Women only used to give birth at home because they could not do so at hospital.

Question 46:

We need to stop focusing so much attention on the dangers of fires. In 2011 there were only 242 deaths due to exposure to smoke, fire and flames, while there were 997 deaths from hernias. We need to think more proportionally as these statistics show that campaigns such as 'fire kills' are unnecessary. In comparison to the risk of death from hernias, clearly shows that fires are not as dangerous as they are perceived to be.

Which of the following statements identify a weakness in the above argument?

1. More people may die in fires if there were no campaigns about their danger and how to prevent them.
2. The smoke of a fire is more dangerous than its flames.
3. There may be more people with hernias than those in fires.

A. 1 only   B. 2 only   C. 2 and 3 only
D. 1 and 2 only   E. 1 and 3 only

Question 47:

School students were surveyed to find out if there was any correlation between the sports students played and the subjects they liked. The findings were as follows: some football players liked maths and some of them liked history. All students liked English. None of the basketball players liked history, but all of them, as well as some rugby players, liked chemistry. All rugby players like geography.

Based on the findings, which one of the below must be true?

A. Some of the footballers liked maths and history.
B. Some of the rugby players liked three subjects.
C. Some rugby players liked history.
D. Some of the footballers liked English but did not like maths and history.
E. Some basketball players like more than 3 subjects.

Question 48:

The control of illegal drug use is becoming increasingly difficult. New 'legal highs' are being manufactured which have a slightly different molecular structure to illegal compounds, so they are not technically illegal. These new 'legal highs' are being brought onto the street at a rate of at least one per week, and so the authorities cannot keep up. Some health professionals therefore believe that knowledge of the legal classification of street drugs is less important than knowledge of the potentially dangerous side effects. The fact that these new compounds are legal may however mean that the public are not aware of their equally high risks.

Which of the following are implied by the argument?
1. Some health professionals believe there is no value in making drugs illegal.
2. The major problem in controlling illegal drug use is the rapid manufacture of new drugs that are not classified as illegal.
3. The general public are not worried about the risks of legal or illegal highs.
4. There is no longer a good correlation between risk of drug taking and the legal status of the drug.

A. 1 only
B. 2 only
C. 1 and 4
D. 2 and 4
E. 2 and 3
F. 1, 2, 3 and 4

Question 49:

WilderTravel Inc. is a company which organises wilderness travel holidays, with activities such as trekking, mountain climbing, safari tours and wilderness survival courses. These activities carry inherent risks, so the directors of the company are drawing up a set of health regulations with the aim of minimising the risks by ensuring that nobody with medical conditions participates in activities that might put their health in jeopardy. They consider the following guidelines:

'Persons with pacemakers, asthma or severe allergies are at significant risk of heart attack in low oxygen environments'. People undertaking mountain climbing activities with WilderTravel frequently encounter environments with low oxygen levels. The directors therefore decide that in order to ensure the safety of customers on WilderTravel holidays, one step that must be taken is to bar those with pacemakers, asthma or allergies from partaking in mountain climbing.

Which of the following best illustrates a flaw in this reasoning?
A. Participants should be allowed to assess the safety risks themselves, and should not be barred from activities if they decide the risk is acceptable.
B. They have assumed that all allergies carry an increased risk of heart attack, when the guidelines only say this applies to those with severe allergies.
C. The directors have failed to consider the health risks of people with these conditions taking part in other activities.
D. People with these conditions could partake in mountain climbing with other holiday organisers, and thus be exposed to danger of heart attack.

Question 50:

St John's Hospital in Northumbria is looking to recruit a new consultant cardiologist, and interviews a series of candidates. The interview panel determines that 3 candidates are clearly more qualified for the role than the others, and they invite these 3 candidates for a second interview. During this second interview, and upon further examination of their previous employment records, it becomes apparent that candidate 3 is the most proficient at surgery of the 3, whilst candidate 1 is the best at patient interaction and explaining the risks of procedures. Candidate 2, meanwhile, ranks between the others in both these aspects of care.

The hospital director tells the interviewing team that the hospital already has a well-renowned team dedicated to patient interaction, but the surgical success record at the hospital is in need of improvement. The director issues instructions that, therefore, it is more important that the new candidate is proficient at surgery, and skills in patient interaction are less of a concern.

Which of the following is a conclusion that can be drawn from the Director's comments?

A. The interviewing team should hire candidate 2, in order to achieve a balance of good patient relations with good surgical records.
B. The interviewing team should hire candidate 1, in order to ensure good patient interactions, as these are a vital part of a doctor's work.
C. The interviewing team should ignore the hospital director and assess the candidates further to see who would be the best fit.
D. The interviewing team should hire candidate 3, in order to ensure that the new candidate has excellent surgical skills, to boost the hospital's success in this area.

Question 51:

Every winter in Britain, there are thousands of urgent callouts for ambulances in snowy conditions. The harsh conditions mean that ambulances cannot drive quickly, and are delayed in reaching patients. These delays cause many injuries and medical complications, which could be avoided with quicker access to treatment. Despite this, very few ambulances are equipped with winter tyres or special tyre coverings to help the ambulances deal with snow. Clearly, if more ambulances were fitted with winter tyres, then we could avoid many medical complications that occur due to delays each winter.

Which of the following is an assumption made in this passage?

A. Fitting winter tyres would allow ambulances to reach patients more quickly.
B. Ambulance trusts have sufficient funding to equip their vehicles with winter tyres.
C. Many medical complications could be avoided with quicker access to medical care.
D. There are no other alternatives to winter tyres that would allow ambulances to reach patients more quickly in snowy conditions.

Question 52:

Vaccinations have been one of the most outstanding and influential developments in medical history. Despite the huge successes, there is a strong anti-vaccination movement active in some countries, particularly the USA, which claims that vaccines are harmful and ineffective.

There have been several high-profile events in recent years where anti-vaccination campaigners have been refused permission to enter countries for campaigns, or have had venues refuse to host them due to the nature of their campaigns. Many anti-vaccination campaigners have claimed this is an affront to free speech, and that they should be allowed to enter countries and obtain venues without hindrance. However, although free speech is desirable, an exception must be made here because the anti-vaccination campaign spreads misinformation, causing vaccination rates to drop.

When this happens, preventable infectious diseases often begin to increase, causing avoidable deaths, particularly in children. Thus, in order to protect people, we must continue to block the anti-vaccine campaigners from spreading misinformation freely by pressuring venues not to host anti-vaccination campaign events.

Which of the following best illustrates the principle that this argument follows?
A. Free speech is always desirable and must not be compromised under any circumstances.
B. The right to protection from infectious diseases by vaccination is more important than the right to freedom of speech.
C. The right of free speech does not apply when the party speaking is lying or spreading misinformation.
D. Public health programmes that achieve significant success in reducing the incidence of disease should be promoted.

Question 53:

In order for a tumour to grow larger than a few centimetres, it must first establish its own blood supply by promoting angiogenesis. Roger has a tumour in his abdomen, which is investigated at the Royal General Hospital. During the tests, they detect newly formed blood vessels in the tumour, showing that it has established its own blood supply. Thus, we should expect the tumour to grow significantly, and become larger than a few centimetres. Action must be taken to deal with this.

Which of the following best illustrates a flaw in this reasoning?
A. It assumes that the tumour in Roger's abdomen has established its own blood supply.
B. It assumes that a blood supply is necessary for a tumour to grow larger than a few centimetres.
C. It assumes that nothing can be done to stop the tumour once a blood supply has been established.
D. It assumes that a blood supply is sufficient for the tumour to grow larger than a few centimetres.

Question 54:

In this year's Great North Run, there are several dozen people running to raise money for the Great North Air Ambulance (GNAA), as part of a large national fundraising campaign. If the runners raise £500,000 between them, then the GNAA will be able to add a new helicopter to its fleet. However, the runners only raise a total of £420,000. Thus, the GNAA will not be able to get a new helicopter.

Which of the following best illustrates a flaw in this passage?

A. It has assumed that the GNAA will not be able to acquire a new helicopter without the runners raising £500,000.

B. It has assumed that the GNAA wishes to add a new helicopter to its fleet.

C. It has assumed that the GNAA does not have better things to spend the money on.

D. It has assumed that only people running in the Great North Run are raising money for the GNAA.

Question 55:

Many courses, spanning universities, colleges, apprenticeship institutions and adult skills courses should be subsidised by the government. This is because they improve the skills of those attending them. It has been well demonstrated that the more skilled people are, the more productive they are economically. Thus, government subsidies of many courses would increase overall economic productivity, and lead to increased growth.

Which of the following would most weaken this argument?

A. The UK already has a high level of growth and does not need to accelerate this growth.

B. Research has demonstrated that higher numbers of people attending adult skills courses results in increased economic growth.

C. Research has demonstrated that the cost of many courses (to those taking them) has little effect on the number of people undertaking the courses.

D. Employers often seek to employ those with greater skill-sets, and appoint them to higher positions.

Question 56:

Pluto was once considered the 9th planet in the solar system. However, further study of the planet led to it being reclassified as a dwarf planet in 2006. One key factor in this reclassification was the discovery of many objects in the solar system with similar characteristics to Pluto, which were also placed into this new category of 'dwarf planet'. Some astronomers believe that Pluto should remain classified as a planet, along with the many entities similar to Pluto that have now been discovered. Considering all of this, it is clear that if we were to reclassify Pluto as a planet, and maintain consistency with classification of astronomical entities, then the number of planets would significantly increase.

Which of the following best illustrates the main conclusion of this passage?

A. If Pluto is classified as a planet, then many other entities should also be planets, as they share similar characteristics.

B. Some astronomers believe Pluto should be classified as a planet.

C. Pluto should not be classified as a planet, as this would also require many other entities to be classified as planets to ensure consistency.

D. If Pluto is to be classified as a planet, then the number of objects classified as planets should increase significantly.

Question 57:

Two trains depart from Birmingham at 5:30 pm. One of the trains is heading to London, whilst the other is heading to Glasgow. The distance from Birmingham to Glasgow is three times larger than the distance from Birmingham to London, and the train to London arrives at 6:30 pm. Thus, the train to Glasgow will arrive at 8:30pm.

Which of the following is an assumption made in this passage?

A. Both trains depart at the same time.

B. Both trains depart from Birmingham.

C. Both trains travel at the same speed.

D. The train heading to Glasgow has to travel three times as far as the train heading to London.

Question 58:

Carcinogenesis, oncogenesis and tumorigenesis are various names given to the generation of cancer, with the term literally meaning 'creation of cancer'. In order for carcinogenesis to happen, there are several steps that must occur. Firstly, a cell (or group of cells) must achieve immortality, and escape senescence (the inherent limitation of a cell's lifespan). Then they must escape regulation by the body, and begin to proliferate in an autonomous way. They must also become immune to apoptosis and other cell death mechanisms. Finally, they must avoid detection by the immune system, or survive its responses to unfamiliar genetic material. If a single one of these steps fails to occur, then carcinogenesis will not be able to occur.

Which of the following is a conclusion that can be reliably drawn from this passage?

A. Carcinogenesis is a multi-step process that occurs in discrete stages and in a specific order.

B. If all the steps mentioned occur, then carcinogenesis will occur.

C. The immune system is unable to tackle cells that have escaped regulation by the body.

D. There are various mechanisms by which carcinogenesis can occur.

E. Carcinogenesis occurs over a period of months to years, as there are multiple steps.

Question 59:

P53 is one of the most crucial genes in the body, responsible for detecting DNA damage and halting cell replication until repair can occur. If repair cannot take place, P53 will signal for the cell to kill itself. These actions are crucial to prevent carcinogenesis, and a loss of functional P53 is identified in over 50% of all cancers. The huge importance of P53 towards protecting the cell from damaging mutations has led to it deservedly being known as 'the guardian of the genome'. The implications of this name are clear – any cell that has a mutation in P53 is at serious risk of developing a potentially dangerous mutation.

Which of the following **CANNOT** be reliably concluded from this passage?
A. P53 is responsible for detecting DNA damage.
B. Most cancers have lost functional P53.
C. P53 deserves its name 'guardian of the genome'.
D. A cell that has a mutation in P53 will develop damaging mutations.
E. None of the above.

Question 60:

Sam is buying a new car, and deciding whether to buy a petrol or a diesel model. He knows he will drive 9,000 miles each year. He calculates that if he drives a petrol car, he will spend £500 per 1,000 miles on fuel, but if he buys a diesel model, he will only spend £300 per 1,000 miles on fuel. He calculates, therefore, that if he purchases a diesel car, then this year he would make a saving of £1800, compared to if he bought the petrol car.

Which of the following is **NOT** an assumption that Sam has made?
A. The price of diesel will not fluctuate relative to that of petrol.
B. The cars will have the same initial purchase cost.
C. The cars will have the same costs for maintenance and garage expenses.
D. The cars will use the same amount of fuel.
E. All of the above are assumptions.

Question 61:

In the UK, cannabis is classified as a Class B drug, with a maximum penalty of up to 5 years imprisonment for possession, or up to 14 years for possession with intent to supply. The justification for drug laws in the UK is that classified drugs are harmful, addictive, and destructive to people's lives. However, plentiful medical evidence indicates that cannabis is relatively safe, non-addictive and overall harmless. In particular, it is certainly shown to be less dangerous than alcohol, which is commonly sold and advertised across the UK. The fact that alcohol can be widely sold and advertised, but cannabis, a less harmful drug, is banned highlights the gross inconsistencies in UK drugs policy.

Which of the following best illustrates the main conclusion of this passage?
A. Cannabis is a less dangerous drug than alcohol, so it should be made more widely available.
B. In order to ensure consistency in the UK drug policy, we should either ban alcohol, or make cannabis widely available.
C. Cannabis is considered harmful and addictive, which is why it is a Class B drug.
D. The UK government's policy on drugs is grossly inconsistent.
E. Alcohol can be advertised in the UK, whereas cannabis cannot.

Question 62:

Every year in Britain, there are thousands of accidents in the home such as burns, broken limbs and severe cuts, which cause a large number of deaths and injuries. Despite this, very few households maintain a sufficient first aid kit equipped with bandages, burn treatments, splints and saline to clean wounds. If more households stocked sufficient first aid supplies, many of these accidents could be avoided.

Which of the following best illustrates a flaw in this argument?
A. It ignores the huge cost associated with maintaining good first aid supplies, which many households cannot afford.
B. It implies that presence of first aid equipment will lead to fewer accidents.
C. It ignores the many accidents that could not be treated even if first aid supplies were readily available.
D. It neglects to consider the need for trained first aid persons in order for first aid supplies to help in reducing the severity of injuries caused by accidents.

Question 63:

Researchers at SmithJones Inc., an international pharmaceuticals firm, are investigating a well-known historic compound, which is thought to reduce levels of DNA replication by inhibiting DNA polymerases. It is proposed that this may be able to be used to combat cancer by reducing the proliferation of cancer cells, allowing the immune system to combat them before they spread too far and cause significant damage. Old experiments have demonstrated the effectiveness of the compound via monitoring DNA levels with a dye that stains DNA red, thus monitoring the levels of DNA present in cell clusters. They report that the compound is observed to reduce the rate at which DNA replicates. However, it is known that if researchers use the wrong solutions when carrying out these experiments, then the amount of red staining will decrease, suggesting DNA replication has been inhibited, even if that is not the case. As several researchers previously used this wrong solution, we can conclude that these experiments are flawed, and do not reflect what is actually happening.

Which of the following best illustrates a flaw in this argument?
A. From the fact that the compound inhibits DNA replication, it cannot be concluded that it has potential as an anticancer drug.
B. From the fact that the wrong solutions were used, it cannot be concluded that the experiments may produce misleading results.
C. From the fact that the experiments are old, it cannot be concluded that the wrong solutions were used.
D. From the fact that the compound is old, it cannot be concluded that it is safe.

Question 64:

Rotherham football club are currently top of the league, with 90 points. Their closest competitors are South Shields football club, with 84 points. Next week, the teams will play each other, and after this, they each have two games left before the end of the season. Each win is worth 3 points, a draw is worth 1 point, and a loss is worth 0 points. Thus, if Rotherham beat South Shields, they will win the league (as they will then be 9 points clear, and South Shields would only be able to earn 6 more points).

In the match of Rotherham vs. South Shields, Rotherham are winning until the 85th minute, when Alberto Simeone scores an equaliser for South Shields, and South Shields then go on to win the match. Thus, Rotherham will not win the league.

Which of the following best illustrates a flaw in this passage's reasoning?
A. It has assumed that Alberto Simeone scored the winning goal for South Shields.
B. It has assumed that beating South Shields was necessary for Rotherham to win the league, when in fact it was only sufficient.
C. Rotherham may have scored an equaliser later in the game, and not lost the match.
D. It has failed to consider what other teams might win the league.

Question 65:

Oakville Supermarkets is looking to build a new superstore, and a meeting of its directors has been convened to decide where the best place to build the supermarket would be. The Chair of the Board suggests that the best place would be Warrington, a town that does not currently have a large supermarket, and would thus give them an excellent share of the grocery shopping market.
However, the CEO notes that the population of Warrington has been steadily declining for several years, whilst Middlesbrough has recently been experiencing high population growth. The CEO therefore argues that they should build the new supermarket in Middlesbrough, as they would then be within range of more people, and so it would have more potential customers.

Which of the following best illustrates a flaw in the CEO's reasoning?
A. Middlesbrough may already have other supermarkets, so the new superstore may get a lower share of the town's shoppers.
B. Despite the recent population changes, Warrington may still have a larger population than Middlesbrough.
C. Middlesbrough's population is projected to continue growing, whilst Warrington's is projected to keep falling.
D. Many people in Warrington travel to Liverpool or Manchester, two nearby major cities, in order to do their shopping.

Question 66:

Global warming is a key challenge facing the world today, and the changes in weather patterns caused by this phenomenon have led to the destruction of many natural habitats, causing many species to become extinct. Recent data shows that extinction events have been occurring at a faster rate over the last 40 years than at any other point in the earth's history, exceeding the great Permian mass extinction, which wiped out 96% of life on earth. If this rate continues, over 50% of species on earth will be extinct by 2100. It is clear that in the face of this huge challenge, conservation programmes will require significantly increased levels of funding in order to prevent most of the species on earth from becoming extinct.

Which of the following are assumptions in this argument?

1. The rate of extinction events seen in the last 40 years will continue to occur without a step-up in conservation efforts.
2. Conservation programmes cannot prevent further extinctions without increased funding.
3. Global warming has caused many extinction events, directly or indirectly.

A. I only   C. 3 only   E.   I and 3
B. 2 only   D. I and 2

Question 67:

After an election in Britain, the new government is debating what policy to adopt on the railway system, and whether it should be entirely privatised, or whether public subsidies should be used to supplement costs and ensure that sufficient services are run. Studies in Austria, which has high public funding for railways, have shown that the rail service is used by many people, and is rated highly by the population. However, this is clearly down to the fact that Austria has many mountainous and high-altitude areas, which experience significant amounts of snow and ice. This makes many roads impassable by car.. Thus, rail is often the only way to travel, explaining the high passenger numbers and approval ratings. Thus, the high public subsidies clearly have no effect.

Which of the following, if true, would weaken this argument?

1. France also has high public subsidy of railways, but does not have large areas where travel by road is difficult. The French railway also has high passenger numbers and approval ratings.
2. Italy also has high public subsidy of railways, but the local population dislike using the rail service, and it has poor passenger numbers.
3. There are many reasons affecting the passenger numbers and approval ratings of a given country's rail service.

A. I only   C. 3 only   E.   I and 3
B. 2 only   D. I and 2

Question 68:

In 2001-2002, 1,019 patients were admitted to hospital due to obesity. This figure was more than 11 times higher by 2011-12 when there were 11,736 patients admitted to hospital with the primary reason for admission being obesity. Data has shown higher percentages of both men and women were either obese or overweight in 2011 compared to 1993, with the percentage of overweight men climbing from 58% to 65%, and female rates of obesity increasing from 49% to 58%. Rates of adult obesity have increased even more steeply within the period from 2001-2012 – 13% to 24% for men and 16% to 26% for women.

Studies in 2011 found that nearly a third of children between 2 – 15 years were either overweight or obese, although this was not significantly higher than in 2008. Lifestyles are also becoming less healthy, with a decline in both children and adults eating the recommended portions of fruit and vegetables each day or taking the recommended amount of exercise each week. The ease and availability of cheap fast-food outlets may be partly to blame for the rising number of obese people. Education is required to teach people the importance of a healthy lifestyle, however people must take some personal responsibility for their health.

Using only information from the passage, which of the following statements is correct?
A. In 2011, there was a higher proportion of obese men than women.
B. Obesity rates are rising steeply for both males and females of all age groups.
C. A combination of education and personal responsibility is needed to improve the population's health
D. The main reason people eat fast food is because it's cheaper than healthy alternatives.

Question 69:

Tobacco companies sell cigarettes despite being fully aware that cigarettes cause significant harm to the health and wellbeing of those that smoke them. Diseases caused or aggravated by smoking cost billions of pounds for the NHS to treat each year and have an enormous impact on the individuals affected by smoking related diseases. This is extremely irresponsible behaviour from the tobacco companies. Tobacco companies should be taxed, and the money raised put towards funding the NHS.

Which of the following conclusions **CANNOT** be drawn from the above?
A. There is a connection between lung cancer and smoking.
B. People who smoke are more likely to also drink.
C. There is a connection between oral cancer and smoking.
D. All smokers drink excessively.
E. All of the above.

Question 70:

Investigations in the origins of species suggest that humans and the great apes have the same ancestors. This is suggested by the high degree of genetic similarity between humans and chimpanzees (estimated at 99%). At the same time there is an 84% homology between the human genome and that of pigs. This raises the interesting question of whether it would be possible to use pig or chimpanzee organs for the treatment of human disease.

Which conclusion can be reasonably drawn from the above article?

A. Pigs and chimpanzees have a common ancestor.

B. Pigs and humans have a common ancestor.

C. It can be assumed that chimpanzees will develop into humans if given enough time to evolve.

D. There seems to be great genetic homology across a variety of species.

E. Organs from pigs or chimpanzees present a good alternative for human organ donation.

Question 71:

Poor blood supply to a part of the body can cause damage to the affected tissue. There are a variety of known risk factors for vascular disease. Diabetes is a major risk factor. Other risk factors are more dependent on the individual as they represent individual choices such as smoking, poor dietary habits as well as little to no exercise. In some cases infarction of the limbs and in particular the feet can become very bad and extensive with patches of tissue dying. This is known as necrosis and is marked by the skin of the affected area of the body (often fingers or toes) turning black. Necrotic tissue is usually removed in surgery.

Which of the following statements **CANNOT** be concluded from the information in the above passage?

A. Smoking causes vascular disease.

B. Diabetes causes vascular disease.

C. Vascular disease always leads to infarctions.

D. Necrotic tissue must be removed surgically.

E. Necrotic tissue only occurs following severe infarction.

F. All of the above.

Question 72:

People who can afford to pay for private education should not have access to the state school system. This would allow more funding to be reserved for educating students from lower income backgrounds. More funding would provide better resources for students from lower income backgrounds, and will help to bridge the gap in educational attainment between students from higher income and lower income backgrounds.

Which of the following statements, if true, would most strengthen the above argument?

A. Educational attainment is a significant factor in determining future prospects.
B. Providing better resources for students has been demonstrated to lead to an increase in educational attainment.
C. Most people who can afford to do so choose to purchase private education for their children.
D. A significant gap exists in educational attainment between students from high and low income backgrounds.
E. Most schools currently receive funding according to the number of students in the school.

Question 73:

Increasing numbers of people are choosing to watch films on DVD in recent years. In the past few years, cinemas have lost customers, causing them to close down. Many cinemas have recently closed, removing an important focal point for many local communities and causing damage to the local economy. Therefore, we should ban DVDs in order to help local communities.

Which of the following best states an assumption made in this argument?

A. Cinemas have closed because of reduced profits due to people choosing to watch DVDs instead.
B. Cinemas being forced to close causes damage to local communities.
C. DVDs are improving local communities by allowing people to meet up and watch films together.
D. Sales of DVDs have increased due to economic growth.
E. Local communities have called for DVDs to be banned.

Question 74:

Aeroplanes are the fastest form of transport available. An aeroplane can travel a given distance in less time than a train or a car. John needs to travel from Glasgow to Birmingham. If he wants to arrive as soon as possible, he should travel by aeroplane.

Which of the following best illustrates a flaw in this argument?

A. One day, cars that travel as fast as aeroplanes could be developed
B. Travelling by air is often more expensive.
C. It ignores the time taken to travel to an airport and check in to a flight, which may mean he will arrive later if travelling by aeroplane.
D. John may not own a car, and thus may not have any option.
E. John may not be legally allowed to make the journey.

Question 75:

During autumn, spiders frequently enter homes to escape the cold weather. Many people dislike spiders and seek ways to prevent them from entering properties, leading to spider populations falling as they struggle to cope with the cold weather. Studies have demonstrated that when spider populations fall, the population of flies rises. Higher numbers of flies are associated with an increase in food poisoning cases. Therefore, people must not seek to prevent spiders from entering their homes.

Which of the following best illustrates the main conclusion of this argument?

A. People should not dislike spiders being present in their homes.

B. People should seek alternative methods to prevent flies from entering their homes.

C. People should actively encourage spiders to occupy their homes to increase biodiversity.

D. People should accept the presence of spiders in their homes to reduce the incidence of food poisoning.

E. Spiders should be cultivated and used as a biological pest control to combat flies.

Question 76:

Each year, thousands of people acquire infections during prolonged stays in hospital. Concurrently, bacteria are becoming resistant to antibiotics at an ever-increasing rate. In spite of this, progressively fewer pharmaceutical companies are investing in research into new antibiotics, and the number of antibiotics coming onto the market is decreasing. As a result, the number of antibiotics that can be used to treat infections is falling. If pharmaceutical companies were pressured into investing in new antibiotic research, many lives could be saved.

Which of the following best illustrates a flaw in this argument?

A. It assumes the infections acquired during stays at hospital are resulting in deaths.

B. It ignores the fact that many people never have to stay in hospital.

C. It does not take into account the fact that antibiotics do not produce much profit for pharmaceutical companies.

D. It ignores the fact that some hospital-acquired infections are caused by organisms that cannot be treated by antibiotics, such as viruses.

E. It assumes that bacterial resistance to antibiotics has not been happening for some time.

Question 77:

Katherine has shaved her armpits most of her adult life, but has now decided to stop. She explains her reasons for this to John, saying she does not like the pressures society puts on women to be shaven in this area. John listens to her reasons, but ultimately responds 'just because you explain why I should find your hairiness attractive, it does not mean I will. I find you unattractive, as I do not like girls with hair on their armpits.'

What assumption has John made?

A. That just because he finds Katherine unattractive, he would find other girls with unshaven armpits unattractive.
B. That Katherine is trying to make John find her armpit hair attractive.
C. That Katherine will never conceal her armpit hair.
D. Katherine must be wrong, because she is a woman.
E. That Katherine thinks women should stop shaving.

Question 78:

Medicine and the availability of powerful drugs have improved significantly over the last century. Better medical practice results in a reduction in the death rate from all causes. However, as people age, they are more likely to suffer from infectious diseases.

Many developing countries have a high rate of deaths from infectious diseases. Sunita argues that this is a result of better medical practices in developing countries. Better medicine has given rise to an ageing population, which explains why there is a higher rate of death from infectious diseases.

However, this cannot be the case. In developing countries, most people do not live to old age. In fact, it is common for people in developing countries to die from infectious diseases at a young age. Therefore, an ageing population cannot be the reason behind the high rate of death from infectious disease in developing countries. Since better medicine reduces the death rate from all causes, it is clear that better medicine would lead to a reduction in the death rate from infectious disease in developing countries.

Which of the following best states the main conclusion of this argument?

A. We can expect that improvements in medicine seen over the last century will improve.
B. Better medicine is not responsible for the increased prevalence of infectious disease in third world countries.
C. Better medicine has caused the overall death rate of third world countries to increase.
D. Better medicine will cause a decrease in the rate of death from infectious disease in third world countries.
E. As people get older, they suffer from infectious disease more commonly.

Question 79:

Bristol and Cardiff are two cities with similar demographics, and are located in a similar area of the country. Bristol has higher demand for housing than Cardiff. Therefore, a house in Bristol will cost more than a similar house in Cardiff.

Which of the following best illustrates an assumption in the statement above?

A. House prices will be higher if demand for housing is higher.

B. People can commute from Cardiff to Bristol.

C. Supply of housing in Cardiff will not be lower than in Bristol.

D. Bristol is a better place to live.

E. Cardiff has sufficient housing to provide for the needs of its communities.

Question 80:

Jellicoe Motors is a small motor company in Sheffield, employing three people. The company is hiring a new mechanic and interviews several candidates. New research into production lines has indicated that having employees with a good ability to work as part of a team boosts the productivity and profits of a company. Therefore, Jellicoe motors should hire a candidate with good team-working skills.

Which of the following best illustrates the main conclusion of this argument?

A. Jellicoe Motors should not hire a new mechanic.

B. Jellicoe motors should hire a candidate with good team-working skills in order to boost their productivity and profits.

C. Jellicoe motors should hire several new candidates in order to form a good team, and boost their productivity.

D. If Jellicoe motors does not hire a candidate with good team-working skills, they may struggle to be profitable.

E. Jellicoe motors should not listen to the new research.

Question 81:

Research into new antibiotics is seldom profitable for pharmaceutical firms. As a consequence, many firms are not investing in antibiotic research, and very few new antibiotics are being produced. However, with bacteria becoming increasingly resistant to current antibiotics, new drugs are desperately needed to avoid running the risk of thousands of deaths from bacterial infections. Therefore, the UK government must provide financial incentives for pharmaceutical companies to invest in research into new antibiotics.

Which of the following best expresses the main conclusion of this argument?

A. If bacteria continue to become resistant to antibiotics, there could be thousands of deaths from bacterial infections.

B. Pharmaceutical firms are not investing in new antibiotic research due to a lack of potential profit.

C. If the UK government invests in research into new antibiotics, thousands of lives will be saved.

D. The pharmaceutical firms should invest in areas of research that are profitable and ignore antibiotic research.

E. The UK government must provide financial incentives for pharmaceutical firms to invest into antibiotic research if it wishes to avoid risking thousands of deaths from bacterial infections.

Question 82:

People in developing countries use far less water per person than those in developed countries. It is estimated that at present, people in the developing world use an average of 30 litres of water per person per day, whilst those in developed countries use on average 70 litres of water per person per day. It is estimated that for the current world population, an average water usage of 60 litres per person per day would be sustainable, but any higher than this would be unsustainable.

The UN has set development targets such that in 20 years, people living in developing countries will be using the same amount of water per person per day as those living in developed countries. Assuming the world population stays constant for the next 20 years, if these targets are met the world's population will be using water at an unsustainable rate.

Which of the following, if true, would most weaken the argument above?

A. The prices of water bills are dropping in developed countries like the UK.

B. The level of water usage in developed countries is falling and may be below 60 litres per person per day in 20 years.

C. The total population of all developing countries is less than the total population of all developed countries.

D. Climate change is likely to decrease the amount of water available for human use over the next 20 years.

E. The UN's development targets are unlikely to be met.

Question 83:

An advert for a senior management post states "we need someone who can keep a cool head in a crisis and react quickly to events". The applicant says he suffers from a phobia of flying, and panics especially when an aircraft is landing and that therefore he would prefer not to travel abroad on business if it could be avoided. The interview panel conclude he is obviously a very nervous type of person who would clearly go to pieces and panic in an emergency and fail to provide the leadership qualities necessary for the job. Therefore, they decide this person is not a suitable candidate for the post.

Which of the following highlights the biggest flaw in the argument above?

A. It falsely assumes that phobias are untreatable or capable of being eliminated.
B. It falsely assumes that the person appointed to the job will need to travel abroad.
C. It falsely assumes that a specific phobia indicates a general tendency to panic.
D. It falsely assumes that people who stay cool in a crisis will be good leaders.
E. It fails to take into account other qualities the person might have for the post.

Question 84:

There are significant numbers of people attending university every year: as many as 45% of 18-year-olds. As a result, there are many more graduates entering the workforce with better skills and better earning potential. Going to university makes economic sense and we should encourage as many young people to attend as possible.

Which of the following highlights the biggest flaw in the argument above?

A. There are no more university places left.
B. Students can succeed without going to university.
C. Not all degrees equip students with the skills needed to earn higher salaries.
D. Some universities are better than others.

Question 85:

Young people spend too much time watching television, which is bad for them. Watching excessive amounts of TV is linked to obesity, social exclusion and can cause eye damage. If young people were to spend just one evening a week playing sport or going for a walk, the benefits would be signficant. They would lose weight, feel better about themselves and it would be a sociable activity. Exercise is also linked to strong performance at school and so young people would be more likely to perform well in their exams.

Which of the following highlights the biggest flaw in the argument above?

A. Young people can watch sport on television.
B. There are many factors that affect exam performance.
C. Television does not necessarily have any damaging effect.
D. Television and sport are not linked.

Question 86:

Campaigners pushing for the legalisation of cannabis present many arguments to support their cause. Most claim there is little evidence of any adverse affects to health caused by cannabis use, that many otherwise law-abiding people are users of cannabis and that in any case, prohibition of drugs does not reduce their usage. Legalising cannabis would also reduce crime associated with drug trafficking and would provide an additional revenue stream for the government.

Which of the following best represents the conclusion of the passage?

A. Regular cannabis users are unlikely to have health problems.
B. Legalising cannabis would be good for cannabis users.
C. There are multiple reasons to legalise cannabis.
D. Prohibition is an effective measure to reduce drugs usage.
E. Drug associated crime would reduce if cannabis was legal.

Question 87:

Mohan has been offered a new job in Birmingham, starting in several months with a fixed salary. In order to ensure he can afford to live in Birmingham on his new salary, Mohan compares the prices of some houses in Birmingham. He finds that a 2-bedroom house will cost £200,000. A 3-bedroom house will cost £250,000. A 4-bedroom house with a garden will cost £300,000.

Mohan's bank tells him that if he is earning the salary of the job he has been offered, they will grant him a mortgage for a house costing up to £275,000. After a month of deliberation, Mohan accepts the job and decides to move to Wolverhampton. He begins searching for a house to buy. He reasons that he will not be able to purchase a 4-bedroomed house.

Which of the following is NOT an assumption that Mohan has made?

A. A house in Wolverhampton will cost the same as a similar house in Birmingham.
B. A different bank will not offer him a mortgage for a more expensive house on the same salary.
C. The salary for the job could increase, allowing him to purchase a more expensive house.
D. A 4-bedroom house without a garden will not cost less than a 4-bedroom house with a garden.
E. House prices in Birmingham will not have fallen in the time between now and Mohan purchasing a house.

Question 88:

We should teach the Holocaust in schools. It is important that young people see what it was like for Jewish people under Nazi rule. If we expose the harsh realities to impressionable people, then this will help improve tolerance of other races. It will also prevent other such terrible events happening again.

Which is the best conclusion?

A. We should teach about the Holocaust in schools.

B. The Holocaust was a tragedy.

C. The Nazis were evil.

D. We should not let terrible events happen again.

E. Educating people is the best solution to the world's problems.

Question 89:

The popular series 'Game of Thrones' should not be allowed on television because it shows scenes of a disturbing nature, in particular scenes of rape. Children may find themselves watching the programme on TV, and then going on to commit the terrible crime of rape, mimicking what they have watched.

Which of the following best illustrates a flaw in this argument?

A. Children may also watch the show on DVD.

B. Adults may watch the show on television.

C. Watching an action does not necessarily lead to recreating the action yourself.

D. There are lots of non-violent scenes in the show.

Question 90:

The TV series 'House of Cards' teaches us all a valuable lesson: the world is not a place that rewards kind behaviour. The protagonist of the series, Frank Underwood, uses intrigue and guile to achieve his goals, and through clever political tactics he is able to climb in rank. If he were to be kinder to people, he would not be able to be so successful. Success is dependent on his refusal to conform to conventional morality. The TV series should be shown to small children in schools, as it could teach them how to achieve their dreams.

Which of the following is an assumption made in the argument?

A. Children pay attention to school lessons.

B. The TV series is sufficiently entertaining.

C. One cannot both obey a moral code and succeed.

D. Frank Underwood is a likable character.

Question 91:

Freddy makes lewd comments about a female passer-by's body to his friend, Neil, loud enough for the woman in question to hear. Neil is uncomfortable with this, and states that it is inappropriate for Freddy to make these comments, and that Freddy is being sexist. Freddy refutes this, and Neil retorts that Freddy would not make these comments about a man's body. Freddy replies by saying 'it is not sexist, I am a feminist, I believe in equality for men and women.'

Which of the following describes a flaw made in Freddy's logic?
A. A self-proclaimed feminist could still say a sexist thing.
B. The female passer-by in question felt uncomfortable.
C. Neil, too, considers himself a feminist.
D. It would still not be OK to make lewd comments at male passers-by.
E. Lewd comments are always inappropriate.

Question 92:

The release of CO_2 from consumption of fossil fuels is the main reason behind global warming, which is causing significant damage to many natural environments across the globe. One significant source of CO_2 emissions is cars, which release CO_2 as they use up petrol. In order to tackle this problem, many car companies have begun to design cars with engines that do not use as much petrol. However, engines which use less petrol are not as powerful, and less powerful cars are not attractive to the public. If a car company produces cars which are not attractive to the public, they will not be profitable.

Which of the following best illustrates the main conclusion of this argument?
A. Car companies which produce cars that use less petrol will not be profitable.
B. The public prefer more powerful cars.
C. Car companies should prioritise profits over helping the environment.
D. Car companies should seek to produce engines that use less petrol but are still just as powerful.
E. The public are not interested in helping the environment.

Question 93:

Automobiles have grown to become an environmental hazard. Previously a prized possession, there are now over 1.4 billion vehicles worldwide, with the number unlikely to decrease anytime soon. In town centres especially, cars have caused many problems for local councils, the public and the air we breathe. With increasing awareness of the damage to the environment we are causing through fossil fuel use, the currently unrestricted use of cars must now be restrained. How many more lung problems and worrying environmental studies will emerge before significant changes are brought in?

Which of the following best illustrates the main conclusion of this argument?
A. The general health of the world population is at risk because so many people use cars.
B. We need to place more restrictions on parking in town centres.
C. The price of automobiles needs to be increased.
D. It is in everyone's best interests to reduce car usage worldwide.
E. We need to be more aware of the environmental issues caused by cars.

Question 94:

If Schweikart do not raise lawyers' salaries, then the morale of their lawyers will fall, leading to a decline in productivity. This means less work is done, so less money is made, and eventually the whole firm could go bankrupt. Higher salaries could help make the firm bigger and more successful, as some case studies from other companies have shown.

Which of the following best illustrates the main conclusion of this argument?
A. The employers will have to accept some decrease in productivity.
B. Fall in productivity could mean the end of Schweikart.
C. The morale of the lawyers is dangerously low right now.
D. If salaries are not raised, the firm could go bankrupt.
E. If salaries are raised, the firm will be bigger and more successful.

Question 95:

Bushfires in Australia this year emitted 900 million tonnes of CO_2 into the atmosphere. Some scientific studies have suggested that the whole planet must release under 100 million tonnes of CO_2 every year from now on if we are to avoid further global warming. When forest vegetation burns, the amount of CO_2 released into the atmosphere can be taken back up again by the plants as they regrow over many years. Bushfires in Australia burn most fiercely in seasons when the air is drier, as it was this year. Rain arrived in Australia recently, and the number of new fires has dropped significantly.

Which of the following is a conclusion that can be drawn from this passage?
A. The Australian bushfires will probably not release as much carbon dioxide next time they occur, as they were so big and devastating this year.
B. Further global warming and climate change would not occur if bushfires could be prevented or controlled in a better way.
C. As a result of these bushfires, it is likely that some carbon dioxide released into the atmosphere will remain there for some time.
D. The bushfires in Australia will prevent the world targets for carbon emissions being reached.
E. Australia is the world's biggest contributor of carbon emissions.

Question 96:

If we are not the only life in the universe, other life must exist on planets of a similar size and with similar terrain to Earth. The hypothetical planet would need to orbit at the right distance from a star to make the climate tolerable for living organisms. Until now, technology has not been able to discover planets like these. China built a new bubble-scope so technologically advanced that for the first time in history astronomers will be able to see if there are any Earth-sized planets in the habitable zones around stars - the region where the temperature is right for liquid water to exist on the surface. If the bubble-scope finds that these planets exist, we can say that there is life on planets other than Earth.

Which of the following is the best statement of the flaw in the above argument?

A. It assumes that it is necessary to have technologically advanced telescopes to see other planets.
B. It assumes that the presence of liquid water is sufficient for life to exist.
C. It assumes that planets of a similar size to Earth will have life similar to that on Earth.
D. It assumes that the Bubble-scope is good enough to see Earth-sized planets in the habitable zones.
E. It assumes life must exist elsewhere in the universe.

Question 97:

The government was criticised for failing to prepare for heavy snowfall last winter. The economy faced a downturn because workers were simply unable to get to work. Others felt that these occasional economic costs should be accepted. Since the probability of heavy snow in the UK is very low, it could be argued that the massive cost of investing in preventative measures would not be a wise use of a finite government budget. Governments must make a judgement on the probability of potential events occurring, assessing risk, the cost of preventative measures and the cost of the event happening. Sometimes, the cost of preventative measures is too high when the risk is low, so it is not worth the investment. The recent extreme weather is an example of this.

Which of the following best illustrates the main conclusion of this argument?

A. The Government should compensate businesses for the money lost and help to bring the economy back to its previous level.
B. The Government should have spent more on preventative measures for the extreme snowfall.
C. Heavy snowfall in the UK is unusual.
D. The Government was right not to take the preventative measures necessary to get us through extreme snowfall.
E. The Government deserves its criticism for the handling of recent extreme weather.

Question 98:

Countries with a thriving arts sector (including architecture, film, literature and music) tend to be less authoritarian, fairer and more economically powerful. They also tend to have happier and more psychologically balanced citizens. If we want to live in a less authoritarian, fairer and more economically powerful society, we should request that the Government financially support the arts.

Which of the following illustrates a flaw in the above argument?

A. A thriving arts sector may positively influence the mental health of citizens.
B. There may be other more important uses of Government spending.
C. Some of the traditional arts are outdated and have no place in modern society.
D. Some of the arts may be seen as totalitarian.
E. A robust economy may be a prerequisite of a thriving traditional arts sector.

Question 99:

When GCSE results come out again next summer, there will inevitably be controversy and criticism about the standard of education in this country. If the results are slightly above average, people will say that the GCSEs are too easy nowadays – not that people are smarter. Alternatively, if the results are slightly below average, then people will raise the issue of state funding for education – rather than arguing it shows a natural fluctuation in ability. In conclusion, there will be negative stories in the media either way, so we should not pay attention to these news stories.

Which of the following is the best statement of the flaw in the above argument?
A. It does not establish that GCSEs are a varying level of difficulty year upon year.
B. The options are either slightly above average or below – the results could also stay roughly the same.
C. It makes an undeserved attack on journalism.
D. The fact that a negative story is inevitable does not mean that it should be ignored.
E. It makes a future prediction without any hard evidence.

Question 100:

Infant road deaths and serious injuries have decreased by 29.8% in the past decade. Nonetheless, we should not assume that guidance for road safety is no longer essential for students. One study shared its results from 2009: almost 1900 young boys and 900 girls were killed or seriously injured in road traffic accidents as pedestrians. On top of this, 750 young cyclists were killed or seriously injured, more than 450 of whom were boys.

Assuming that 2009 is a representative year for road accidents involving infant pedestrians and cyclists, which one of the following is a conclusion that can be drawn from the passage?
A. Boys are more than twice as likely as girls to be killed or seriously injured as cyclists.
B. Boys are more than twice as likely as girls to be killed or seriously injured as pedestrians.
C. Girls are usually supervised as adults, whereas boys are more reckless and thus are subject to more accidents on the road.
D. There are more boys than girls who ride bicycles and walk on the streets.
E. Lessons in road safety specifically designed for boys would be beneficial in reducing these worrying numbers.

Question 101:

The conduct of the public has been a major contributing factor to the disinclination of doctors to work outside of regular office hours. When doctors only dealt with actual medical emergencies on call, the workload was under control. The state has contributed to making the public think they should now be entitled to medical care 24 hours a day, no matter how insignificant or non-urgent the problem is. Only some of these minor problems should be dealt with at GP surgeries or by NHS 111.

Which of the following best illustrates the main conclusion of this argument?

A. The State should encourage doctors to offer separate urgent and non-urgent surgeries outside of office hours.

B. The public are having difficulty drawing a line between which medical problems require hospital assistance and medical care, and which do not.

C. Doctors should turn away patients who come with non-urgent problems to their surgeries outside of office hours.

D. The problems with providing medical care outside of office hours are partly due to the general public.

E. There needs to be a change in the rules for what defines an urgent medical problem.

Question 102:

Vast increases in the cost of bringing a case to court introduced last week are an outrage on the general public's access to legal representation. It means that justice is treated as a commodity. Fees are set to increase by more than 700% for claims of a quarter of a million pounds or more. This deters small businesses and ordinary people from taking cases to court, for fear of the financial loss if they lose, bankrupting anyone who tries to get justice. All civil settlements are affected. If justice is not accessible for most ordinary people, the government must change things. The civil courts are the cornerstone of a just and fair society.

Which of the following best illustrates the main conclusion of this argument?

A. The civil courts are crucial to fairness and justice in society.

B. The government needs to rethink its policies, or justice will simply be inaccessible to most due to cost.

C. Capitalism is corrupting the justice system.

D. The planned increase in court fees aids people's ability to seek justice.

E. Small businesses and individuals will not use civil courts anymore to resolve matters.

Question 103:

Studies reveal that families from more affluent socio-economic backgrounds tend to have children who score highly in IQ tests. It seems unlikely that there is a direct relationship between money and IQ. These studies did not record whether the link between family income and children's IQ scores was affected by the parents' profession. A much more likely explanation is that intelligence tends to result in higher income, since a certain level of intelligence is needed for higher paid and skilled careers such as medicine, law, banking or engineering. If these studies found that the children of high earners in sport, music and entertainment did not generally have high IQ scores, we could conclude that the intelligence level of children is largely genetically inherited from their parents.

Which of the following is the main assumption underlying this argument?

A. High IQ scores are not found in children of parents in the sports, entertainment and music industries.
B. Children of affluent families are likely to have a much better education than others.
C. To get into professions that are higher paid and more skilled, all that is needed is a high level of intelligence.
D. IQ is a representative test for intelligence.
E. Careers in entertainment, music and sport do not require high levels of intelligence.

Question 104:

One cohort of medical researchers have a unique chance to analyse human brains due to the recent contributions of brain post-mortems. Dissection of these brains and the analysis of synapses may add evidence to the hypothesis of cognitive reserve. This states that when a person gets older, they gradually lose some brain synapses, leading to declining cognitive function. Those who have led an active lifestyle with respect to their brain are thought to have created more synapses, creating a resistance to the natural degradation of synapses as they age. Having a healthy lifestyle when younger should make it feasible to maintain a higher standard of cognitive ability into old age. This sort of lifestyle includes eating a healthy diet, reading and exercising regularly.

Which of the following can be reliably be concluded from this passage??

A. Reading, eating well and keeping active will ensure quality of life in old age.
B. Declining brain performance in old age is not understood very well because of the limited resources scientists have.
C. Scientists should encourage the public to donate their brains for medical research after death.
D. The decline in brain synapses in old age is due to poor lifestyle choices made in youth.
E. On average, those older people who have lived an active lifestyle are expected to have a higher number of synapses in their brain.

Question 105:

Everyone who lives in the UK is entitled to free healthcare. It is paid for by everyone who pays tax. Some people who use a lot of drugs and drink a lot of alcohol use this free healthcare service a lot more than others who lead healthy lives. This is unfair, as healthy people should not have to pay for the healthcare of people who use drugs and drink lots of alcohol. These people who use the healthcare system a lot more than others should therefore pay for their treatments.

Which of the following best illustrates the principle underlying this argument?
A. People who earn more should get their treatment faster than others.
B. People who drive more should pay more road tax.
C. All channels on television should be provided in one bundle, not on separate subscriptions.
D. If people have leftover mobile data on their monthly plan, it should rollover onto the next month.
E. People should have to empty their own bins in designated disposal sites, so those with less waste have less to do.

Question 106:

A journalist says; "we should impose a tariff on imported vegetables so that they cost consumers more than domestic vegetables. Otherwise, growers from other countries who can grow vegetables more cheaply will put domestic vegetable growers out of business. This will result in farmland being converted for more lucrative industrial uses and the consequent vanishing of a unique way of life."

Which of the following principles does the journalist's recommendation most closely conform to?
A. Government intervention sometimes creates more economic efficiency than free markets.
B. A country should put the interests of its own citizens ahead of those of citizens of other countries.
C. Social concerns should sometimes take precedence over economic efficiency.
D. The interests of producers should always take precedence over those of consumers.
E. A country should put its own economic interest over that of other countries.

Question 107:

A columnist said that; "there should be complete freedom of thought and expression. That means there is nothing wrong with exploiting unsavoury popular tastes for the sake of financial gain."

Which of the following judgments conforms most closely to the principle the columnist is expressing?
A. There should be no laws restricting which books are published, but publishing books that pander to people with depraved tastes is not necessarily morally acceptable.
B. The public have the freedom to purchase whatever recordings are produced, but that does not mean that the government may not limit the production of recordings.
C. People who produce depraved movies have the freedom to do so, but they should be discouraged.
D. The government should grant artists the right to create whatever works of art they want to create so long as no one considers those works to be depraved.
E. If we are all free to say what we like, we are free to say things which are offensive to others if it benefits us.

Question 108:

Lions do not tolerate an attack by one lion on another if the latter demonstrates submission by baring its throat. The same is true of tigers and domesticated cats. So, it would be erroneous to deny that animals have rights on the grounds that only human beings are capable of obeying moral rules.

What is the underlying structure of this argument?
A. It provides counterexamples to refute the premise on which this particular conclusion is based.
B. It establishes inductively that all animals possess some form of morality.
C. It casts doubt on the principle that being capable of obeying moral rules is a necessary condition for having rights.
D. It establishes a claim by showing that the denial of that claim entails a logical contradiction.
E. It provides evidence suggesting that the concept of morality is often applied too broadly.

Question 109:

When a nation is on the brink of financial crisis, its government must violate free market principles in order to prevent economic collapse by limiting the extent to which foreign investors and lenders can withdraw their money. After all, the right to free speech does not include the right to shout "Fire!" in a crowded theatre, and the harm done when investors and lenders rush madly to get their money out before anyone else does can be just as real as the harm resulting from a stampede in a theatre.

What is the underlying structure of this argument?
A. It uses an analogy to show that a set of principles should not be adhered to in every circumstance.
B. It makes a claim by arguing that the truth of that claim best explains the observed facts.
C. It presents numerous experimental results as evidence for a general principle.
D. It attempts to demonstrate that the explanation of a phenomenon is flawed by showing that it fails to explain a particular instance of that phenomenon.
E. It applies an empirical generalisation to reach a conclusion about a particular case.

Question 110:

The most advanced kind of moral motivation is based solely on abstract principles. This form of motivation is in contrast to calculated self-interest or the desire to adhere to societal norms and conventions.

The actions of which of the following individuals exhibit the most advanced kind of moral motivation, as described above?

A. Jay gave money to a local charity after walking past a charity collection box at work because he worried that not doing so would make him look stingy.

B. Will gave money to a local charity after walking past a charity collection box at work as he believed that doing so would improve his employer's opinion of him.

C. Harvey's employers engaged in an illegal but profitable practice that caused serious environmental damage. He did not report this to the authorities out of fear his employers would retaliate against him.

D. Rachel's employers engaged in an illegal but profitable practice that caused serious environmental damage. She reported it to the authorities out of a belief that protecting the environment is always more important than money.

E. Lee's employers engaged in an illegal but profitable practice that caused serious environmental damage. He reported it to the authorities because several colleagues pressured him to do so.

Question 111:

José claims that the Northeast Faithville Neighbourhood Federation opposes the new electricity system and uses this as evidence of citywide opposition. The Federation passed a resolution opposing it, but less than 10% of members voted, and 40% of those who voted, voted in favour of the system. The opposing votes represent less than 1% of the population of Faithville. One should not assume that so few votes represent the majority view.

Which of the following most accurately describes the author's form of argument?

A. Attempting to cast doubt on a conclusion by claiming the statistical sample on which the conclusion is based is too small to be dependable.

B. Criticizing a view on grounds that the view is based on evidence that is impossible to dispute.

C. Questioning a conclusion based on vote results, on the grounds that people with certain views are more likely to vote.

D. It is in everyone's best interests to reduce car usage worldwide.

E. Attempting to refute an argument by showing that, contrary to what has been claimed, the truth of the premise does not guarantee the truth of the conclusion.

Question 112:

All sharks in the Atlantic Ocean have flat parts on their bodies, called fins. All dolphins in the Pacific Ocean also have fins. Therefore, they are similar.

Which of the following most closely parallels the reasoning of this argument?
A. Sharks and dolphins are similar; however, they have many differences.
B. All sharks have teeth, dolphins are similar to sharks, therefore they must have teeth.
C. Bats and eagles must be similar because every bat has wings, and so does every eagle.
D. All dogs have eyes, and all cats have eyes.
E. Some rats have fur, and all mammals have fur; therefore, rats are mammals.

Question 113:

If I have promised to keep a secret and someone asks me a question, I cannot answer truthfully without breaking this promise. I cannot keep and break the same promise. So, one cannot be obliged to answer all questions truthfully and to keep all promises.

Which of the following arguments is the most similar in its reasoning to the argument above?
A. If business hours are extended, we will have to hire new employees or have current staff work overtime. Both options would increase labour costs. We cannot afford to do this, so we will have to keep business hours as they are.
B. Some politicians gain votes by making extravagant promises, but this deceives people. Since the only way for some politicians to be popular is by deception, and all politicians need to be popular, it follows that some must deceive.
C. If we put an effort into making a report look good, the client might think we did so because we thought the proposal would not stand on its own merits. But if we do not try to make it look good, they might think we are not serious about our business. So, whatever we do, we risk criticism from the client.
D. It is claimed that we have the right to say whatever we want. We also have an obligation to be civil to others. But civility requires that we are not always able to say what we want. Therefore, it cannot be true that we have the right to say whatever we want while still being civil.
E. If creditors have legitimate claims against a business, and they have resources to pay those debts, then they are obliged to pay them. Also, if this is the case, then a court will enforce it. But the courts did not force this business to pay its debts, so either the creditors did not have sufficient legitimate claims, or the business did not have sufficient resources.

Question 114:

Phone companies frequently request consumer information about human factors, such as whether the phone is comfortable to hold or whether a set of features are easy to use. However, designer interaction with consumers is superior to survey data; the data may tell the designer why a feature on last year's model had a lower rating, but it will not explain how that feature needs to be changed to increase the rating.

Which of the following arguments is the most similar in its reasoning to the argument above?
A. Designers aim to create features that will appeal to specific market niches.
B. A phone will have unappealing features if consumers are not consulted during the design stage.
C. Consumer input affects external rather than internal design components of phones.
D. Getting consumer input for design changes can help contribute to successful product design.
E. Phone companies always conduct extensive post-market surveys.

Question 115:

Someone living in a cold climate buys a winter coat that is stylish but not warm in order to appear sophisticated. People are sometimes willing to sacrifice practicality and comfort for the sake of appearances.

Which of the following situations and explanations are most similar to the above passage?
A. A performer convinces his entertainment company to purchase an expensive outfit so he can impress the audience more.
B. A woman sets her thermostat at a low temperature in the winter because she is concerned about environmental damage caused by using fossil fuels to heat her home.
C. Someone buys a particular wine even though their favourite wine tastes better and is easier to find because they think the other wine will impress their dinner guests more.
D. A parent buys a car seat for their child because it is more colourful and comfortable for the child than other seats on the market, though no safer.
E. A man buys a car to commute to work even though public transport is already quick and reliable.

Question 116:

Chris owns a car dealership which has donated cars to driver education programmes for over 5 years. He finds the statistics on car accidents disturbing and wants to do something to encourage better driving in young drivers. Some people show support by buying cars from Chris' dealership.

Which of the following is best illustrated by the passage?
A. The only way to reduce traffic accidents is through driver education programmes.
B. Altruistic actions sometimes have additional positive consequences for those who perform them.
C. Young drivers are the group most likely to benefit from driver education programs.
D. It is usually in one's best interest to perform actions that benefit others.
E. An action must have broad community support if it is to be successful.

Question 117:

In academia, sources are always cited when used in articles or presentations. In open-source software, the code in which the program is written can be viewed and modified by individual users for their purposes without getting permission. In contrast, the use of proprietary software is kept secret, and modifications can only be made by the producer, for a fee. This shows that open-source software better matches the values embodied in academic scholarship and since scholarship is central to the mission of universities, universities should use only open-source software.

Which of the following most closely conforms to the reasoning above?

A. Whatever software tools are the most advanced and can achieve the goals of academic scholarship are the ones that should be used in universities.

B. Universities should use the type of software technology that is cheapest, as long as it is adequate for their purposes.

C. Universities should choose the type of software technology that best matches the values embodied in the activities central to the mission of universities.

D. The form of software technology that best matches the values embodied in the activities central to the mission of universities is the form of software technology most efficient for universities to use.

E. A university should not pursue any activity that would block the achievement of academic goals at that university.

Question 118:

Every business strives to increase its productivity, as this increases profits for the owners and the probability that the business will survive. Not all efforts to increase productivity are beneficial to the business as a whole. A lot of the time, attempts to increase productivity decrease the number of employees, which clearly harms the dismissed employees as well as the sense of security of the employees kept at the company.

Which of the following best illustrates the main conclusion of this argument?

A. Reducing the number of employees in a business undermines the sense of security of the retained employees.

B. Every business makes efforts to increase productivity.

C. Interests align only if the employees of a business are also its owners, which enables measures that are beneficial to the overall business.

D. Some measures taken by a business to increase productivity fail to be beneficial to the overall business.

E. If an action taken to secure the survival of a business fails to enhance employee welfare, that action cannot be good for the overall business.

Question 119:

A recent report suggests that Saino's pre-packaged meals are lacking nutritional value. But this report was commissioned by Tesoc, Saino's largest corporate rival. Some primary drafts of the report were submitted for approval to Tesoc's public relations department. Due to the obvious bias of this report, it is clear that Saino's pre-packaged meals really are nutritious.

Which of the following best outlines the flaw in this argument?

A. It treats evidence of an apparent bias as actual evidence that the report's claims are false.
B. It draws a conclusion based solely on an unrepresentative sample of Saino's products.
C. It fails to take into account the possibility that Saino's has just as much motivation to create negative publicity for Tesoc as Tesoc has to create negative publicity for Saino's.
D. It fails to provide evidence that Tesoc's pre-packaged meals are not more nutritious than Saino's meals are.
E. It assumes that Tesoc's public relations department would not approve a draft of a report that was hostile to Tesoc's products.

Question 120:

No one with a criminal conviction can be appointed to the board. An undergraduate degree is necessary for appointment to the executive board. Therefore, Jim – who has a master's degree as well as a bachelor's degree – cannot be accepted as the new executive administrator, as he has a criminal conviction.

Which of the following is an assumption of the above argument?

A. Anyone with a bachelor's degree without a criminal conviction is eligible for appointment to the executive board.
B. Only candidates eligible for appointment to the executive board can be accepted for the position of executive administrator.
C. A bachelor's degree is not necessary for acceptance for the role of executive administrator.
D. If Jim did not have a criminal conviction, he would be accepted for the position of executive administrator.
E. The criminal charge on which Jim was convicted is relevant to the duties of the role of executive administrator.

Question 121:

Whenever possible, all scientific experiments should be performed using a double-blind protocol. This helps prevent common misinterpretations of trial results due to expectations and opinions that scientists already hold. Clearly scientists should be extremely diligent in trying to avoid such misinterpretations.

Which of the following best expresses the main conclusion of this argument?
A. Double blind experiments are an effective way of ensuring scientific objectivity.
B. Whenever they can, scientists should refrain from interpreting evidence based on previously formed expectations.
C. The objectivity of scientists may be impeded when interpreting experimental evidence on the basis of expectations and opinions they already have.
D. It is advisable for scientists to use double blind techniques in as many experiments as they can.
E. Scientists sometimes fail to adequately consider the risk of misinterpreting evidence on the basis of prior expectations and opinions.

Question 122:

Aluminium soft drink cans all contain the same amount of aluminium. At dispoal and manufacture, the cans are divided into several groups. Fifty percent of the aluminium in Group B was recycled from cans in Group A, a group of used aluminium soft drink cans. Since all the cans from Group A were recycled into cans in Group B, and since the amount of material other than aluminium in the cans is negligible, it follows that Group B has twice as many cans as Group A.

Which of the following best illustrates the main assumption of this argument?
A. The aluminium of cans in Group B cannot be recycled further.
B. Recycled aluminium is of poorer quality than unrecycled aluminium.
C. All of the aluminium of an aluminium can is recovered when the can is recycled.
D. Aluminium soft drink cans are more easily recycled than other cans made of different materials.
E. None of the soft drinks from Group A had been made from recycled aluminium.

Question 123:

A policy has been developed to avoid many serious cases of influenza. This goal will be met by the annual vaccination of high-risk individuals. This means everyone over 65 years old and/or with a chronic disease will be offered a vaccine. The vaccination produced each year only prevents the strain deemed most prevalent that year. Every year a vaccine will be necessary for all high-risk individuals to prevent serious influenza.

Which of the following best expresses the main assumption of this argument?
A. The number of individuals in the high-risk group will not significantly change every year.
B. The likelihood that a serious influenza epidemic will occur varies every year.
C. No vaccine against the influenza virus protects against more than one strain of the virus.
D. Every year, the strain deemed most prevalent will be one that had not previously been deemed most prevalent.
E. Every year, the vaccine will have fewer side effects than the previous year's vaccine, since technology will improve year by year.

Question 124:

At a recent conference on non-profit management, several computer experts maintained that the most significant threat faced by large institutions such as universities and hospitals is unauthorised access to confidential data. In light of this testimony, we should make the protection of our clients' confidentiality our highest priority.

Which of the following best is a flaw in this argument?
A. It confuses the causes of a problem with the appropriate solutions to it.
B. It relies on the testimony of experts whose expertise is not shown to be sufficient to support their general claim.
C. It assumes a correlation between two phenomena is evidence that one is the cause of the other.
D. It draws a general conclusion about a group, based on data about an unrepresentative sample.
E. It infers that a property belonging to large institutions belongs to all institutions.

Question 125:

My friends say I will have a road accident one day because I drive my sports car recklessly. But I have done some research, and it says minivans and larger sedans have very low accident rates compared to sports cars. So, trading my sports car in for a minivan would reduce my risk of having an accident.

Which of the following best expresses the main flaw of the passage?
A. It infers a cause from a correlation.
B. It relies on a sample that is too narrow.
C. It misinterprets evidence that a result is likely as evidence that the result is certain.
D. It mistakes a condition sufficient for bringing about a result for a condition necessary for doing so.
E. It relies on an unreliable source.

Question 126:

An action is morally right if it would be reasonably expected to increase the overall wellbeing of the people affected by it. An action is morally wrong if and only if it can be reasonably expected to reduce the wellbeing of the people affected by it. Therefore, actions that would be reasonably expected to leave the overall wellbeing of the people affected by them unchanged are also morally right.

Which of the following best the ain assumption of the passage above?

A. Only morally wrong actions can be reasonably expected to reduce the overall wellbeing of the people affected by them.
B. No action is both right and wrong.
C. Any action that is not morally wrong is morally right.
D. There are actions reasonably expected to leave the overall wellbeing of the people affected by them unchanged.
E. Only morally right actions have good consequences.

Question 127:

Many companies have started to decorate their halls with motivational posters in the hope of boosting the motivation and producitivity of their employees. However, almost all of these employees are already motivated to work productively. Therefore, although these companies use motivational posters it is unlikely to achieve their intended purpose.

Which of the following best expresses the main flaw of the passage?

A. It fails to consider whether companies that do not currently use motivational posters would increase their employees' motivation to work productively if they began to use the posters.
B. It takes for granted that companies that decorate their halls with motivational posters are representative of companies in general.
C. It fails to consider whether even if motivational posters do not have one particular benefit for companies, they may have similar effects that are equally beneficial.
D. It does not adequately address the possibility that employee productivity is strongly affected by factors other than their motivation.
E. It fails to consider that even if employees are already motivated to work productively, motivational posters may increase that motivation.

Question 128:

An entomologist observed ants carrying particles to neighbouring ant colonies and inferred that the ants were bringing food to their neighbours. However, further research revealed that the ants were emptying their own colony's dump site. Therefore, the entomologist was wrong.

A. Which of the following best illustrates the main assumption of this argument?

B. Ant colonies do not interact in the same way human societies do.

C. There is only weak evidence for the view that ants have the capacity to make use of objects as gifts.

D. Ant dumping sites do not contain particles that could be used as food.

E. The ants to whom the particles were brought never carried the particles into their own colonies.

Question 129:

Febrooze leaves clothes fluffy and soft to touch – combine this with its fresh odour and it is a delight. We conducted a test with over a hundred consumers to prove it is indeed the best clothes product out there. Each person was given one towel washed with Febrooze and one towel washed without it. 99% of these consumers preferred the Febroozed towel. Therefore, Febrooze is the most effective fabric softener available.

The reasoning in the passage is most vulnerable to criticism because it fails to consider which of the following points?

A. If any of the consumers tested are allergic to fabric softeners.

B. If Febrooze is more or less harmful to the environment than other fabric softeners.

C. If Febrooze is cheaper or more expensive than other fabric softeners.

D. If the consumers tested find the benefits of using fabric softeners worth the expense.

E. If the consumers tested had the chance to evaluate fabric softeners other than Febrooze.

Question 130:

The government pays for individual medical needs through a healthcare system, requiring all citizens to pay for this service through taxes. Since it is individuals who need medical attention that primarily benefit from this, the government should ensure individuals who have the medical attention bear the cost and should impose an end-user fee on medical care.

Which one of the following principles would do most to justify the conclusion of the argument?

A. The government should avoid any actions that might alter the behaviour of taxpayers.

B. Any rational healthcare system must base the amount of the cost on the usage involved.

C. The people who stand to benefit from a service should always be made to bear the cost of a service.

D. Government based medical care for individuals should be provided only when it does not reduce incentives for individuals to seek medical care.

E. The choice of not accepting an offered service should be available, even if there is no charge.

Question 131:

Most people can tell whether a sequence of words in their own dialect is grammatically correct or not, yet few people who can do so are able to specify the relevant grammatical rules.

Which of the following best illustrates the principle underlying the argument above?

A. Some people are able to write coherent and accurate narrative descriptions of events, but these people are not necessarily capable of composing emotionally moving and satisfying poems.

B. Engineers who apply the principles of physics to design buildings and bridges must know a great deal more than the physicists who discovered these principles.

C. Some people are able to tell whether any given piece of music is a waltz, but the majority of these people cannot state the defining characteristics of a waltz.

D. Those travellers who most enjoy their journeys are not always those most capable of vividly describing details of those journeys to others.

E. Quite a few people know the rules of chess, but only a small number of them can play chess well.

Question 132:

A robber comes to your house and steals your computer. You lose all your pictures and music. Later, the robber turns himself in to police. He apologises, says the robbery was a moment of weakness, and repents. The robber is prepared to accept whatever punishment the law thinks appropriate. This is the robber's third crime, and he will go to jail for life if found guilty. Unfortunately, the computer was broken in the robbery, and you can't get any data back. The police come to you and ask if you want to press charges. You say no.

Which of the following sufficiently illustrates the principle underlying the argument above?

A. Anyone who repents of a crime deserves mercy and should not be punished.

B. All convicted criminals must be sent to jail.

C. You should not have mercy for a criminal who does not repent.

D. You should only punish a criminal if they cause irreparable damage.

E. You should only have mercy for a criminal if they repent.

Question 133:

Hospitals, universities, labour unions, and other institutions may well have public purposes and be quite successful at achieving them even though each of their individual staff members does what he or she does only for selfish reasons.

Which of the following best illustrates the principle underlying the argument above?

A. What is true of some social organisations is not necessarily true of all such organisations.

B. An organisation can have a property that not all of its members possess.

C. People often claim altruistic motives for actions that are in fact selfish.

D. Many social institutions have social consequences unintended by those who founded them.

E. Often an instrument created for one purpose will be found to serve another purpose just as effectively.

Question 134:

When drivers are sleep deprived, there are definite behavioural changes, such as slower response times to stimuli and a reduced ability to concentrate, but self-awareness of these changes is poor. Most drivers think they can tell when they are about to fall asleep, but they cannot.

Select the answer option that does not illustrate the same principle as the passage above:

A. People who have been drinking alcohol are not good judges of whether they are too drunk to drive.

B. Primary school students who dislike arithmetic are not good judges of whether multiplication tables should be included in the school's curriculum.

C. Industrial workers who have just been exposed to noxious fumes are not good judges of whether they should keep working.

D. People who have just donated blood and have become faint are not good judges of whether they are ready to walk out of the facility.

E. People who are being treated for schizophrenia are not good judges of whether they should continue their medical treatment.

Question 135:

One of our local television stations has been criticised for its recent coverage of the personal problems of a local politician's nephew, but the coverage was in fact good journalism. The information was accurate. Furthermore, the newscast had significantly more viewers than it normally does, because many people are curious about the politician's nephew's problems.

Which of the following principles, if valid, would most help to justify the reasoning above?

A. Journalism deserves to be criticised if it does not provide information that people want.

B. Any journalism that intentionally misrepresents the facts of a case deserves to be criticised.

C. Any journalism that provides accurate information on a subject about which there is considerable interest is good journalism.

D. Good journalism will always provide people with information that they desire or need.

E. Journalism that neither satisfies the public's curiosity nor provides accurate information can never be considered good journalism.

Question 136:

Because people are generally better at detecting mistakes in others' work than in their own, a prudent principle is that one should always have one's own work checked by someone else.

Which of the following provides the best illustration of the principle above?

A. The best primary school maths teachers are not those for whom maths was always easy. Teachers who had to struggle through maths themselves are better able to explain maths to students.

B. One must make a special effort to clearly explain one's views to someone else; people normally find it easier to understand their own views than to understand others' views.

C. Juries composed of legal novices, rather than panels of lawyers, should be the final arbiters in legal proceedings. People who are not legal experts are in a better position to detect good legal arguments by lawyers than are other lawyers.

D. People should always have their writing proofread by someone else. Someone who does not know in advance what is going to be said is in a better position to spot typographical errors.

E. Two people going out for dinner will have a more enjoyable meal if they order for each other. By allowing someone else to choose, one opens oneself up to new and exciting dining experiences.

Question 137:

All beagles bark a lot when they are not supposed to. All pugs do not bark often. Each of Amy's dogs is a cross between a beagle and a pug, so these dogs are moderate barkers that bark some of the time.

Which of the following parallels the reasoning in the argument above?

A. All of Faith's dresses are very well made. All of Hannah's dresses are very badly made. Half of the dresses in this closet are very well made and half are very badly made. So, half of these dresses must be Faith's and half are Hannah's.

B. All students at Wonn School live in Blue County. All students at Perrie School live in Bong County. Members of the Edwards family attend both Wonn and Perrie School. Therefore, some members of the Edwards family live in Blue County and some live in Bong County.

C. All stenographers know shorthand. All engineers know calculus. Tom has worked as both a stenographer and an engineer, so he knows both shorthand and calculus.

D. All mercury is extremely toxic to humans. All glycols are nontoxic to humans. This household cleaner is a mixture of mercury and glycols, so is moderately toxic.

E. All students who study a lot get good grades. But some students who do not study a lot also get good grades. Michael studies some of the time, so has pretty good grades.

Question 138:

We should accept the proposal to demolish the old railway station because the local historical society, who intensely oppose this, is dominated by people who have no commitment to long-term economic wellbeing. Preserving old buildings blocks the progress of new development, which is crucial for economic health and wellbeing.

Which of the following is the flawed reasoning in the passage above most similar to?

A. We should try to protect works of art that are of national cultural significance. They might not be recognised as such by all taxpayers or critics, nevertheless, we should expend whatever money is needed to procure all such works as they become available.

B. Documents of importance to local heritage should be preserved for future generations. If even one of these documents is lost or damaged, the integrity of the overall historical record will be damaged.

C. You should have your hair cut only once a month. Beauticians often suggest that their customers have haircuts twice a month in order to generate more business for themselves.

D. The committee should endorse the plan to postpone construction of the new motorway. Many local residents who would be affected are opposed to it, and the committee has an obligation to avoid alienating those residents.

E. One should not borrow even small sums of money unless strictly necessary. Once one borrows small sums, the interest starts to accumulate. The longer one takes to repay, the more one owes, and eventually a small debt has become a large one.

Question 139:

It has been scientifically established that all dogs do indeed bark. As a result, any animal that barks is a dog. So, if a person hears an animal bark, that person can safely conclude that the animal is a dog.

Which of the following arguments most closely parallels the flawed reasoning above?

A. Only high interest debt is debt that should be avoided. Debt that is not high interest should not be avoided.

B. All high interest debt should be avoided. Debt that isn't high interest need not be avoided. So, people should prefer low interest debt.

C. All debt that should be avoided is high interest debt because all high interest debt should be avoided. Debt that should be avoided must be high interest debt.

D. High interest debt should sometimes be avoided. As a result, some debt that should be avoided is high interest debt. So, a person can safely conclude that high interest debt should be avoided.

E. If all high interest debt should be avoided, and if some debt is high interest, then some debt should be avoided.

Question 140:

Everyone who thinks the Raiders should win the title thought that Jackie would win the award for Most Valuable Payer, but Jackie did not get this award. Therefore, anyone who believes the Raiders will win the title is wrong.

Which of the following arguments contains similarly flawed reasoning?

A. Anyone who thinks exercising after eating is a good idea has never taken a health class. But Paulo has never taken a health class and knows he should not eat before exercising. Therefore, taking a health class is not necessary for you to know eating before exercise is a bad idea.

B. Anyone who believes seagull migration is based on advanced spatial recognition patterns believes that most bird species have highly developed frontal cortices. But it has been proven that most bird species do not have highly developed frontal cortices. Therefore, the belief that seagulls migrate based on advanced spatial recognition patterns is false.

C. Anyone who thinks animals deserve better treatment believes animals are capable of moral judgment. You do not believe that animals deserve better treatment, so you do not think they are capable of moral judgment.

D. Anyone who thinks chickens are ugly thinks ducks are also ugly. Since there is no reason to think ducks are ugly, there is no reason to think chickens are ugly.

E. If you believe in the tooth fairy, then you do not believe in blood sucking vampires. Since Calum believes in blood sucking vampires, he cannot believe in the tooth fairy.

Question 141:

Winning requires the willingness to cooperate, which in turn requires motivation. So, you will not win if you are not motivated.

Which of the following is most similar to the argument above?

A. Being healthy requires exercise, but exercise involves risk of injury. So, paradoxically, anyone who wants to be healthy will not exercise.

B. Learning requires making some mistakes, and you must learn if you are to improve. So, you will not make mistakes without there being a noticeable improvement in your skills.

C. Our political party will retain its status only if it raises more money, but raising more money requires increased campaigning. So, our party will not retain its status unless it increases campaigning.

D. Getting a ticket requires waiting in line. Waiting in line requires patience. So, if you do not wait in line, you lack patience.

E. You can repair your own bicycle only if you are enthusiastic. And if you are enthusiastic, you will also have mechanical aptitude. So, if you are not able to repair your own bicycle you lack mechanical aptitude.

Question 142:

The national Sooper Plate was held last weekend. In order to win the tournament, a contestant must answer three questions correctly and consecutively. Angela answered three questions correctly and consecutively, so she must have won the Sooper Plate.

Which of the following arguments contains similar reasoning to the passage above?

A. To win the county swim meet, a swimmer needs to win three heats. Dave won the swim meet, so he must have won three heats.

B. Good doctors spend time getting to know their patients as people, not just their medical history. Doctor Smith is a bad doctor, so he must not know his patients on a personal level.

C. Daniel likes to win tournaments. He enters a new tournament every week, even if he is completely unskilled in the tasks involved. He has even won a few trophies.

D. People who win tournaments are confident. People with confidence are usually successful at work. It must follow then, that people who win tournaments are successful at work.

E. When a television station is owned by a large conglomerate, it often has to edit its news to be favourable to its parent company and all of its related products. The local independent station, channel 7, never runs a bad story about SOD Industries. It must be owned by SOD Industries, and so is incapable of fair reporting.

Question 143:

If Max Landsy is healthy, it is highly unlikely the Tokes will win their match-up with the Patties. But in fact, the Tokes did win the match-up, so it's highly unlikely that Max Landsy was healthy.

The pattern of reasoning in the passage is most similar to that in which of the following arguments?

A. If the football cup was not fixed, the winning team would have been highly unlikely to win the whole thing. Thus, since this team was highly unlikely to win, the tournament was probably fixed.

B. If the Prime Minister was a speaker at the event, it's highly unlikely that the Sun would cover the event. It's very unlikely that the Prime Minister did speak, since as it turns out, the Sun did cover the event.

C. If the dice were not loaded, Oliver, who planned to win the roulette 5 times in a row, would have been highly unlikely to win the roulette 5 times in a row. Since he was unlikely to win, the dice were probably loaded.

D. If the star player is raring to go, it is highly likely her team will win their crucial match to win the league title. They did win. It was reported that it was an outstanding team effort to win.

E. If the first-choice keeper has an illness, it is highly unlikely that The United would win their match against The City. The City did in fact win this match, so it follows that the first choice keeper was not playing.

Question 144:

Although birds have long been considered much less intelligent than humans and apes, new research has shown that some species of birds have similar thinking skills to apes. Crows can create and use tools and are socially sophisticated when finding and protecting food. How is a bird with a walnut-sized brain capable of this higher level of cognition? The answer is that both crows and apes have much bigger brains than you would expect from the size of their bodies. The same pattern is found in other intelligent animals, including humans, parrots and chimps.

Which of the following can be drawn as a conclusion from this passage?
A. Apes are not as similar to humans as had been thought.
B. Crows are more intelligent than other species of birds.
C. Animals that cannot create tools are not intelligent.
D. Relative brain size is a better indicator of intelligence than absolute brain size.
E. It could be argued that birds are as intelligent as apes.

Question 145:

There is a higher than average risk of death or injury to young drivers and their passengers. In 2007, 32 per cent of car driver deaths and 40 per cent of car passenger deaths occurred in people aged between 17 and 24. Young male drivers were much more likely to be killed or seriously injured than young female drivers. In order to reduce the number of road accidents and the numbers of people killed or injured, young people should not be allowed to drive until they reach the age of 24

Which of the following is an assumption upon which this argument depends?
A. Young people would not accept the raising of the legal driving age.
B. Most of the accidents involving young people were the fault of the young drivers.
C. The driving test does not effectively test the skill of drivers.
D. The majority of drivers aged between 17 and 24 drive dangerously.
E. Amongst drivers aged between 17 and 24, there are more male drivers than female drivers.

Question 146:

Some disabled people find it difficult to gain access to some of our older public buildings because the entrances have steps. The problem is most often solved by installing ramps. All public buildings should be accessible to everyone therefore they must all install ramps.

Which of the following identifies the flaw in this argument?
A. Disabled people must have access to all buildings not just public ones so all buildings should have ramps.
B. Installing ramps in all public buildings would be extremely expensive.
C. It is unreasonable to suggest that disabled people should be able to access all public buildings.
D. Some older public buildings without ramps may already be accessible to disabled people.
E. Inaccessible public buildings should be replaced by buildings accessible to all.

Question 147:

If more workers worked for only four days each week there would be fewer commuters, and therefore less traffic congestion and less pollution. Fewer people would be unemployed because there would be more work to go around. There is evidence that part-time workers are absent from work less often than full-time workers, so a person working a four-day week is likely to be more productive. Less work means less pressure, which means less stress and happier people.

Which of the following can be drawn as a conclusion from this passage?
A. People choosing to work a four-day week would have to take a 20% pay cut.
B. There would be less pressure on the health services if most workers were on a four-day week.
C. The economy would be more competitive if people worked more productively.
D. The government should enforce a four-day working week.
E. There would be many benefits to working a four-day week.

Question 148:

Some types of migratory birds that are unable to fly long distances without resting have to use the shortest distance over water in their flights to and from Africa, and so they cross at the Strait of Gibraltar. It is essential for these birds, some of which are very rare, that the route remains open. For that reason, it is important that plans to build electricity-generating wind farms on the hills surrounding the Strait of Gibraltar do not go ahead.

Which of the following is an assumption upon which this argument depends?
A. The birds that migrate across the Strait of Gibraltar are close to extinction.
B. Electricity-generating wind farms have to be built on hills.
C. The planned wind farms will make it dangerous for migratory birds to use their usual route.
D. Other species of bird can fly further and can thus use other routes in their migration.
E. There are no plans to build wind farms at other places along the coast.

Question 149:

When mobile phones were first introduced there were concerns about the microwave radiation they produced, and the effects that these could have on the brain, given that phones are held close to the ear while being used. These concerns have ultimately been shown to be unfounded, as mobile phones are more frequently used for sending text messages than for making phone calls. Sending a text message does not require the phone to be anywhere near to the brain so it cannot cause any problems.

Which of the following identifies the flaw in this argument?
A. It ignores research showing that microwaves from the phones cannot penetrate far enough to reach the brain.
B. It ignores evidence suggesting that text messaging is only popular in certain age groups.
C. It does not consider uses of mobile phones other than making phone calls and sending text messages.
D. It does not consider other technology such as wireless internet which could cause similar problems.
E. It ignores the possible effects of the phone calls that are still made.

Question 150:

A comparison is sometimes made between fast-food restaurants and factories. This is because fast-food is a mass-produced, heavily processed product, and restaurant workers' jobs are as routine and boring as those in manufacturing. Not only does fast-food taste the same everywhere, but all workers involved are on low wages and have little power to improve their conditions.

Which of the following best expresses the main conclusion of this argument?

A. Workers who do routine and boring jobs are often poorly paid.
B. Mass production in factories leads to poor working conditions.
C. It is not unrealistic to compare fast-food restaurants with factories.
D. All fast-food tastes the same because it is heavily processed.
E. Working in a fast-food restaurant is no different from working in a factory.

SECTION 1: PROBLEM SOLVING

Section I problem solving questions are arguably the hardest to prepare for. However, there are some useful techniques you can employ to solve some types of questions much more quickly:

Construct Equations
Some of the problems in Section I are quite complex and you'll need to be comfortable with turning prose into equations and then manipulating them. For example, when you read "Mark is twice as old as Jon" – this should immediately register as M = 2J. Once you get comfortable forming equations from the text, you can start to approach some of the harder questions in this book (and past papers) which may require you to form and solve simultaneous equations. Consider the following example:
Nick has a sleigh that contains toy horses and clowns. He counts 44 heads and 132 legs in his sleigh. Given that horses have one head and four legs, and clowns have one head and two legs, calculate the difference between the number of horses and clowns.

A. 0
B. 5
C. 22
D. 28
E. 132
F. More information is needed.

To start with, let C= Clowns and H= Horses.
For Heads: $C + H = 44$; For Legs: $2C + 4H = 132$
This now sets up your two equations that you can solve simultaneously.
$C = 44 - H$ so $2(44 - H) + 4H = 132$
Thus, $88 - 2H + 4H = 132$;
Therefore, $2H = 44$; $H = 22$
Substitute back in to give $C = 44 - H = 44 - 22 = 22$
Thus, the difference between horses and clowns = $C - H = 22 - 22 = 0$

It's important you are able to do these types of questions quickly (and **without resorting to trial & error**) as they are commonplace in Section I of the BMAT.
Diagrams
When a question asks about timetables, orders or sequences, draw out diagrams. By doing this, you can organise your thoughts and help make sense of the question.
"Mordor is West of Gondor but East of Rivendale. Lorien is midway between Gondor and Mordor. Erebus is West of Mordor. Eden is not East of Gondor."

*Which of the following **cannot** be concluded?*
A. Lorien is East of Erebus and Mordor.
B. Mordor is West of Gondor and East of Erebus.
C. Rivendale is west of Lorien and Gondor.
D. Gondor is East of Mordor and East of Lorien
E. Erebus is West of Mordor and West of Rivendale.

Whilst it is possible to solve this in your head, it becomes much more manageable if you draw a quick diagram and plot the positions of each town:

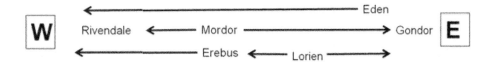

Now, it's a simple case of going through each option and seeing if it is correct according to the diagram. You can now easily see that Option E- Erebus cannot be west of Rivendale.

Don't feel that you have to restrict yourself to linear diagrams like this either – for some questions you may need to draw tables or even Venn diagrams. Consider this example:

Slifers and Osiris are not legendary. Krakens and Minotaurs are legendary. Minotaurs and Lords are both divine. Humans are neither legendary nor divine.

A. Krakens may be only legendary or legendary and divine.
B. Humans are not divine.
C. Slifers are only divine.
D. Osiris may be divine.
E. Humans and Slifers are the same in terms of both qualities.

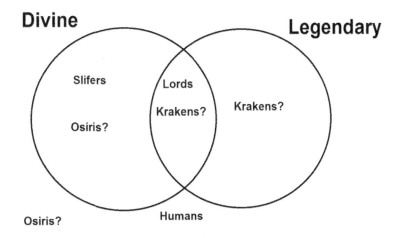

Constructing a Venn diagram allows us to quickly see that the position of Osiris and Krakens aren't certain. Thus, A and D must be true. Humans are neither so B is true. Krakens may be divine so A is true. E cannot be concluded as Slifers are divine but are humans are not. Thus, E is False.

Spatial Reasoning

There are usually 1-2 spatial reasoning questions every year. They usually give nets for a shape or a patterned cuboid and ask which of the answer options are possible from rotations of the original shape. Unfortunately, they are extremely difficult to prepare for because the skills necessary to solve these types of questions can take a very long time to improve. The best thing you can do to prepare is to familiarise yourself with the basics of how cube nets work and what the effect of transformations are e.g. what happens if a shape is reflected in a mirror etc.

It is also a good idea to try to learn to draw basic shapes like cubes from multiple angles if you can't do so already. Finally, remember that if the shape is straightforward like a cube, it might be easier for you to draw a net, cut it out and fold it yourself to see which of the options are possible. Another option is to practice rotating small cube or cuboidal objects such as dice or erasers, to understand the 3D relationships between the 6 faces better.

PROBLEM SOLVING QUESTIONS

Question 151:
Pilbury is south of Westside, which is south of Harrington. Twotown is north of Pilbury and Crewville but not further north than Westside. Crewville is:
A. South of Westside, Pilbury and Harrington but not necessarily Twotown.
B. North of Pilbury, and Westside.
C. South of Westside and Twotown, but north of Pilbury.
D. South of Westside, Harrington and Twotown but not necessarily Pilbury.
E. South of Harrington, Westside, Twotown and Pilbury.

Question 152:
The hospital coordinator is making the rota for the ward for next week; two of Drs Evans, James and Luca must be working on weekdays, none of them on Sundays and all of them on Saturdays. Dr Evans works 4 days a week including Mondays and Fridays. Dr Luca cannot work Monday or Thursday. Only Dr James can work 4 days consecutively, but he cannot do 5.

What days does Dr James work?
A. Saturday, Sunday and Monday.
B. Monday, Tuesday, Wednesday, Thursday and Saturday.
C. Monday, Thursday Friday and Saturday.
D. Tuesday, Wednesday, Friday and Saturday.
E. Monday, Tuesday, Wednesday, Thursday and Friday.

Question 153:
Michael, a taxi driver, charges a call out rate and a rate per mile for taxi rides. For a 4-mile ride he charges £11, and for a 5 mile ride, £13.

How much does he charge for a 9-mile ride?
A. £15   B. £17   C. £19   D. £20   E. £21

Question 154:
Goblins and trolls are not magical. Fairies and goblins are both mythical. Elves and fairies are magical. Gnomes are neither mythical nor magical.

Which of the following is **FALSE**?
A. Elves may be only magical or magical and mythical.
B. Gnomes are not mythical.
C. Goblins are only mythical.
D. Trolls may be mythical.
E. Gnomes and goblins are the same in terms of both qualities.

Question 155:

Jessica runs a small business making bespoke wall tiles. She has just had a rush order for 100 tiles placed that must be ready for today at 7pm. The client wants the tiles packed all together, a process which will take 15 minutes. Only 50 tiles can go in the kiln at any point, and they must be put in the kiln to heat for 45 minutes. The tiles then sit in the kiln to cool before they can be packed, a process which takes 20 minutes. While tiles are in the kiln Jessica is able to decorate more tiles at a rate of 1 tile per minute.

What is the latest time Jessica can start making the tiles?

A. 2:55pm   B. 3:15pm   C. 3:30pm   D. 3:45pm

Question 156:

Pain nerve impulses are twice as fast as normal touch impulses. If Yun touches a boiling hot pan this message reaches her brain, 1 metre away, in 1 millisecond.

What is the speed of a normal touch impulse?

A. 5 m/s   B. 20 m/s   C. 50 m/s   D. 200m/s   E. 500 m/s

Question 157:

A woman has two children, Melissa and Jack. Their birthdays are 3 months apart, both being on the 22nd of the month. The woman wishes to continue the trend of her children's names beginning with the same letter as the month they were born. If her next child, Alina is born on the 22nd 2 months after Jack's birthday, how many months after Alina is born will Melissa have her next birthday?

A. 2 months   B. 4 months   C. 5 months   D. 6 months   E. 7 months

Question 158:

Policemen work in pairs. PC Carter, PC Dirk, PC Adams and PC Bryan must work together but not for more than seven days in a row, which PC Adams and PC Bryan now have. PC Dirk has worked with PC Carter for 3 days in a row. PC Carter does not want to work with PC Adams if it can be avoided.

Who should work with PC Bryan?
A. PC Carter
B. PC Dirk
C. PC Adams
D. Nobody is available under the guidelines above.

Question 159:

My hairdressers charges £30 for a haircut, £50 for a cut and blow-dry, and £60 for a full hair dye. They also do manicures, of which the first costs £15, and includes a bottle of nail polish, but are subsequently reduced by £5 if I bring my bottle of polish. The price is reduced by 10% if I book and pay for the next 5 appointments in advance and by 15% if I book at least the next 10.

I want to pay for my next 5 cut and blow-dry appointments, as well as for my next 3 manicures. How much will it cost?

A. £170   B. £255   C. £260   D. £285   E. £305

Question 160:

Alex, Bertha, David, Gemma, Charlie, Elena and Frankie are all members of the same family consisting of three children, two of whom, Frankie and Gemma are girls. No other assumption of gender based on name can be established. There are also four adults. Alex is a doctor and is David's brother. One of them is married to Elena, and they have two children. Bertha is married to David; Gemma is their child.

Who is Charlie?

A. Alex's daughter   C. Gemma's brother   E. Gemma's sister
B. Frankie's father   D. Elena's son

Question 161:

At 14:30 three medical students were asked to examine a patient's heart. Having already watched their colleague, the second two students were twice as fast as the first to examine. During the 8 minutes break after the final student had finished, they were told by their consultant that they had taken too long and so should go back and do the examinations again. The second time, all the students took half as long as they had taken the first time with the exception of the first student who, instead took the same time as his two colleagues' second attempt. Assuming there was a one-minute change over time between each student and they were finished by 15:15, how long did the second student take to examine the first time?

A. 3 minutes   B. 4 minutes   C. 6 minutes   D. 7 minutes   E. 8 minutes

Question 162:

I pay for 2 chocolate bars that cost £1.65 each with a £5 note. I receive 8 coins change, only 3 of which are the same.

Which **TWO** coins do I not receive in my change?
A. 1p   C. 20p   E. £2
B. 2p   D. 10p

Question 163:

Two 140m long trains are running at the same speed in opposite directions. If they cross each other in 14 seconds, then what is the speed of each train?

A. 10 km/hr　　　　B. 18 km/hr　　　　C. 32 km/hr　　　　D. 36 km/hr　　　　E. 42 km/hr

Question 164:

Anil has to refill his home's swimming pool. He has four hoses which all run at different speeds. Alone, the first would completely fill the pool with water in 6 hours, the second in two days, the third in three days and the fourth in four days.

Using all the hoses together, how long will it take to fill the pool to the nearest quarter of an hour?

A. 4 hours 15 minutes　　　　C. 4 hours 45 minutes　　　　E. 5 hours 15 minutes
B. 4 hours 30 minutes　　　　D. 5 hours

Question 165:

An ant is stuck in a 30 cm deep ditch. When the ant reaches the top of the ditch, he will be able to climb out straight away. The ant is able to climb 3 cm upwards during the day, but falls back 2 cm at night.

How many days does it take for the ant to climb out of the ditch?

A. 27　　　　B. 28　　　　C. 29　　　　D. 30　　　　E. 31

Question 166:

When buying his ingredients, a chef gets a discount of 10% when he buys 10 or more of each item, and a 20% discount when he buys 20 or more. On one order he bought 5 sausages and 10 oranges, and paid £8.50. On another, he bought 10 sausages and 10 apples and paid £9, on a third he bought 30 oranges and paid £12.

How much would an order of 2 oranges, 13 sausages and 12 apples cost?

A. £12.52　　　　B. £12.76　　　　C. £13.52　　　　D. £13.76　　　　E. £13.80

Question 167:

My hairdressers encourage all of their clients to become members. By paying an annual membership fee, the cost of haircuts decreases. VIP membership costs £125 annually with a £10 reduction on each haircut. Executive VIP membership costs £200 for the year with a £15 reduction per haircut. At the moment I am not a member and pay £60 per haircut. I know how many haircuts I have a year, and I work out that by becoming a member on either programme it would work out cheaper, and I would save the same amount of money per year on either programme.

How much will I save this year by buying membership?

A. £10　　　　B. £15　　　　C. £25　　　　D. £30　　　　E. £50

Question 168:

If criminals, thieves and judges are represented below:

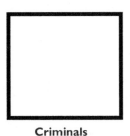

Criminals

Thieves

Judges

Assuming that judges must have a clean record, all thieves are criminals and all those who are guilty are convicted of their crimes, which of one of the following best represents their interaction?

A.

B.

C.

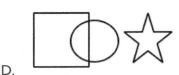

D.

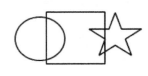

E.

Question 169:

The months of the year have been made into number codes. The code is comprised of three factors, including two of these being related to the letters that make up the name of the month. No two months would have the same first number. But some such as March, which has the code 3513, have the same last number as others, such as May, which has the code 5313. October would be coded as 10715 while February is 286.

What would be the code for April?

A. 154   B. 441   C. 451   D. 514   E. 541

Question 170:

A mother gives yearly birthday presents of money to her children, based on their age and their exam results. She gives them £5 each, plus £3 for every year they are older than 5, and a further £10 for every A they achieved in their results. Josie is 16 and gained 9 As in her results. Although Josie's brother Carson is 2 years older, he receives £44 less a year for his birthday.

How many more As did Josie get than Carson?

A. 2   B. 3   C. 4   D. 5   E. 10

Question 171:

Apples are more expensive than pears, which are more expensive than oranges. Peaches are more expensive than oranges. Apples are less expensive than grapes.

Which **two** of the following must be true?
A. Grapes are less expensive than oranges.
B. Peaches may be less expensive than pears.
C. Grapes are more expensive than pears.
D. Pears and peaches are the same price.
E. Apples and peaches are the same price.

Question 172:

What is the minimum number of straight cutting motions needed to slice a cylindrical cake into 8 equally sized pieces?
A. 2   B. 3   C. 4   D. 5   E. 6   F. 8

Question 173

Three friends, Mark, Russell and Tom had agreed to meet for lunch at 12 PM on Sunday. Daylight saving time (GMT+1) had started at 2 AM the same day, where clocks should be put forward by one hour. Mark's phone automatically changes the time, but he does not realise this so when he wakes up he puts his phone forward an hour and uses his phone to time his arrival to lunch. Tom puts all of his clocks forward one hour at 7 AM. Russell forgets that the clocks should go forward, wakes at 10 AM, and doesn't change his clocks. All of the friends arrive on time as far as they are concerned.

Assuming that none of the friends realise any errors before arriving, which **TWO** of the following statements are **FALSE**?
A. Tom arrives at 12 PM (GMT +1).
B. All three friends arrive at the same time.
C. There is a 2-hour difference between when the first and last friend arrive.
D. Mark arrives late.
E. Mark arrives at 1 PM (GMT+3).

Question 174:

A class of young students has a pet spider. Deciding to play a practical joke on their teacher, one day during morning break one of the students put the spider in their teacher's desk. When first questioned by the headteacher, Mr Jones, the five students who were in the classroom during morning break all lied about what they saw. Realising that the students were all lying, Mr Jones called all 5 students back individually and, threatened with suspension, all the students told the truth. Unfortunately, Mr Jones only wrote down the student's statements not whether they had been told in the truthful or lying questioning.

The students' two statements appear below:

Archie: "It wasn't Edward. "
"It was Bella."

Darcy: "It was Charlotte"
"It was Bella"

Edward: "It was Darcy"
"It wasn't Archie"

Charlotte: "It was Edward."
"It wasn't Archie"

Bella: "It wasn't Charlotte."
"It wasn't Edward."

Who put the spider in the teacher's desk?

A. Edward   C. Darcy   E. More information needed.
B. Bella   D. Charlotte

Question 175:

Dr Massey wants to measure out 0.1 litres of solution. Unfortunately, the lab assistant dropped the 200 ml measuring cylinder, and so the scientist only has a 300 ml and a half litre-measuring beaker. Assuming he cannot accurately use the beakers to measure anything less than their full capacity, what is the minimum volume he will have to use to be able to ensure he measures the right amount?

A. 100 ml   C. 300 ml   E. 600 ml
B. 200 ml   D. 400 ml

Question 176:

Francis lives on a street with houses all consecutively numbered evenly. When one adds up the value of all the house numbers it totals 870.

In order to determine Francis' house number:
1. The relative position of Francis' house must be known.
2. The number of houses in the street must be known.
3. At least three of the house numbers must be known.

A. 1 only   B. 2 only   C. 3 only   D. 1 and 2   E. 2 and 3

Question 177:

There were 20 people exercising in the cardio room of a gym. Four people were about to leave when suddenly a man collapsed on one of the machines. Fortunately, a doctor was on the machine beside him. Emerging from his office, one of the personal trainers called an ambulance. In the 5 minutes that followed before the two paramedics arrived, half of the people who were leaving, left upon hearing the commotion, and eight people came in from the changing rooms to hear the paramedics pronouncing the man dead.

How many living people were left in the room?
A. 25   B. 26   C. 27   D. 28   E. 29

Question 178:

A man and woman are in an accident. They both suffer the same trauma, which causes both of them to lose blood at a rate of 0.2 litres/minute. At normal blood volume the man has 8 litres and the woman 7 litres, and people collapse when they lose 40% of their normal blood volume.

Which **TWO** of the following are true?
A. The man will collapse 2 minutes before the woman.
B. The woman collapses 2 minutes before the man.
C. The total blood loss is 5 litres.
D. The woman has 4.2 litres of blood in her body when she collapses.
E. The man's blood loss is 4.8 litres when he collapses.

Question 179:

Jenny, Helen and Rachel have to run a distance of 13 km. Jenny runs at a pace of 8 kmph, Helen at a pace of 10 kmph, and Rachel 11 kmph.

If Jenny sets off 15 minutes before Helen, and 25 minutes before Rachel, what order will they arrive at the destination?
A. Jenny, Helen, Rachel.   C. Helen, Jenny, Rachel.   E. Jenny, Rachel, Helen.
B. Helen, Rachel, Jenny.   D. Rachel, Helen, Jenny.

Question 180:

On a specific day at a GP surgery 150 people visited the surgery, and common complaints were recorded as a percentage of total patients. Each patient could use their appointment to discuss up to 2 complaints. The split of complaints was as follows: 56% flu-like symptoms, 48% pain, 20% diabetes, 40% asthma/COPD and 30% high blood pressure.

Which statement **must** be true?

A. A minimum of 8 patients complained of pain and flu-like symptoms.
B. No more than 45 patients complained of high blood pressure and diabetes.
C. There were a minimum of 21 patients who did not complain about flu-like symptoms or high blood pressure.
D. There were actually 291 patients who visited the surgery.
E. None of the above.

Question 181:

The prices of all products in a store were marked up by 15%. The products were subsequently reduced in a sale, with quoted savings of 25% from the higher price. What is the true reduction from the original price?

A. 5%
B. 10%
C. 13.75%
D. 18.25%
E. 20%

Question 182:

A recipe states it makes 12 pancakes and requires the following ingredients: 2 eggs, 100g plain flour, and 300ml milk. Steve is cooking pancakes for 15 people and wants to have sufficient mixture for 3 pancakes each.

What quantities should Steve use to ensure this whilst using whole eggs?

A. 2½ eggs, 125g plain flour, 375ml milk
B. 3 eggs, 150g plain flour, 450 ml milk
C. 7½ eggs, 375g plain flour, 1125 ml milk
D. 8 eggs, 400g plain flour, 1200 ml milk
E. 12 eggs, 600g plain flour, 1800 ml milk

Question 183:

Spring Cleaning cleaners buy industrial bleach from a warehouse and dilute it twice before using it domestically. The first dilution is by 9:1 and then the second, 4:1.

If the cleaners require 6 litres of diluted bleach, how much warehouse bleach do they require?

A. 30 ml
B. 120 ml
C. 166 ml
D. 666 ml
E. 1,200 ml

Question 184:

During a GP consultation in 2015, Ms Smith tells the GP about her grandchildren. Ms Smith states that Charles is the middle grandchild and was born in 2002. In 2010, Bertie was twice the age of Adam, and in 2015 there are 5 years between Bertie and Adam. Charles and Adam are separated by 3 years.

How old are the 3 grandchildren in 2015?
A. Adam = 16, Bertie = 11, Charles = 13
B. Adam = 5, Bertie = 10, Charles = 8
C. Adam = 10, Bertie = 15, Charles = 13
D. Adam = 10, Bertie = 20, Charles = 13
E. Adam = 11, Bertie = 10, Charles = 8

Question 185:

Kayak Hire charges a fixed flat rate and then an additional half-hourly rate. Peter hires a kayak for 3 hours and pays £14.50, and his friend Kevin hires 2 kayaks for 4 hrs 30 mins each and pays £41.

How much would Tom pay to hire one kayak for 2 hours?
A. £8
B. £10.50
C. £15
D. £33.20
E. £35.70

Question 186:

A ticketing system uses a common digital display of numbers 0 – 9. The number 7 is showing. However, a number of the light elements are not currently working.

Which set of digits is possible given the configuration of a common digital display?
A. 3, 4, 7   B. 0, 1, 9   C. 2, 7, 8   D. 0, 5, 9   E. 3, 8, 9

Question 187:

A team of 4 builders take 12 days of 7 hours' work to complete a house. The company decides to recruit 3 extra builders.

How many 8 hour days will it take the new workforce to build a house?
A. 2 days
B. 6 days
C. 7 days
D. 10 days
E. 12 days

Question 188:

All astragalus are fabacaea as are all gummifer. Acacia are not astragalus. Which of the following statements is true?

A. Acacia are not fabacaea.
B. No astragalus are also gummifer.
C. All fabacae are astragalus or gummifer.

D. Some acacia may be fabacaea.
E. Gummifer are all acacia.

Question 189:

The Smiths want to reupholster both sides of their seating cushions (dimensions shown on diagram). The fabric they are using costs £10/m, can only be bought in whole metre lengths and has a standard width of 1m. Each side of a cushion must be made from a single piece of fabric. The seamstress changes a flat rate of £25 per cushion. How much will it cost them to reupholster 4 cushions?

A. £ 20
B. £ 80
C. £ 110
D. £ 130
E. £ 150

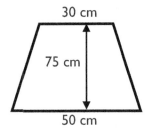

30 cm

75 cm

50 cm

Negligible Thickness

Question 190:

Lisa buys a cappuccino from either Milk or Beans Coffee shops each day. The quality of the coffee is the same but she wishes to work out the relative costs once the loyalty scheme has been taken into account. At Milk, a regular cappuccino is £2.40, and at Beans, £2.15. However, the loyalty scheme in Milk gives Lisa a free cappuccino for every 9 drinks she buys, whereas Beans use a points system of 10 points per full pound spent (each point is worth 1p) which can be used to cover the cost of a full cappuccino.

If Lisa buys a cappuccino each day of September, which coffee shop would work out cheaper, and by how much?

A. Milk, by £4.60
B. Beans by £6.30

C. Beans, by £4.60
D. Beans, by £2.45

E. Milk, by £2.45

Question 191:

Paula needs to be at a meeting in Notting Hill at 11 am. The route requires her to walk 5 minutes to the 283 bus, which takes 25 minutes, and then change to the 220 bus which takes 14 minutes. Finally, she walks for 3 minutes to her meeting. If the 283 bus comes every 10 minutes, and the 220 bus at 0 minutes, 20 minutes and 40 minutes past the hour, what is the latest time she can leave and still be at her meeting on time?

A. 09.45   B. 09.58   C. 10.01   D. 10.05   E. 10.10

Question 192:

Two trains, a high-speed train A and a slower local train B, travel from Manchester to London. Train A travels the first 20 km at 100 km/hr and then at an average speed of 150km/hr. Train B travels at a constant average speed of 90 km/hr. If train B leaves 20 minutes before train A, at what distance will train A pass train B?

A. 75 km   B. 90 km   C. 100 km   D. 120 km   E. 150 km

Question 193:

The university gym has an upfront cost of £35 with no contracted monthly fee, but classes are charged at £3 each. The local gym has no joining fee and is £15 per month including all classes. What is the minimum number of classes I need to attend in a 12 month period to make the local gym cheaper than the university gym?

A. 40   B. 48   C. 49   D. 50   E. 55   F. 60

Question 194:

"All medicines are drugs, but not all drugs are medicines", goes a well-known saying. If we accept this statement as true, and consider that all antibiotics are medicines, but no herbal drugs are medicines, then which of the following is definitely **FALSE**?

A. Some herbal drugs are not medicines.   D. Some medicines are antibiotics.
B. All antibiotics are drugs.   E. None of the above.
C. Some herbal drugs are antibiotics.

Question 195:

Sonia has been studying the routes taken by various trains travelling between London and Edinburgh on the East coast. Trains can stop at the following stations: Newark, Peterborough, Doncaster, York, Northallerton, Darlington, Durham and Newcastle. She notes the following:

- All trains stop at Peterborough, York, Darlington and Newcastle.
- All trains that stop at Northallerton also stop at Durham.
- Each day, 50% of the trains stop at both Newark *and* Northallerton.
- All designated "fast" trains make less than 5 stops. All other trains make 5 stops or more.
- On average, 16 trains run each day.

Which of the following can be reliably concluded from these observations?
A. All trains, which are not designated "fast" trains, must stop at Durham.
B. No more than 8 trains on any given day will stop at Northallerton.
C. No designated "fast" trains will stop at Durham.
D. It is possible for a train to make 5 stops, including Northallerton.
E. A train which stops at Newark will also stop at Durham.

Question 196:

Rakton is 5 miles directly north of Blueville. Gallford is 8 miles directly south of Haston. Lepstone is situated 5 miles directly east of Blueville, and 5 miles directly west of Gallford.

Which of the following **CANNOT** be reliably concluded from this information?

A. Lepstone is South of Rakton
B. Haston is North of Rakton
C. Gallford is East of Rakton
D. Blueville is East of Haston
E. Haston is North of Lepstone

Question 197:

The Eastminster Parliament is undergoing an election. There are 600 seats up for election, each of which will be elected separately by the people living in that constituency. Six parties win at least one seat in the election, the Blue Party, the Red Party, the Orange Party, the Yellow Party, the Green Party and the Purple Party. In order to form a government, a party (or coalition) must hold *over* 50% of the seats. After the election, a political analysis committee produces the following report:

- No party has gained more than 45% of the seats, so no party is able to form a government alone.
- The Red and the Blue party each gained over 40% of the seats.
- No other party gained more than 4% of the seats.
- The Yellow party did not win the fewest seats

The Red party work out that if they collaborate with the Green party and the Orange party, between the 3 of them, they will have enough seats to form a coalition government.

What is the minimum number of seats that the Green party could have?

A. 5   B. 6   C. 13   D. 14   E. 23   F. 24

Questions 198-202 are based on the following information:

A grandmother wants to give her five grandchildren £100 between them for Christmas this year. She wants to grade the money she gives to each grandchild exactly so that the older children receive more than the younger ones. She wants to share the money such that she will give the 2nd youngest child as much more than the youngest, as the 3rd youngest gets than the 2nd youngest, as the 4th youngest gets from the 3rd youngest and so on. The result will be that the two youngest children together will get seven times less money than the three oldest.

M is the amount of money the youngest child receives, and *D* the difference between the amount the youngest and 2nd youngest children receive.

Question 198:
What is the expression for the amount the oldest child receives?

A. M

B. M + D

C. 2M

D. 4M²

E. M + 4D

F. None of the above.

Question 199:
What is the correct expression for the total money received?

A. 5M = £100

B. 5D + 10M = £100

C. $D = \frac{M}{100}$

D. 5M + 10D = £100

E. $M = \frac{2D}{11}$

Question 200:
"The two youngest children together will get seven times less money than the three oldest."
Which one of the following best expresses the above statement?

A. 7(3M + 9D) = 2M + D

B. 7D = M

C. 7(2M + D) = 3M + 9D

D. 2(7M + D) = 3M + 9D

E. None of the above

Question 201:
Using the statement in the previous question, what is the correct expression for *M*?

A. $\frac{2D}{11}$

B. $\frac{2}{11}$

C. $\frac{10D}{11}$

D. $\frac{120}{11}$

E. None of the above

Question 202:
Express £100 in terms of D.

A. $£100 = \frac{120D}{11}$

B. $£100 = \frac{120D}{10}$

C. $£100 = \frac{120}{11D}$

D. $£100 = 21D$

E. $£100 = 5M + 10D$

Question 203:
Four young girls entered a local baking competition. Though a bit burnt, Ellen's carrot cake did not come last. The girl who baked a Madeira sponge had practised a lot, and so came first, while Jaya came third with her entry. Aleena did better than the girl who made the tiramisu, and the girl who made the Victoria sponge did better than Veronica.

Which **TWO** of the following were **NOT** results of the competition?

A. Veronica made a tiramisu

B. Ellen came second

C. Aleena made a Victoria sponge

D. The Victoria sponge came in 3rd place

E. The carrot cake came 3rd

Question 204:

In a young children's football league, the 5 teams were Celtic Changers, Eire Lions, Nordic Nesters, Sorten Swipers and the Whistling Winners. One of the boys playing in the league, after being asked by his parents, said that while he could remember the other teams' total points, he could not remember the score for his own team, the Eire Lions. He said that all the teams played each other and when teams lost, they were given 0 points, when they drew, 1 point, and 3 points for a win. He remembered that the Celtic Changers had a total of 2 points; the Sorten Swipers had 5; the Nordic Nesters had 8, and the Whistling Winners 1.

How many points did the boy's team score?

A. 1

B. 4

C. 8

D. 10

E. 11

Question 205:

T is the son of Z, Z and J are sisters, R is the mother of J and S is the son of R.

Which one of the following statements is correct?

A. T and J are cousins

B. S and J are sisters

C. J is the maternal uncle of T

D. S is the maternal uncle of T

E. R is the grandmother of Z.

Question 206:

John likes to shoot bottles off a wall. In the first round, he places 16 bottles on the wall and knocks off 8 bottles. 3 of the knocked off bottles are damaged and can no longer be used, whilst 1 bottle is lost. He puts the undamaged bottles back on the wall before continuing. In the second round he shoots six times and misses 50% of these shots. He damages two bottles with every shot that does not miss. 2 bottles also fall off the wall at the end. He puts up 2 new bottles before continuing. In the final round, John misses all his shots and in frustration, knocks over 50% of the remaining bottles.

How many bottles were left on the wall after the final round?

A. 2

B. 3

C. 4

D. 5

E. 6

Questions 207 - 213 are based on the information below:

All train lines are named after a station they serve, apart from the Oval and Rectangle lines, which are named for their recognisable shapes. Trains run in both directions.

- There are express trains that run from end to end of the St Mark's and Straightly lines taking 5 and 6 minutes respectively.
- It takes 2 minutes to change between St Mark's and both Oval and Rectangle lines, 1 minute between Rectangle and Oval.
- It takes 3 minutes to change between the Straightly and all other lines, except for the St Mark's line which only takes 30 seconds to change between.
- The Straightly line is a fast line and takes only 2 minutes between stops apart from to and from Keyton, which only takes 1 minute, and to and from Lime St which takes 3 minutes.
- The Oval line is much slower and takes 4 minutes between stops, apart from between Baxton and Marven, and also Archite and West Quays, where travelling between stops takes 5 minutes.
- The Rectangle line is a reliable line; it never runs late but as a consequence is much slower taking 6 minutes between stops.
- The St Mark's line is fast and takes 2 and half minutes between stations.
- If a passenger reaches the end of the line, it takes three minutes to change onto a train travelling back in the opposite direction.

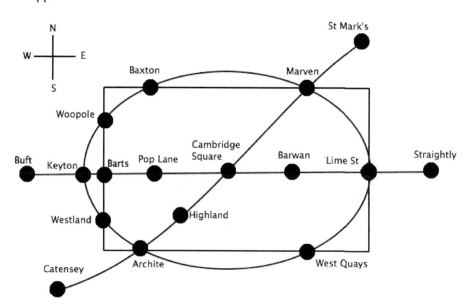

Question 207:

Assuming all lines are running on time, how long does it take to go from St Mark's to Archite on the St Mark's line?

A. 5 minutes

B. 6 minutes

C. 7.5 minutes

D. 10 minutes

E. 12.5 minutes

Question 208:

Assuming all lines are running on time, what's the shortest time it will take to go from Buft to Straightly?

A. 6 minutes

B. 10 minutes

C. 12 minutes

D. 14 minutes

E. 16 minutes

Question 209:

What is the shortest time it will take to go from Baxton to Pop Lane?

A. 11 minutes

B. 12 minutes

C. 13 minutes

D. 14 minutes

E. 15 minutes

Question 210:

Which station, even allowing for the quickest journey time, takes the longest to reach from Cambridge Square?

A. Catensey

B. Buft

C. Woopole

D. Westland

Questions 211-213 use this additional information:

On a difficult day there are signal problems whereby all lines except the reliable line are delayed, such that train travel times between stations are doubled. These delays have caused overcrowding at the platforms which means that while changeover times between lines are still the same, passengers always have to wait an extra 5 minutes on all of the platforms before catching the next train.

Question 211:

At best, how long will it now take to go from Westland to Marven?

A. 25 minutes

B. 29 minutes

C. 30 minutes

D. 33 minutes

E. 35 minutes

Question 212:

There is a bus that goes from Baxton to Archite and takes 27-31 minutes. Susan lives in Baxton and needs to get to her office in Archite as quickly as possible. With all the delays and lines out of service,

How should you advise Susan to enable her to get to work most quickly?

A. Baxton to Archite via Barts using the Rectangle line.

B. Baxton to Woopole on the Rectangle line, then Oval to Archite via Keyton.

C. It is not possible to tell between the fastest two options.

D. Baxton to Woopole on the Rectangle line, then Oval to Archite via Keyton.

E. Baxton to Archite on the Oval line.

F. Baxton to Archite using the bus.

Question 213:

In addition to the delays, the Oval line signals fail completely, so the line falls out of service. How long will the fastest journey now take to go from St Mark's to West Quays?

A. 35 minutes

C. 33 minutes

E. 30.5 minutes

B. 30 minutes

D. 29 minutes

F. None of the above.

Question 214:

In an unusual horserace, only 4 horses competed. Each had different racing colours and numbers. Simon's horse wore number 1. Lila's horse was painted neither yellow nor blue, and the horse that wore number 3, which was wearing red, beat the horse that came in third place. Only one horse wore the same number as the position it finished in. Arthur's horse beat Simon's horse, whereas Celia's horse beat the horse that wore number 1. The horse wearing green, Celia's, came second, and the horse wearing blue wore number 4. Which one of the following must be true?

A. Simon's horse was yellow and placed 3rd.

D. Arthur's horse was blue.

B. Celia's horse was red.

A. Lila's horse wore number 4.

C. Celia's horse was in third place.

Question 215:

Jessie plants a tree with a height of 40 cm. The information leaflet states that the plant should grow by 20% each year for the first 2 years, and then 10% each year thereafter.

What is the expected height at 4 years?

A. 58.08 cm

C. 69.696 cm

E. 82.944 cm

B. 64.89 cm

D. 89.696 cm

Question 216:

A company is required to pay each employee 10% of their wages into a pension fund if their annual total wage bill is above £200,000. However, there is a legal loophole stating that if the company splits over two sites, the £200,000 limit applies to the total wages for each individual site. The company therefore decides to have an east site and a west site.

Name	Annual Salary (£)
Luke	47,000
John	78,400
Emma	68,250
Nicola	88,500
Victoria	52,500
Daniel	63,000

Which employees should be grouped at the same site to minimise the cost to the company?

A. John, Nicola, Luke   C. Nicola, Daniel, Luke   E. Luke, Victoria, Emma

B. Nicola, Victoria, Daniel   D. John, Daniel, Emma

Question 217:

A bus takes 24 minutes to travel from White City to Hammersmith with no stops. Each time the bus stops to pick up and/or drop off passengers, it takes approximately 90 seconds. This morning, the bus picked up passengers from 5 stops, and dropped off passengers at 7 stops.

What is the minimum journey time from White City to Hammersmith this morning?

A. 28 minutes   C. 34.5 minutes   E. 37.5 minutes

B. 34 minutes   D. 36 minutes   F. 42 minutes

Question 218:

Sally is making a Sunday roast for her family and is planning her schedule regarding cooking times. The chicken takes 15 minutes to prepare, 75 minutes to cook, and needs to stand for exactly 5 minutes after cooking. The potatoes take 18 minutes to prepare, 5 minutes to boil, then 50 minutes to roast. The potatoes must be roasted immediately after boiling, and then served immediately after roasting. The vegetables require only 5 minutes preparation time and 8 minutes boiling time before serving, and can be kept warm to be served at any time after cooking. Given that the cooker can only be cooking two items at any given time and Sally can prepare only one item at a time, what should Sally's schedule be if she wishes to serve dinner at 4pm and wants to start cooking each item as late as possible?

A. Chicken 2.25, potatoes 2.47, vegetables 2.42
B. Chicken 2.25, potatoes 2.47, vegetables 3.47
C. Chicken 2.35, potatoes 3.47, vegetables 2.47
D. Chicken 2.35, potatoes 2.47, vegetables 3.47
E. Chicken 2.45, potatoes 3.47, vegetables 2.47
F. Chicken 2.45, potatoes 2.47, vegetables 3.47

Question 219:

The Smiths have 4 children with a total age of 80. Paul is double the age of Jeremy. Annie is exactly halfway between the ages of Jeremy and Paul, and Rebecca is 2 years older than Paul. How old are each of the children?

A. Paul 23, Jeremy 12, Rebecca 26, Annie 19.
B. Paul 22, Jeremy, 11, Rebecca 24, Annie 16.
C. Paul 24, Jeremy 12, Rebecca 26, Annie 18.

D. Paul 28, Jeremy 14, Rebecca 30, Annie 21.
E. More information needed.

Question 220:

Sarah has a jar of spare buttons that are a mix of colours and sizes. The jar contains the following assortment of buttons:

	10mm	25mm	40mm
Cream	15	22	13
Red	6	15	7
Green	9	19	8
Blue	20	6	15
Yellow	4	8	26
Black	17	16	14
Total	71	86	83

Sarah wants to use a 25mm diameter button but doesn't mind if it is cream or yellow. What is the maximum number of buttons she will have to remove in order to guarantee picking a suitable button on the next attempt?

A. 210   C. 219   E. None of the above

B. 218   D. 239

Question 221:

Ben wants to optimise his score with one throw of a dart. 50% of the time he hits a segment to either side of the one he is aiming at. With this in mind, which of the following segments should he aim for? [Ignore all double/triple modifiers]

A. 15

B. 16

C. 17

D. 18

E. 19

Question 222:

Victoria is completing her weekly shop and the total cost of the items is £8.65. She looks in her purse and sees that she has a £5 note and a large amount of change, including all types of coins. She uses the £5 note and pays the remainder using the maximum number of coins possible in order to remove some weight from the purse. However, the store has certain rules she must follow when paying:

- No more than 20p can be paid in "bronze" change (the name given to any combination of 1p pieces and 2p pieces)
- No more than 50p can be paid using any combination of 5p pieces and 10p pieces.
- No more than £1.50 can be paid using any combination of 20p pieces and 50p pieces.

Victoria pays the exact amount and does not receive any change. Under these rules, what is the *maximum* number of coins that Victoria can have paid with?

A. 30   B. 31   C. 36   D. 41   E. 46

Question 223:

I look at the clock on my bedside table, and I see the following digits:

However, I also see that there is a glass of water between me and the clock, which is in front of 2 adjacent figures. I know that this means these 2 figures will appear reversed. For example, 10 would appear as 01, and 20 would appear as 05 (as 5 on a digital clock is a reversed image of a 2). Some numbers, such as 3, cannot appear reversed because there are no numbers which look like the reverse of 3.

Which of the following could be the actual time shown on the clock?

A. 15:52   B. 21:25   C. 12:55   D. 12:22   E. 21:52

Question 224:

Slavica has invaded Worsid, whilst Nordic has invaded Lorkdon. Worsid, spotting an opportunity to bolster its amount of land and natural resources, invades Nordic. Each of these countries is either a dictatorship or a democracy. Slavica is a dictatorship, but Lorkdon is a democracy. 10 years ago, a treaty was signed which guaranteed that no democracy would invade another democracy. No dictatorship has both invaded another dictatorship *and* been invaded by another dictatorship.

Assuming the aforementioned treaty has been upheld, what style of government is practised in Worsid?
A. Worsid is a dictatorship.
B. Worsid is a democracy.
C. Worsid does not practice either of these forms of government.
D. It is impossible to tell.

Question 225:

Sheila is on a shift at the local supermarket. Unfortunately, the till has developed a fault, meaning it cannot tell her how much change to give each customer. A customer is purchasing the following items, at the following costs:

- A packet of grated cheese priced at £3.25
- A whole cucumber, priced at 75p
- 750g of carrots, priced at 60p/kg
- 3 DVDs, each priced at £3.00

Sheila knows there is an offer on DVDs in the store at present, in which 3 DVDs bought together will be a third off. The customer pays with a £50 note.

How much change will Sheila need to give the customer?
A. £36.00   B. £36.40   C. £36.55   D. £39.40   E. £39.55

Question 226:

Ryan is cooking breakfast for several guests at his hotel. He is frying most of the items using the same large frying pan, to get as much food prepared in as little time as possible. Ryan is cooking bacon, sausages, and eggs in this pan. He calculates how much room is taken up in the pan by each item. He calculates the following:

- Each rasher of bacon takes up 7% of the available space in the pan
- Each sausage takes up 3% of the available space in the pan.
- Each egg takes up 12% of the available space in the pan.

Ryan is cooking 2 rashers of bacon, 4 sausages and 1 egg for each guest. He decides to cook all the food for each guest at the same time, rather than cooking all of each item at once.

How many guests can he cook for at once?
A. 1   B. 2   C. 3   D. 4   E. 5

Question 227:

SafeEat Inc. is a national food development testing agency. The Manchester-based laboratory has a system for recording all of the laboratory employees' birthdays, and presenting them with cake on their birthday, in order to keep staff morale high. Certain amounts of petty cash are set aside each month in order to fund this. 40% of the staff have their birthday in March, and the secretary works out that £60 is required to fund the birthday cake scheme during this month.

If all birthdays cost £2 to provide a cake for, how many people work at the laboratory?

A. 45   B. 60   C. 75   D. 100   E. 150

Question 228:

Many diseases, such as cancer, require specialist treatment, and thus cannot be treated by a general practitioner. Instead, these diseases must be referred to a specialist after an initial, more generalised, medical assessment. Bob has had a biopsy on the 1st of August on a lump in his abdomen. The results show that it is a cancerous tumour, with a slight chance of spreading to other parts of the body, so he is referred to a waiting list for specialist radiotherapy and chemotherapy. The average waiting time in the UK for such treatment is 3 weeks, but in Bob's local district, high demand means that it takes 50% longer for each patient to receive treatment. As he is a lower risk case (due to a low risk of the disease spreading), his waiting time is extended by another 20%.

How many weeks will it be before Bob receives specialist treatment?

A. 4.5   B. 4.6   C. 5.0   D. 5.1   E. 5.4   F. 5.6

Question 229:

In a class of 30 seventeen-year-old students, 40% drink alcohol at least once a month. Of those who drink alcohol at least once a month, 75% drink alcohol at least once a week. 1 in 3 of the students who drink alcohol at least once a week also smoke marijuana. 1 in 3 of the students who drink alcohol less than once a month also smoke marijuana.

How many of the students in total smoke marijuana?

A. 3   B. 4   C. 6   D. 9   E. 10   F. 15

Question 230:

Complete the following sequence of numbers: 1, 4, 10, 22, 46, …

A. 84   B. 92   C. 94   D. 96   E. 100

Question 231:

If the mean of 5 numbers is 7, the median is 8 and the mode is 3, what must the two largest numbers in the set of numbers add up to?

A. 14   C. 24   E. 35
B. 21   D. 26   F. More information needed.

Question 232:

Ahmed buys 1kg bags of potatoes from the supermarket. By law, 1kg bags have to weigh between 900 and 1100 grams. In the first week, there are 10 potatoes in the bag. The next week, there are only 5. Assuming that the potatoes in the bag in week 1 are all the same weight as each other, and the potatoes in the bag in week 2 are all the same weight as each other, what is the maximum possible difference between the heaviest and lightest potato in the two bags?

A. 50g   B. 70g   C. 90g   D. 110g   E. 130g

Question 233:

A football tournament involves a group stage, then a knockout stage. In the group stage, groups of four teams play in a round robin format (i.e. each team plays every other team once) and the team that wins the most matches in each group proceeds through to a knockout stage. In addition, the single best performing second place team across all the groups gains a place in the knockout stage. In the knockout stage, sets of two teams play each other and the one that wins proceeds to the next round until there are two teams left, who play the final.

If we start with 60 teams, how many matches are played altogether?

A. 75   B. 90   C. 100   D. 105   E. 165

Question 234:

The last 4 digits of my card number are 2 times my Personal Identification Number (PIN), plus 200. The last 4 digits of my husband's card number are the last four digits of my card number doubled, plus 200. My husband's PIN is 2 times the last 4 digits of his card number, plus 200. Given that all these numbers are 4 digits long, whole numbers, and cannot begin with 0, what is the largest number my PIN can be?

A. 1,074   C. 2,348   E. 9,999
B. 1,174   D. 4,096

Question 235:

All women between the age 50 to 70 in the UK are invited for breast cancer screening every 3 years. Patients at Doddinghurst Surgery are invited for screening for the first time at any point between their 50th and 53rd birthday. If they ignore an invitation, they are sent reminders every 5 months. We can assume that a woman is screened exactly 1 month after she is sent the invitation or reminder that she accepts. The next invitation for screening is sent exactly 3 years after the previous screening.

If a woman accepts the screening on the second reminder each time, what is the youngest she can be when she has her 4th screening?

A. 60   B. 61   C. 62   D. 63   E. 64

Question 236:

Ellie gets a pay rise of k thousand pounds on every anniversary of joining the company, where k is the number of years she has been at the company. She currently earns £40,000, and she has been at the company for 5.5 years. What was her salary when she started at the company?

A. £25,000   C. £28,000   E. £31,000

B. £27,000   D. £30,000

Question 237:

Northern Line trains arrive at Kings Cross station every 8 minutes, Piccadilly Line trains every 5 minutes and Victoria Line trains every 2 minutes. If trains from all 3 lines arrived at the station exactly 15 minutes ago, how long will it be before they do so again?

A. 24 minutes   C. 40 minutes   E. 65 minutes

B. 25 minutes   D. 60 minutes

Question 238:

If you do not smoke or drink alcohol, your risk of getting Disease X is 1 in 12. If you smoke, you are half as likely to get Disease X as someone who does not smoke. If you drink alcohol, you are twice as likely to get Disease X. A new drug is released that halves anyone's total risk of getting Disease X for each tablet taken. How many tablets of the drug would someone who drinks alcohol have to take to reduce their risk to the same level as someone who smoked but did not take the drug?

A. 0   B. 1   C. 2   D. 3   E. 4

Questions 239 – 241 refer to the following information:

There are 20 balls in a bag. 1/2 are red. 1/10 of those balls that are not red are yellow. The rest are green except 1, which is blue.

Question 239:

If I draw 2 balls from the bag (without replacement), what is the most likely combination to draw?

A. Red and green   C. Red and red

B. Red and yellow   D. Blue and yellow

Question 240:

If I draw 2 balls from the bag (without replacement), what is the least likely (without being impossible) combination to draw?

A. Blue and green   C. Yellow and yellow

B. Blue and yellow   D. Yellow and green

Question 241:

How many balls do you have to draw (without replacement) to guarantee getting at least one of at least three different colours?

A. 5 B. 12 C. 13 D. 17 E. 19

Question 242:

A general election in the UK resulted in a hung parliament, with no single party gaining more than 50% of the seats. Thus, the main political parties are engaged in discussion over the formation of a coalition government. The results of this election are shown below:

Political Party	Seats won
Conservatives	260
Labour	270
Liberal Democrats	50
UKIP	35
Green Party	20
Scottish National Party	17
Plaid Cymru	13
Sinn Fein	9
Democratic Unionist Party (DUP)	11
Other	14 (14 other parties won 1 seat each)

There are a total of 699 seats, meaning that in order to form a government, any coalition must have at least 350 seats between them. Several of the party leaders have released statements about who they are and are not willing to form a coalition with, which are summarised as follows:

- The Conservative party and Labour party are not willing to take part in a coalition together.
- The Liberal Democrats refuse to take part in any coalition which also involves UKIP.
- The Labour party will only form a coalition with UKIP if the Green party are also part of this coalition.
- The Conservative party are not willing to take part in any coalition with UKIP unless the Liberal Democrats are also involved.

Considering this information, what is the minimum number of parties required to form a coalition government?

A. 2 B. 3 C. 4 D. 5 E. 6

Question 243:

On Tuesday, 360 patients attend appointments at Doddinghurst Surgery. Of the appointments that are booked in, only 90% are attended. Of the appointments that are booked in, 1 in 2 are for male patients, the remaining appointments are for female patients. Male patients are three times as likely to miss their booked appointment as female patients.

How many male patients attend appointments at Doddinghurst Surgery on Tuesday?

A. 30   B. 60   C. 130   D. 150   E. 170

Question 244:

Every A Level student at Greentown Sixth Form studies maths. Additionally, 60% study biology, 50% study economics and 50% study chemistry. The other subject on offer at Greentown Sixth Form is physics. Assuming every student studies 3 subjects and that there are 60 students altogether, how many students study physics?

A. 15   C. 30   E. 60

B. 24   D. 40

Question 245:

100,000 people are diagnosed with chlamydia each year in the UK. An average of 0.6 sexual partners are informed per diagnosis. Of these, 80% have tests for chlamydia themselves. Half of these tests come back positive.

Assuming that each of the people diagnosed has had an average of 3 sexual partners (none of them share sexual partners or have sex with each other) and that the likelihood of having chlamydia is the same for those partners who are tested and those who are not, how many of the sexual partners who were not tested (whether they were informed or not) have chlamydia?

A. 120,000   C. 136,000   E. 240,000

B. 126,000   D. 150,000

Question 246:

In how many different positions can you place an additional tile to make a straight line of 3 tiles?

A. 6
B. 7
C. 8
D. 9
E. 10
F. 11
G. 12

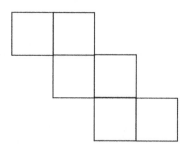

Question 247:

Harry is making orange squash for his daughter's birthday party. He wants to have a 200ml glass of squash for each of the 20 children attending and a 300ml glass of squash for him and each of 3 parents who are helping him out. He has 1,040ml of the concentrated squash.

What ratio of water: concentrated squash should he use in the dilution to ensure he has the right amount to go around?

A. 2:1   B. 3:1   C. 4:1   D. 5:1   E. 6:1

Question 248:

4 children, Alex, Beth, Cathy and Daniel are each sitting on one of the 4 swings in the park. The swings are in a straight line. One possible arrangement of the children is, left to right, Alex, Beth, Cathy, Daniel.

How many other possible arrangements are there?

A. 5   B. 12   C. 23   D. 24   E. 64

Question 249:

A delivery driver is looking to make deliveries in several towns. He is given the following map of the various towns in the area. The lines indicate roads between the towns, along with the lengths of these roads, where m is miles.

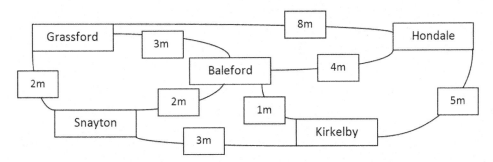

The delivery driver's vehicle has a black box which records the distance travelled and locations visited. At the end of the day, the black box recording shows that he has travelled a total of 14 miles. It also shows that he has visited one town twice, but has not visited any other town more than once. Which of the following is a possible route the driver could have taken?

A. Snayton → Baleford→ Grassford → Snayton→ Kirkelby

B. Baleford → Kirkelby→ Hondale → Grassford→ Baleford→ Snayton

C. Kirkelby → Hondale→ Baleford →Grassford→ Snayton

D. Baleford → Hondale→ Grassford → Baleford→ Hondale→ Kirkelby

E. None of the above.

Question 250:

Ellie, her brother Tom, her sister Georgia, her mum and her dad line up in height order from shortest to tallest for a family photograph. Ellie is shorter than her dad but taller than her mum. Georgia is shorter than both her parents. Tom is taller than both his parents.

If 1 is shortest and 5 is tallest, what position is Ellie in the line?

A. 1   B. 2   C. 3   D. 4   E. 5

Question 251:

Miss Briggs is trying to arrange the 5 students in her class into a seating plan. Ashley must sit on the front row because she has poor eyesight. Danielle disrupts anyone she sits next to apart from Caitlin, so she must sit next to Caitlin and no-one else. Bella needs to have a teaching assistant sat next to her. The teaching assistant must be sat on the left-hand side of the row. Emily does not get on with Bella, so they need to be sat apart from one another. The teacher has 2 tables which each sit 3 people, which are arranged 1 behind the other.

Who is sitting in the front right seat?

A. Ashley   C. Caitlin   E. Emily
B. Bella   D. Danielle

Question 252:

My aunt runs the dishwasher twice a week as standard. She also runs the dishwasher an additional time for each person who is living in the house that week. When her son is away at university, she buys a new pack of dishwasher tablets every 6 weeks, but when her son is home, she must buy a new one every 5 weeks. How many people are living in the house when her son is home?

A. 2   B. 3   C. 4   D. 5   E. 6

Question 253:

Dates can be written in an 8-digit form, for example 26-12-2014. How many days after 26-12-2014 would be the next time that the 8 digits were made up of exactly 4 different integers?

A. 6   B. 8   C. 10   D. 16   E. 24

Question 254:

Redtown is 4 miles east of Greentown. Bluetown is 5 miles north of Greentown. If every town is due North, South, East or West of at least two other towns, and the only other town is Yellowtown, how many miles away from Yellowtown is Redtown, and in what direction?

A. 4 miles east of Yellowtown.   D. 4 miles west of Yellowtown.
B. 5 miles south of Yellowtown.   E. 5 miles west of Yellowtown.
C. 5 miles north of Yellowtown.

Question 255:

Jenna pours wine from two 750ml bottles into glasses. The glasses hold 250ml, but she only fills them to 4/5 of capacity, except the last glass, into which she pours whatever liquid she has left. How full is the last glass compared to its capacity?

A. 1/5   B. 2/5   C. 3/5   D. 4/5   E. 5/5

Question 256:

There are 30 children in Miss Smith's class. Two thirds of the girls in Miss Smith's class have brown eyes, and two thirds of the class as a whole have brown hair. Given that the class is half boys and half girls, what is the difference between the minimum and maximum number of girls that could have brown eyes and brown hair?

A. 0   C. 5   E. 10
B. 2   D. 7

Question 257:

A biased die with the numbers 1 to 6 on it is rolled twice. The resulting numbers are multiplied together, and then their sum subtracted from this result to get the 'score' of the dice roll. If the probability of getting a negative (non-zero) score is 0.75, what is the probability of rolling a 1 on a third throw of the die?

A. 0.1   C. 0.3   E. 0.5
B. 0.2   D. 0.4

Questions 258 - 260 are based on the following information:

Fares on the number 11 bus are charged at a number of pence per stop that you travel, plus a flat rate. Emma, who is 21, travels 15 stops and pays £1.70. Charlie, who is 43, travels 8 stops and pays £1.14. Children (under 16) pay half the adult flat rate plus a quarter of the adult charge "per stop".

Question 258:

How much does 17-year-old Megan pay to travel 30 stops to college?

A. £0.85   C. £2.90   E. More information needed.
B. £2.40   D. £3.40

Question 259:

How much does 14-year-old Alice pay to travel 25 stops to school?

A. £0.50   C. £1.25   E. More information needed.
B. £0.75   D. £2.50

Question 260:

James, who is 24, wants to get the bus into town. The town stop is the 25th stop along a straight road from his house, but he only has £2.

Assuming he has to walk past the stop nearest his house, how many stops will he need to walk past before he gets to the stop he can afford to catch the bus from?

A. 4   B. 6   C. 7   D. 8   E. 9

Questions 261 -263 are based on the following information:

Emma mounts and frames paintings. Each painting needs a mount which is 2 inches bigger in each dimension than the painting, and a wooden frame which is 1 inch bigger in each dimension than the mount. Mounts are priced by multiplying 50p by the largest dimension of the mount, so a mount which is 8 inches in one direction and 6 in the other would be £4. Frames are priced by multiplying £2 by the smallest dimension of the frame, so a frame which is 8 inches in one direction and 6 in the other would be £12.

Question 261:

How much would mounting and framing a painting that is 10 x 14 inches cost?

A. £8   B. £26   C. £27   D. £34   E. £42

Question 262:

How much more would mounting and framing a 10 x 10 inch painting cost than mounting and framing an 8 x 8 inch painting?

A. £ 3.00   B. £ 4.00   C. £ 5.00   D. £ 6.00   E. £ 7.00

Question 263:

What is the largest square painting that can be framed for £40?

A. 12 inches   C. 14 inches   E. 16 inches
B. 13 inches   D. 15 inches

Question 264:

The word 'CREATURES' is coded as 'FTEAWUTEV', which would be coded for a second time as 'HWEAYUWEX'. What would be the second coding of the word 'MAGICAL'?

A. QCKIGAN   C. PAJIFAN   E. RCIMGEP
B. OCIIEAN   D. RALIHAQ

Question 265:

Jane's mum has asked Jane to go to the shops to get some items that they need. She tells Jane that she will pay her per kilometre that she cycles on her bike to get to the shop, plus a flat rate payment for each shop she goes to. Jane receives £6 to go to the grocers, a distance of 5 km, and £4.20 to go the supermarket, a distance of 3km.

How much would she earn if she then cycles to the library to change some books, a distance of 7 km?

A. £7.50   B. £7.70   C. £7.80   D. £8.00   E. £8.10

Question 266:

In 2001-2002, 1,019 patients were admitted to hospital due to obesity. This figure was more than 11 times higher by 2011-12 when there were 11,736 patients admitted to hospital with the primary reason for admission being obesity.

If the rate of admissions due to obesity continues to increase at the same linear rate as it has from 2001/2 to 2011/12, how many admissions would you expect in 2031/32?

A. 22,453   C. 33,170   E. 269,928
B. 23,437   D. 134,964

Question 267:

A shop puts its dresses on sale at 20% off the normal selling price. During the sale, the shop makes a 25% profit over the price at which they bought the dresses. What is the percentage profit when the dresses are sold at full price?

A. 36%   C. 56.25%   E. 77%
B. 42.5%   D. 64%

Question 268:

The 'Keys MedSoc committee' is made up of 20 students from each of the 6 year groups of medical students at the university. However, the president and vice-president are sabbatical roles which require students to take a year out from studying. There must be at least two general committee students from each year group as well as the specialist roles. Also, the social and welfare officers must be pre-clinical students (years 1-3) but not first years, and the treasurer must be a clinical student (years 4-6).

Which **TWO** of the following statements must be true?

1. There can be a maximum of 13 preclinical (years 1-3) students on the committee.
2. There must be a minimum of 6 2nd and 3rd years.
3. There is an unequal distribution of committee members over the different year groups.
4. There can be a maximum of 10 clinical (years 4-6) students on the committee.
5. There can be a maximum of 2 first year students on the committee.
6. General committee members are equally spread across the 6 years.

A. 1 and 4   C. 2 and 4   E. 4 and 5
B. 2 and 3   D. 3 and 6

Question 269:

Friday the 13th is superstitiously considered an 'unlucky' day. If 13th January 2012 was a Friday, when would the next Friday the 13th be?

A. March 2012   C. July 2012   E. January has the only
B. April 2012   D. August 2012   Friday 13th in 2012.

Question 270:

A farmer has 18 sheep, 8 of which are male. Unfortunately, 9 sheep die, of which 5 were female. The farmer decides to breed his remaining sheep in order to increase the size of his herd. Assuming every female gives birth to two lambs, how many sheep does the farmer have after all the females have given birth once?

A. 10   B. 14   C. 15   D. 16   E. 19

Question 271:

Piyanga writes a coded message for Nishita. Each letter of the original message is coded as a letter a specific number of characters further on in the alphabet (the specific number is the same for all letters). Piyanga's coded message includes the word "PJVN". What could the original word say?

A. CAME   B. DAME   C. FAME   D. GAME   E. LAME

Question 272:
A number of people get on the bus at the station, which is considered the first stop. At each subsequent stop, 1/2 of the people on the bus get off and then 2 people get on. Between the 4th and 5th stop after the station, there are 5 people on the bus.

How many people got on at the station?

A. 4   B. 6   C. 20   D. 24   E. 30

Question 273:
I recently moved into a new house, and I am looking to paint my new living room. The price of several different colours of paint is displayed in the table below. A small can contains enough to paint 10 m² of wall. A large can contains enough to paint 25 m² of wall.

Colour	Cost for a Small Can	Cost for a Large Can
Red	£4	£12
Blue	£8	£15
Black	£3	£9
White	£2	£13
Green	£7	£15
Orange	£5	£20
Yellow	£10	£12

I decide to paint my room a mixture of blue and white, and I purchase some small cans of blue paint and white paint. The cost of blue paint accounts for 50% of the total cost. I paint a total of 100 m² of wall space.
I use up all the paint. How many m² of wall space have I painted blue?

A. 10 m²   B. 20 m²   C. 40 m²   D. 50 m²   E. 80 m²

Question 274:
Cakes usually cost 42p at the bakers. The bakers want to introduce a new offer where the amount in pence you pay for each cake is discounted by the square of the number of cakes you buy. For example, buying 3 cakes would mean each cake costs 33p. Isobel says that this is not a good offer from the baker's perspective as it would be cheaper to buy several cakes than just 1. How many cakes would you have to buy for the total cost to fall below 40p?

A. 2   B. 3   C. 4   D. 5   E. 6

Question 275:

The table below shows the percentages of students in two different universities who take various courses. There are 800 students in University A and 1200 students in University B. Biology, chemistry and physics are counted as "sciences".

	University A	University B
Biology	23.50	13.25
Economics	10.25	14.5
Physics	6.25	14.75
Mathematics	11.50	17.25
Chemistry	30.25	7.00
Psychology	18.25	33.25

Assuming each student only takes one course, how many more students in University A than University B study a "science"?

A. 10   B. 25   C. 60   D. 250   E. 600

Question 276:

Traveleasy Coaches charge passengers at a rate of 50p per mile travelled, plus an additional charge of £5.00 for each international border crossed during the journey. Europremier Coaches charge £15 for every journey, plus 10p per mile travelled, with no charge for crossing international borders. Sonia is travelling from France to Germany, crossing 1 international border. She finds that both companies will charge the same price for this journey.

How many miles is Sonia travelling?
A. 10   B. 20   C. 25   D. 35   E. 40

Question 277:

Lauren, Amy and Chloe live in different cities across England. They decide to meet up together in London and have a meal together. Lauren departs from Southampton at 2:30pm and arrives in London at 4pm. Amy's journey lasts twice as long as Lauren's journey and she arrives in London at 4:15pm. Chloe departs from Sheffield at 1:30pm, and her journey lasts an hour longer than Lauren's journey.

Which of the following statements is definitely true?
A. Chloe's journey took the longest time.
B. Amy departed after Lauren.
C. Chloe arrived last.
D. Everybody travelled by train.
E. Amy departed before Chloe.

Question 278:

Emma is packing to go on holiday by aeroplane. On the aeroplane, she can take a case with dimensions of 50cm by 50cm by 20cm, which, when fully packed, can weigh up to 20kg. The empty suitcase weighs 2kg. In her suitcase, she needs to take 3 books, each of which is 0.2m by 0.1m by 0.05m in size, and weighs 1000g. She would also like to take as many items of clothing as possible. Each item of clothing has volume 1500cm³ and weighs 400 g.

Assuming each item of clothing can be squashed so as to fill any shape gap, how many items of clothing can she take in her case?

A. 28   B. 31   C. 34   D. 37   E. 40

Question 279:

Alex is buying a new bed and mattress. There are 5 bed shops Alex can buy the bed and mattress he wants from, each of which sells the bed and mattress for a different price as follows:

- **Bed Shop A:** Bed £120, mattress £70
- **Bed Shop B:** All beds and mattresses £90 each
- **Bed Shop C:** Bed £140, mattress £60. Mattress half price when you buy a bed and mattress together.
- **Bed Shop D:** Bed £140, mattress £100. Get 33% off when you buy a bed and mattress together.
- **Bed Shop E:** Bed £175. All beds come with a free mattress.

Which is the cheapest place for Alex to buy the bed and mattress from?

A. Bed Shop A   C. Bed Shop C   E. Bed Shop E
B. Bed Shop B   D. Bed Shop D

Question 280:

In Joseph's sock drawer, there are 21 socks. 4 are blue, 5 are red, 6 are green and the rest are black. How many socks does he need to take from the drawer in order to guarantee he has a matching pair?

A. 3   B. 4   C. 5   D. 6   E. 7

Question 281:

Printing a magazine uses 1 sheet of card and 25 sheets of paper. It also uses ink. Paper comes in packs of 500 sheets and card comes in packs of 60 sheets. A pack of card is twice the price of a pack of paper. Each ink cartridge prints 130 sheets of either paper or card. A pack of paper costs £3. Ink cartridges cost £5 each.

How many complete magazines can be printed with a budget of £300?

A. 210   B. 220   C. 230   D. 240   E. 250

Question 282:

Rebecca went swimming yesterday. After a while she had covered one fifth of her intended distance. After swimming six more lengths of the pool, she had covered one quarter of her intended distance. How many lengths of the pool did she intend to complete?

A. 40   B. 72   C. 80   D. 100   E. 120

Question 283:

As a special treat, Sammy is allowed to eat five sweets from his very large jar. The jar contains 3 flavours of sweets – lemon, orange and strawberry. He wants to eat his five sweets in such a way that no two consecutive sweets have the same flavour.

In how many ways can he do this?

A. 32   B. 48   C. 72   D. 108   E. 162

Question 284:

Granny and her granddaughter Gill both had their birthday yesterday. Today, Granny's age in years is an even number and 15 times that of Gill. In 4 years' time Granny's age in years will be the square of Gill's age in years.

How many years older than Gill is Granny today?

A. 42   B. 49   C. 56   D. 60   E. 64

Question 285:

Pierre said, "Just one of us is telling the truth". Qadr said, "What Pierre says is not true". Ratna said, "What Qadr says is not true". Sven said, "What Ratna says is not true". Tanya said, "What Sven says is not true".

How many of them were telling the truth?

A. 0   B. 1   C. 2   D. 3   E. 4

Question 286:

Two entrants in a school's sponsored run adopt different tactics. Angus walks for half the time and runs for the other half, whilst Bruce walks for half the distance and runs for the other half. Both competitors walk at 3 mph and run at 6 mph. Angus takes 40 minutes to complete the course.

How many minutes does Bruce take?

A. 30   B. 35   C. 40   D. 45   E. 50

Question 287:

Dr Song discovers two new alien life forms on Mars. Species 8472 have one head and two legs. Species 24601 have four legs and one head. Dr Song counts a total of 73 heads and 290 legs in the area. How many members of Species 8472 are present?

A. 0
B. 1

C. 72
D. 73

E. 145

Question 288:

A restaurant menu states that:

"All chicken dishes are creamy, and all vegetable dishes are spicy. No creamy dishes contain vegetables."

Which of the following **must** be true?

A. Some chicken dishes are spicy.
B. All spicy dishes contain vegetables.
C. Some creamy dishes are spicy.
D. Some vegetable dishes contain tomatoes.
E. None of the above

Question 289:

Simon and his sister Lucy both cycle home from school. One day, Simon is kept back in detention, so Lucy sets off for home first. Lucy cycles the 8 miles home at 10 mph. Simon leaves school 20 minutes later than Lucy. How fast must he cycle in order to arrive home at the same time as Lucy?

A. 10 mph
B. 14 mph
C. 17 mph
D. 21 mph
E. 24 mph

Question 290:

Dr. Whu buys 2000 shares in a company at a rate of 50p per share. He then sells the shares for 58p per share. Subsequently he buys 1000 shares at 55p per share then sells them for 61p per share. There is a charge of £20 for each transaction of either buying or selling shares. What is Dr. Whu's total profit?

A. £140
B. £160
C. £180
D. £200
E. £220

Question 291:

Jina is playing darts. A dartboard is composed of equal segments, numbered from 1 to 20. She takes three throws, and each of the darts lands in a numbered segment. None land in the centre or in double or triple sections. What is the probability that her total score with the three darts is odd?

A. $1/4$
B. $1/3$
C. $1/2$
D. $3/5$
E. $2/3$

Question 292:

John Morgan invests £5,000 in a savings bond paying 5% interest per annum. What is the value of the investment in 3 years' time?

A. £5,250   C. £5,750   E. £5,788
B. £6,125   D. £6,442

Question 293:

Joe is 12 years younger than Michael. In 5 years, the sum of their ages will be 62. How old was Michael two years ago?

A. 20   B. 24   C. 26   D. 30   E. 32

Question 294:

A book has 500 pages. Vicky tears every page out that is a multiple of 4. She then tears out a fifth of the remaining pages. If the book measures 15 cm x 30cm and is made from paper of weight 110 gm^{-2}, how much lighter is the book now than at the start?

A. 990 g   B. 1,010 g   C. 1,250 g   D. 1,485 g   E. 1,590 g

Question 295:

A farmer is fertilising his crops. The more fertiliser is used, the more the crops grow. Fertiliser costs 80p per kilo. Fertilising at a rate of 0.2 kgm^{-2} increases the crop yield by £1.30 m^{-2}. For each additional 100g of fertiliser above 200g, the extra yield is 30% lower than the linear projection of the stated rate. At what rate of fertiliser application is it no longer cost effective to increase the dose?

A. 0.5 kgm^{-2} B. 0.6 kgm^{-2} C. 0.7 kgm^{-2} D. 0.8 kgm^{-2} E. 0.9 kgm^{-2}

Question 296:

Pet-Star, Furry Friends and Creature Cuddles are three pet shops, which each sell food for various types of pets.

Type of pet food	Amount of food required per week	Price per Kg in:		
		Pet-star	Furry Friends	Creature Cuddles
Guinea Pig	3 Kg	£2	£1	£1.50
Cat	6 Kg	£4	£6	£5
Rabbit	4 Kg	£3	£1	£2.50
Dog	8 Kg	£5	£8	£6
Chinchilla	2 Kg	£1.50	£0.50	£1

Given the information above, which of the following statements can we state is definitely *not* true?

A. Regardless of which of these shops you use, the most expensive animal to provide food for will be a dog.

B. If I own a mixture of cats and rabbits, it will be cheaper for me to shop at Pet-star.

C. If I own 3 cats and a dog, the cheapest place for me to shop is at Pet-star

D. Furry Friends sells the cheapest food for the type of pet requiring the most food

E. If I only have one pet, Creature Cuddles will not be the cheapest place to shop regardless of which type of pet I have.

Question 297:

I record my bank balance at the start of each month for six months to help me see how much I am spending each month. My salary is paid on the 10th of each month. At the start of the year, I earn £1000 a month but from March inclusive I receive a pay rise of 10%.

Date	Bank balance
January 1st	1,200
February 1st	1,029
March 1st	1,189
April 1st	1,050
May 1st	925
June 1st	1,025

In which month did I spend the most money?

A. January   B. February   C. March   D. April   E. May

Question 298:

Amy needs to travel from Southtown station to Northtown station, which are 100 miles apart. She can travel by 3 different methods: train, aeroplane or taxi. The tables below show the different times for these 3 methods. The taxi takes 1 minute to cover a distance of 1 mile. Aeroplane passengers must be at the airport 30 minutes before their flight. Southtown airport is 10 minutes travelling time from Southtown station and Northtown airport is 30 minutes travelling time from Northtown station.

If Amy wants to arrive by 1700 and wants to set off as late as possible, what method of travel should she choose and what time will she leave Southtown station?

Train	Departs Southtown station	1400	1500	1600
	Arrives Northtown station	1615	1650	1715
Flights	Departs Southtown airport	1610		
	Arrives Northtown airport	1645		

A. Flight, 1530   C. Taxi, 1520   E. Flight, 1610
B. Train, 1600   D. Train, 1500

Question 299:

In the multiplication grid below, a, b, c and d are all integers. What does d equal?

A. 18   B. 24   C. 30   D. 40   E. 45

	c	d
a	168	720
b	119	510

Question 300:

A sixth form college has 1,500 students. 48% are girls. 80 of the girls are mixed race.

If an equal proportion of boys and girls are mixed race, how many mixed-race boys are there in the college to the nearest 10?

A. 50   B. 60   C. 70   D. 80   E. 90

SECTION 2

Section 2 is undoubtedly the most time-pressured section of the BMAT. This section tests concepts from GCSE level biology, chemistry, physics and maths. You have 27 questions to answer in 30 minutes. The questions can be quite difficult and it's easy to trip up when scientific concepts are presented in unfamiliar ways. However, Section 2 is also the section in which you can improve the most quickly in so it's well worth spending time practicing for it.

Although the vast majority of questions in section 2 aren't particularly difficult, the intense time pressure of having to do one question every minute makes this section the hardest for many students sitting the BMAT. As with section 1, the trick is to identify and complete the easier questions first whilst leaving the harder ones for the end.

In general, the biology and chemistry questions in the BMAT require the least amount of time per question whilst the maths and physics are more time consuming as they usually consist of multi-step calculations.

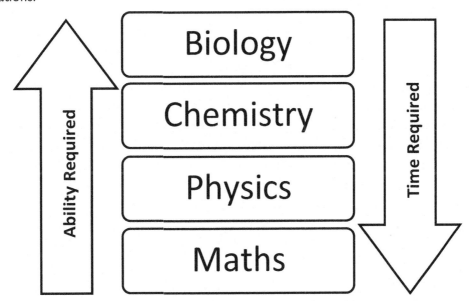

Gaps in Knowledge

The BMAT only tests GCSE level content. However, there is a large variation in content between the GCSE exam boards meaning that you may not have covered some topics that are examinable. This is more likely if you didn't carry on with biology or physics to AS level (e.g. Newtonian mechanics and parallel circuits in physics; hormones and stem cells in biology). If you fall into this category, you are highly advised to go through the BMAT specification and ensure that you have covered all examinable topics. An electronic copy of this can be obtained from the official BMAT website at www.admissionstestingservice.org/bmat.

It is worth noting that although the content (i.e. concepts and scientific facts) tested in the BMAT is set at GCSE level, very often the questions will required *application* of this content to an unfamiliar scenario. This is designed to test your understanding as well as recall of the material, so it's paramount that you brush up on any weak areas.

The questions in this book will help highlight any particular areas of weakness or gaps in your knowledge that you may have. Upon discovering these, make sure you take some time to revise these topics before carrying on – there is little to be gained by attempting section 2 questions with huge gaps in your knowledge.

Maths

Being confident with maths is extremely important for section 2. You won't have access to a calculator, so need to be confident manipulating numbers in your head or using pen and paper. Many students find that improving their numerical and algebraic skills usually results in big improvements in their section 1 and 2 scores. Remember that maths concepts tested in section 2 will also span physics (manipulating equations and standard form) and chemistry (mass calculations). So, if you find yourself consistently running out of time in section 2, spending a few hours on brushing up your basic maths skills (such as long division, converting between percentages and decimals…) may do wonders for you.

SECTION 2: BIOLOGY

Thankfully, the biology questions tend to be fairly straightforward and require the least amount of time compared to other questions in Section 2. You should be able to do the majority of these well within 60 seconds. This means that you should be aiming to make up time by quickly completing biology questions, leaving more time for calculations etc. Unlike the other topics in section 2, the majority of biology questions will simply test your recall, so the trick really is to ensure that there are no obvious gaps in your knowledge.

Before going onto to do the practice questions in this book, ensure you are comfortable with the following commonly tested topics:

- Structure of animal, plant and bacterial cells
- Osmosis, diffusion and active transport
- Cell Division (mitosis + meiosis)
- Family pedigrees and Inheritance
- DNA structure and replication
- Gene technology & stem Cells
- Enzymes – function, mechanism and examples of digestive enzymes
- Aerobic and anaerobic respiration
- The central vs. peripheral nervous system
- The respiratory cycle including movement of ribs and diaphragm
- The cardiac cycle
- Hormones
- Basic immunology
- Food chains and food webs
- The carbon and nitrogen cycles

Top tip! If you find yourself getting less than 50% of biology questions correct in this book, make sure you revisit the syllabus before attempting more questions as this is the best way to maximise your efficiency. In general, there is no reason why you shouldn't be able to get the vast majority of biology questions correct (and in well under 60 seconds) with sufficient practice.

BIOLOGY QUESTIONS

Question 301:

In relation to the human genome, which of the following statements are correct?

1. The genome is encoded by 4 different bases in DNA.
2. The sugar backbone of the DNA strand is formed of glucose.
3. DNA is found in the nucleus of bacteria.

A. 1 only
B. 2 only
C. 3 only
D. 1 and 2
E. 1 and 3

Question 302:

Animal cells contain organelles that take part in vital metabolic processes. Which of the following is true?

1. The majority of energy production by animal cells occurs in the mitochondria.
2. The cell wall protects the animal cell membrane from outside pressure differences.
3. Chloroplasts are not present in animal cells.

A. 1 only
B. 2 only
C. 3 only
D. 1 and 3
E. 2 and 3

Question 303:

With regards to animal mitochondria, which of the following is correct?

A. Mitochondria are not necessary for aerobic respiration.
B. Mitochondria are enveloped by a double membrane.
C. Mitochondria are more abundant in skeletal muscle than fat cells.
D. The majority of DNA replication happens inside mitochondria.
E. The majority of protein synthesis occurs in mitochondria.

Question 304:

In relation to bacteria, which of the following is **FALSE**?

A. Bacteria always lead to disease.
B. Bacteria contain plasmid DNA.
C. Bacteria do not contain mitochondria.
D. Bacteria have a cell wall and a plasma membrane.
E. Some bacteria are susceptible to antibiotics.

Question 305:

In relation to bacterial replication, which of the following is correct?

A. Bacteria undergo sexual reproduction.
B. Bacteria have a nucleus.
C. Bacteria carry genetic information on circular plasmids.
D. Bacterial genomes are formed of RNA instead of DNA.
E. Bacteria require gametes to replicate.

Question 306

Which of the following statements are correct regarding active transport?

A. ATP is necessary and sufficient for active transport.
B. ATP is not necessary but is sufficient for active transport.
C. The relative concentrations of the material being transported have little impact on the rate of active transport.
D. Transport proteins are necessary and sufficient for active transport.
E. Active transport relies on transport proteins that are powered by an electrochemical gradient.

Question 307:

Concerning mammalian reproduction, which of the following is **FALSE**?

A. Fertilisation involves the fusion of two gametes.
B. Reproduction is sexual and the offspring display genetic variation.
C. Reproduction relies upon the exchange of genetic material.
D. Mammalian gametes are diploid cells produced via meiosis.
E. Embryonic growth requires carefully controlled mitosis.

Question 308:

Which of the following apply to Mendelian inheritance?

1. It only applies to plants.
2. It treats different traits as either dominant or recessive.
3. Heterozygotes have a 25% chance of expressing a recessive trait.

A. I only
B. 2 only
C. 3 only
D. I and 2
E. I and 3

Question 309:
Which of the following statements are correct?

A. Hormones are secreted into the blood stream and act over long distances at specific target organs.
B. Hormones are substances that almost always cause muscles to contract.
C. Hormones have no impact on the nervous or enteric systems.
D. Hormones are always derived from food and never synthesised.
E. Hormones act rapidly to restore homeostasis.

Question 310:
With regard to neuronal signalling in the body, which of the following are true?

1. Neuronal transmission can be caused by both electrical and chemical stimulation.
2. Synapses ultimately result in the production of an electrical current for signal transduction.
3. All synapses in humans are electrical and unidirectional.

A. 1 only   C. 3 only   E. 1 and 3
B. 2 only   D. 1 and 2

Question 311:
What is the **primary** reason that pH is controlled so tightly in the human body?

A. To allow rapid protein synthesis.
B. To allow for effective digestion throughout the GI tract.
C. To ensure ions can function properly in neural signalling.
D. To ensure enzymes are able to function properly.
E. To prevent changes in core body temperature.

Question 312:
Which of the following statements are correct regarding the bacterial cell wall?

1. It confers bacteria protection against external environmental stimuli.
2. It is an evolutionary remnant and now has little functional significance in most bacteria.
3. It is made up primarily of glucose in bacteria.

A. Only 1   C. Only 3   E. 2 and 3
B. Only 2   D. 1 and 2

Question 313:

Which of the following statements are correct regarding mitosis?

1. It is important in sexual reproduction.
2. A single round of mitosis results in the formation of 2 genetically distinct daughter cells.
3. Mitosis is vital for tissue growth, as it is the basis for cell multiplication.

A. Only 1
B. Only 2

C. Only 3
D. 1 and 2

E. 2 and 3

Question 314:

Which of the following is the best definition of a mutation?

A. A mutation is a permanent change in DNA.
B. A mutation is a permanent change in DNA that is harmful to an organism.
C. A mutation is a permanent change in the structure of intra-cellular organelles caused by changes in DNA/RNA.
D. A mutation is a permanent change in chromosomal structure caused by DNA/RNA changes.

Question 315:

In relation to mutations, which of the following statements are correct?

1. Mutations always lead to discernible changes in the phenotype of an organism.
2. Mutations are central to natural processes such as evolution.
3. Mutations play a role in cancer.

A. Only 1
B. Only 2

C. Only 3
D. 1 and 2

E. 2 and 3

Question 316:

Which of the following is the most accurate definition of an antibody?

A. An antibody is a molecule that protects red blood cells from changes in pH.
B. An antibody is a molecule produced only by humans and has a pivotal role in the immune system.
C. An antibody is a toxin produced by a pathogen to damage the host organism.
D. An antibody is a molecule that is used by the immune system to identify and neutralize foreign objects and molecules.
E. Antibodies are small proteins found in red blood cells that help increase oxygen carriage.

Question 317:

Which of the following statements about the kidney are correct?

1. The kidneys filter the blood and remove waste products from the body.
2. The kidneys are involved in the digestion of food.
3. In a healthy individual, the kidneys produce urine that contains high levels of glucose.

A. Only 1   C. Only 3   E. 2 and 3

B. Only 2   D. 1 and 2

Question 318:

Which of the following statements are correct?

1. Hormones are slower acting than nerves.
2. Hormones act for a very short time.
3. Hormones act more generally than nerves.
4. Hormones are released when you are frightened.

A. 1 only   C. 2 and 4 only   E. 1, 2, 3 and 4

B. 1 and 3 only   D. 1, 3 and 4 only

Question 319:

Which statements about homeostasis are correct?

1. Homeostasis is about ensuring the inputs within your body exceed the outputs to maintain a constant internal environment.
2. Homeostasis is about ensuring the inputs within your body are less than the outputs to maintain a constant internal environment.
3. Homeostasis is about balancing the inputs within your body with the outputs to ensure your body fluctuates with the needs of the external environment.
4. Homeostasis is about balancing the inputs within your body with the outputs to maintain a constant internal environment.

A. 1 only   C. 3 only   E. 1 and 3 only

B. 2 only   D. 4 only

Question 320:

Which of the following statements regarding the food chain is true?

A. There is more energy and biomass each time you move up a trophic level.
B. There is less energy and biomass each time you move up a trophic level.
C. There is more energy but less biomass each time you move up a trophic level.
D. There is less energy but more biomass each time you move up a trophic level.
E. There is no difference in the energy or biomass when you move up a trophic level.

Question 321:

Which of the following statements are true about asexual reproduction?

1. There is no fusion of gametes.
2. There are two parents.
3. There is no mixing of chromosomes.
4. Binary fission is an example of asexual reproduction.

A. 1, 3 and 4 only   C. 1 and 4 only   E. All are true
B. 1 and 3 only   D. 3 and 4 only

Question 322:

Put the following components of a stimulus-reponse arc in the order in which they are activated when Jonas sees a bowl of chicken and moves towards it.

1. Retina   3. Sensory neuron   5. Muscle
2. Motor neuron   4. Brain

A. 1 - 3 - 4 - 5 - 2   C. 5 - 1 - 3 - 2 - 4   E. 1 - 3 - 4 - 2 - 5
B. 1 - 2 - 3 - 4 - 5   D. 1 - 3 - 2 - 4 - 5

Question 323:

The following statements relate to the flow of blood through the heart.

1. The right-hand side of the heart contains deoxygenated blood.
2. The aorta receives oxygenated blood from the left atrium.
3. The heart pumps deoxygenated blood into the pulmonary vein.
4. Valves are present to prevent flow of blood from the left ventricle to the right ventricle.

Which of these statements is / are correct?

A. 3 only
B. 1 and 3
C. 2 and 4
D. 1 only
E. None of the above

Question 324:

Which of the following statements are true about animal cloning?

1. Animals cloned from embryo transplants are genetically identical.
2. The genetic material is removed from an unfertilised egg during adult cell cloning.
3. Cloning can cause a reduced gene pool.
4. Cloning is only possible with mammals.

A. 1 only   C. 1 and 2 only   E. 1, 2 and 3 only
B. 2 only   D. 4 only

Question 325:

Which of the following statements are true with regards to evolution?

1. Individuals within a species show variation because of differences in their genes.
2. Beneficial mutations will accumulate within a population.
3. Gene differences are caused by sexual reproduction and mutations.
4. Species with similar characteristics never have similar genes.

A. 1 only   C. 2 and 3 only   E. 1, 2 and 3 only
B. 1 and 4 only   D. 2 and 4 only

Question 326:
Which of the following statements about genetics are correct?

1. Alleles are a similar version of different cells.
2. If you are homozygous for a trait, you have three alleles the same for that particular gene.
3. If you are heterozygous for a trait, you have two different alleles for that particular gene.
4. To show the characteristic that is caused by a recessive allele, both carried alleles for the gene have to be recessive.

A. 1 only
B. 2 only
C. 3 only
D. 4 only
E. 3 and 4 only

Question 327:
Which of the following statements are correct about meiosis?

1. The DNA content of a gamete is half that of a human red blood cell.
2. Meiosis requires ATP.
3. Meiosis only takes place in reproductive tissue.
4. In meiosis, a diploid cell divides in such a way so as to produce two haploid cells.

A. 1 only
B. 3 only
C. 1 and 2 only
D. 2 and 3 only
E. 2 and 4 only

Question 328:
Put the following statements in the correct order of events for when there is too little water in the blood.

1. Urine is more concentrated
2. Pituitary gland releases ADH
3. Blood water level returns to normal
4. Hypothalamus detects too little water in blood
5. Kidney affects water level

A. 1 - 2 - 3 - 4 - 5
B. 5 - 4 - 3 - 2 - 1
C. 4 - 2 - 5 - 1 - 3
D. 3 - 2 - 4 - 1 - 5
E. 5 - 2 - 3 - 4 - 1

Question 329:
The pH of venous blood is 7.35. Which of the following is the likely pH of arterial blood?

A. 5.2
B. 6.5
C. 7.0
D. 7.4
E. 8.0

Question 330:

Which of the following statements are true of the cytoplasm?

1. The vast majority of the cytoplasm is made up of water.
2. All contents of animal cells are contained in the cytoplasm.
3. The cytoplasm contains electrolytes and proteins.

A. 1 only   C. 3 only   E. 1 and 3 only

B. 2 only   D. 1 and 2 only

Question 331:

ATP is produced in which of the following organelles?

1. The cytoplasm
2. Plasmids
3. The mitochondria
4. The nucleus

A. 1 only   C. 3 only   E. 1 and 2

B. 2 only   D. 1 and 3

Question 332:

The cell membrane:

A. Is made up of a phospholipid bilayer which only allows active transport across it.
B. Is not found in bacteria.
C. Is a semi-permeable barrier to ions and organic molecules.
D. Consists purely of enzymes.

Question 333:

Cells of the *Polyommatus atlantica* butterfly of the Lycaenidae family have 446 chromosomes. Which of the following statements about a *P. atlantica* butterfly are correct?

1. Mitosis will produce 2 daughter cells each with 223 pairs of chromosomes
2. Meiosis will produce 4 daughter cells each with 223 chromosomes
3. Mitosis will produce 4 daughter cells each with 446 chromosomes
4. Meiosis will produce 2 daughter cells each with 223 pairs of chromosomes

A. 1 and 2 only   C. 2 and 3 only   E. 1, 2 and 3 only

B. 1 and 3 only   D. 3 and 4 only   F. 1, 2, 3 and 4

Questions 334-336 are based on the following information:
Assume that hair colour is determined by a single allele. The R allele is dominant and results in black hair. The r allele is recessive for red hair. Mary (red hair) and Bob (black hair) are having a baby girl.

Question 334:
What is the probability that the baby will have red hair?

A. 0% only

B. 25% only

C. 50% only

D. 0% or 25%

E. 0% or 50%

F. 25% or 50%

Question 335:
Mary and Bob have a second child, Tim, who is born with red hair. What does this confirm about Bob?

A. Bob is heterozygous for the red hair allele.
B. Bob is homozygous dominant for the red hair allele.
C. Bob is homozygous recessive for the red hair allele.
D. Bob does not have the red hair allele.

Question 336:
Mary and Bob go on to have a third child. What are the chances that this child will be born homozygous for black hair?

A. 0%   B. 25%   C. 50%   D. 75%   E. 100%

Question 337:
Why does air flow into the chest on inspiration?

1. Atmospheric pressure is less than intra-thoracic pressure during inspiration.
2. Atmospheric pressure is greater than intra-thoracic pressure during inspiration.
3. Anterior and lateral chest expansion decreases absolute intra-thoracic pressure.
4. Anterior and lateral chest expansion increases absolute intra-thoracic pressure.

A. 1 only

B. 2 only

C. 2 and 3

D. 1 and 4

E. 1 and 3

Question 338:
Which of the following components of a food chain represent the largest biomass?

A. Producers

B. Decomposers

C. Primary consumers

D. Secondary consumers

E. Tertiary consumers

Question 339:

Concerning the nitrogen cycle, which of the following are true?

1. The majority of the Earth's atmosphere is nitrogen.
2. Most of the nitrogen in the Earth's atmosphere is inert.
3. Bacteria are essential for nitrogen fixation.
4. Nitrogen fixation occurs during lightning strikes.

A. 1 and 2

B. 1 and 3

C. 2 and 3

D. 2 and 4

E. 1, 2, 3 and 4

Question 340:

Which of the following statement are correct regarding mutations?

1. Mutations always cause proteins to lose their function.
2. Mutations always change the structure of the protein encoded by the affected gene.
3. Mutations always result in cancer.

A. Only 1

B. Only 2

C. Only 3

D. 1 and 2

E. None of
   the above

Question 341:

Which of the following is not a function of the central nervous system?

A. Coordination of movement

B. Decision making and executive functions

C. Control of heart rate

D. Cognition

E. Memory

Question 342:

Which of the following control mechanisms is / are involved in modulating the amount of blood that is pumped per unit time?

1. Voluntary control.
2. Sympathetic control to decrease heart rate.
3. Parasympathetic control to increase heart rate.

A. Only 1

B. Only 2

C. 2 and 3

D. 1, 2 and 3

E. None of
   the above

Question 343:

Vijay goes to see his GP with fatty, smelly stools that float in water. Which of the following enzymes is most likely to be malfunctioning?

A. Amylase
B. Lipase

C. Protease
D. Sucrase

E. Lactase

Question 344:

Which of the following statements concerning the cardiovascular system is correct?

A. Oxygenated blood from the lungs flows to the heart via the pulmonary artery.
B. All arteries carry oxygenated blood.
C. All animals have a double circulatory system.
D. The superior vena cava contains oxygenated blood
E. None of the above.

Question 345:

In which part of the GI tract is there the least enzymatic activity for digestion?

A. Mouth
B. Stomach

C. Small intestine
D. Large intestine

E. Rectum

Question 346:

Oge touches a hot stove and immediately moves her hand away. Which of the following components are **NOT** involved in this reflex reaction?

1. Thermo-receptor
2. Brain

3. Spinal Cord
4. Sensory nerve

5. Motor nerve
6. Muscle

A. 1 only
B. 2 only

C. 3 only
D. 1 and 2 only

E. 1, 2 and 3 only

Question 347:

Which of the following represents a scenario with an appropriate description of the mode of transport?

1. Osmosis = water moving from a hypotonic solution outside of a potato cell, across the cell wall and cell membrane and into the hypertonic cytoplasm of the potato cell.
2. Active transport = carbon dioxide moving across a respiring cell's membrane and dissolving in blood plasma.
3. Diffusion = reabsorption of amino acids against a concentration gradient in the glomeruluar apparatus.

A. 1 only
B. 2 only
C. 3 only
D. 1 and 2 only
E. 2 and 3 only

Question 348:

Which of the following equations represents anaerobic respiration in animal cells?

1. Carbohydrate + Oxygen $\rightarrow$ Energy + Carbon dioxide + Water
2. Carbohydrate $\rightarrow$ Energy + Lactic acid + Carbon dioxide
3. Carbohydrate $\rightarrow$ Energy + Lactic acid
4. Carbohydrate $\rightarrow$ Energy + Ethanol + Carbon dioxide

A. 1 only
B. 2 and 4
C. 3 and 4 only
D. 4 only
E. 3 only

Question 349:

Which of the following statements regarding respiration in animal cells are correct?

1. Mitochondria are the centres of both aerobic and anaerobic respiration.
2. The cytoplasm is the main site of anaerobic respiration.
3. In aerobic respiration, every two moles of glucose results in the liberation of 12 moles of CO_2.
4. Anaerobic respiration is more efficient than aerobic respiration.

A. 1 and 2
B. 1 and 4
C. 2 and 3
D. 2 and 4
E. 3 and 4

Question 350:

Which of the following statements are true?

1. The nucleus contains the cell's chromosomes.
2. The cytoplasm consists purely of water.
3. The plasma membrane is a single phospholipid layer.
4. The cell wall prevents plants cells from lysis due to osmotic pressure.

A. 1 and 2   C. 1, 3 and 4   E. 1, 2 and 4
B. 1 and 4   D. 1, 2 and 3

Question 351:

Which of the following statements are true about osmosis?

1. If a medium is more concentrated than the cell cytoplasm, the cell will gain water through osmosis.
2. If a medium is less concentrated than the cell cytoplasm, the cell will gain water through osmosis.
3. If a medium is less concentrated than the cell cytoplasm, the cell will lose water through osmosis.
4. If a medium is more concentrated than the cell cytoplasm, the cell will lose water through osmosis.
5. The medium's tonicity has no impact on the movement of water.

A. 1 only   B. 2 only   C. 1 and 3   D. 2 and 4   E. 5 only

Question 352:

Which of the following statements are true about stem cells?

1. Stem cells have the ability to differentiate into other mature types of cells.
2. Stem cells are unable to maintain their undifferentiated state.
3. Stem cells can be classified as embryonic stem cells or adult stem cells.
4. Stem cells are only found in embryos.

A. 1 and 3   B. 3 and 4   C. 2 and 3   D. 1 and 2   E. 2 and 4

Question 353:

Which of the following are **NOT** examples of natural selection?

1. Giraffes growing longer necks to eat taller plants.
2. Antibiotic resistance developed by certain strains of bacteria.
3. Pesticide resistance among locusts in farms.
4. Breeding of horses to make them run faster.

A. 1 only   B. 4 only   C. 1 and 3   D. 1 and 4   E. 2 and 4

Question 354:

Which of the following statements are true?

1. Enzymes stabilise the transition state and therefore lower the activation energy.
2. Enzymes distort substrates in order to lower activation energy.
3. Enzymes decrease temperature to slow down reactions and lower the activation energy.
4. Enzymes provide alternative pathways for reactions to occur.

A. I only   B. I and 3   C. I and 4   D. 2 and 4   E. I, 2 and 4

Question 355:

Which of the following are examples of negative feedback?

1. Salivating whilst waiting for a meal.
2. Throwing a dart.
3. The regulation of blood pH.
4. The regulation of blood pressure.

A. I only   B. I and 2   C. 3 and 4   D. 2, 3, and 4   E. I, 2, 3 and 4

Question 356:

Which of the following statements about the immune system are true?

1. White blood cells defend against bacterial and fungal infections.
2. Red blood cells are involved in the process of phagocytosis.
3. White blood cells use antibodies to fight pathogens.
4. Antibodies are produced by bone marrow stem cells.

A. I and 3   C. 2 and 3   E. I, 2, and 3
B. I and 4   D. 2 and 4

Question 357:

The cardiovascular system does **NOT**:

A. Deliver vital nutrients to peripheral cells.
B. Oxygenate blood and transport it to peripheral cells.
C. Act as a mode of transportation for hormones to reach their target organ.
D. Facilitate thermoregulation.
E. Respond to exercise by increasing the heart rate.

Question 358:
Which of the following statements is correct?

A. Adrenaline can sometimes decrease heart rate.
B. Adrenaline is rarely released during flight or fight responses.
C. Adrenaline causes peripheral vasoconstriction.
D. Adrenaline only affects the cardiovascular system.
E. Adrenaline travels primarily in lymphatic vessels.
F. None of the above.

Question 359:
Which of the following statements is true?

A. Protein synthesis occurs solely in the nucleus.
B. Each amino acid is coded for by three DNA bases.
C. Each protein is coded for by three amino acids.
D. Red blood cells can create new proteins to prolong their lifespan.
E. Protein synthesis isn't necessary for mitosis to take place.

Question 360:
A solution of amylase and carbohydrate is present in a beaker, where the pH of the contents is 6.3. Assuming amylase is saturated, which of the following will increase the rate of production of the product?

1. Add sodium bicarbonate
2. Add carbohydrate
3. Add amylase
4. Increase the temperature to 100° C

A. I only
B. 2 only
C. I and 3
D. 4 only
E. I and 2

Question 361:
Celestial necrosis is a newly discovered autosomal recessive disorder. A female carrier and a male with the disease produce two sons. What is the probability that neither son's genotype contains the celestial necrosis allele?

A. 100%
B. 75%
C. 50%
D. 25%
E. 0%

Question 362:
Which of the following organs has **no** endocrine function?

A. The thyroid
B. The ovary

C. The pancreas
D. The testes

E. None of the above.

Question 363:
Which of the following statements are true?

1. Increasing levels of insulin cause a decrease in blood glucose levels.
2. Increasing levels of glycogen cause an increase in blood glucose levels.
3. Increasing levels of adrenaline decrease the heart rate.

A. 1 only
B. 2 only
C. 3 only
D. 1 and 2
E. 2 and 3

Question 364:
Which of the following rows is correct?

	Oxygenated Blood		Deoxygenated Blood	
A.	Left atrium	Left ventricle	Right atrium	Right ventricle
B.	Left atrium	Right atrium	Left ventricle	Right ventricle
C.	Left atrium	Right ventricle	Right atrium	Right ventricle
D.	Right atrium	Right ventricle	Left atrium	Left ventricle
E.	Left ventricle	Right atrium	Left atrium	Right ventricle

Questions 365-367 are based on the following information:

The pedigree below shows the inheritance of a newly discovered disease that affects connective tissue called Nafram syndrome. Individual 1 is a normal homozygote (disease-free).

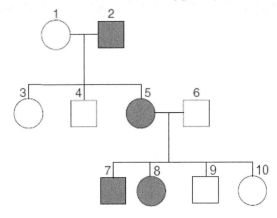

Question 365:

Based on the pedigree, what is the pattern of inheritance for Nafram syndrome?

A. Autosomal dominant

B. Autosomal recessive

C. X-linked dominant

D. X-linked recessive

E. Cannot be determined

Question 366:

Which individuals in the pedigree must be heterozygous for Nafram syndrome?

A. 1 and 2

B. 8 and 9

C. 2 and 5

D. 5 and 6

E. 6 and 8

Question 367:

Taking N to denote a disease-conferring allele and n to denote a normal allele, which of the following are **NOT** possible genotypes for 6's parents?

1. NN x NN

2. NN x Nn

3. Nn x nn

4. Nn x Nn

5. nn x nn

A. 1 and 2

B. 1 and 3

C. 2 and 3

D. 2 and 5

E. 3 and 4

Question 368:

Which of the following correctly describes the passage of urine through the body?

	1st	2nd	3rd	4th
A	Kidney	Ureter	Bladder	Urethra
B	Kidney	Urethra	Bladder	Ureter
C	Urethra	Bladder	Ureter	Kidney
D	Ureter	Kidney	Bladder	Urethra

Question 369:

Which of the following best describes the passage of blood from the body, through the heart, and back to the body?

A. Aorta → Left Ventricle → Left Atrium → Inferior Vena Cava → Right Atrium → Right Ventricle → Lungs → Aorta

B. Inferior vena cava → Left Atrium → Left Ventricle → Lungs → Right Atrium → Right Ventricle → Aorta

C. Inferior vena cava → Right Ventricle → Right Atrium → Lungs → Left Atrium → Left Ventricle → Aorta

D. Aorta → Left Atrium → Left Ventricle → Lungs → Right Atrium → Right Ventricle → Inferior Vena Cava

E. None of the above.

Question 370:

Which of the following best describes the events during inspiration?

	Intrathoracic Pressure	Intercostal Muscles	Diaphragm
A	Increases	Contract	Contracts
B	Increases	Relax	Contracts
C	Increases	Contract	Relaxes
D	Increases	Relax	Relaxes
E	Decreases	Contract	Contracts

Questions 371-372 are based on the following information:

DNA is made up of four nucleotide bases: adenine, cytosine, guanine and thymine. A triplet of bases (codon) is a sequence of three nucleotides which code for an amino acid. While there are only 20 amino acids there are 64 different combinations of the four DNA nucleotide bases. This means that more than one combination of 3 DNA nucleotide sequences code for the same amino acid.

Question 371:

Which property of the genetic code is described above?

A. The code is unambiguous.
B. The code is universal.
C. The code is non-overlapping.
D. The code is degenerate.
E. The code is preserved.

Question 372:

Which type of mutation does the described property protect against the most?

A. An insertion - where a single nucleotide is inserted.
B. A point mutation - where a single nucleotide is replaced for another.
C. A deletion - where a single nucleotide is deleted.
D. A repeat expansion - where a repeated trinucleotide sequence is added.
E. A duplication - where a piece of DNA is abnormally copied.

Question 373:

Which row of the table below describes what happens when the temperature decreases in the external environment?

	Temperature Change Detected by	Sweat Gland Secretion	Cutaneous Blood Flow
A	Hypothalamus	Increases	Increases
B	Hypothalamus	Increases	Decreases
C	Hypothalamus	Decreases	Increases
D	Hypothalamus	Decreases	Decreases
E	Cerebral Cortex	Increases	Increases

Question 374:

Which of the following processes involve active transport?

1. Reabsorption of glucose in the kidney.
2. Movement of carbon dioxide into the alveoli of the lungs.
3. Movement of chemicals in a synapse.

A. 1 only
B. 2 only
C. 3 only
D. 1 and 2
E. 1 and 3

Question 375:

Which of the following statements is correct about enzymes?

A. All enzymes are made up of amino acids only.
B. Enzymes can sometimes slow the rate of reactions.
C. Enzymes are heat sensitive but resistant to changes in pH.
D. Enzymes are unspecific in their substrate use.
E. None of the above.

SECTION 2: CHEMISTRY

Most students don't struggle with the content of BMAT chemistry as they'll be studying it at A level. However, there are certain questions that even very good students tend to find difficult under time pressure e.g. balancing equations and mass calculations. It is essential that you do enough practice so that you're able to do these questions quickly and accurately as they have the potential to really slow your progress through the Section 2 questions.

Balancing Equations

For some reason, most students are rarely shown how to formally balance equations – including those studying it at A-level. Balancing equations intuitively or via trial and error will only get you so far in the BMAT as the equations you'll have to work with will be fairly complex. To avoid wasting valuable time, it is essential you learn a clear method that will allow you to consistently solve these in less than 60 seconds. The method shown below is the simplest way and requires you to be able to do quick mental arithmetic (which is something you should be aiming for anyway). The easiest way to do learn it is through an example:

The following equation shows the reaction between iodic acid (HIO_3), hydrochloric acid (HCl) and copper iodide (CuI_2):

$$\textbf{a } HIO_3 + \textbf{b } CuI_2 + \textbf{c } HCl \rightarrow \textbf{d } CuCl_3 + \textbf{e } ICl + \textbf{f } H_2O$$

What values of **a**, **b**, **c**, **d**, **e** and **f** are needed in order to balance the equation?

Step 1: Pick an element and see how many atoms there are on the left and right sides.

Step 2: Form an equation to represent this. For Cu: b = d

Step 3: See if any of the answer options given **don't** satisfy b=d. In this case, for option E, b is 8 and d is 10. This allows us to eliminate option E immediately.

	a	b	c	d	e	f
A	5	4	25	4	13	15
B	5	4	20	4	8	15
C	5	6	20	6	8	15
D	2	8	10	8	8	15
E	6	8	24	10	16	15
F	6	10	22	10	16	15

Once you've eliminated as many options as possible using this method, go back to step 1 and pick another element.

For Hydrogen (H): a + c = 2. Then see if any of the answer options **don't** satisfy a + c = 2f.

- Option A: 5 + 25 is equal to 2 x 15
- Option B: 5 + 20 is not equal to 2 x 15
- Option C: 5 + 20 is not equal to 2 x 15
- Option D: 2 + 10 is not equal to 2 x 15

This allows us to eliminate option B, C and D. E has already been eliminated. Thus, the only solution possible is A.

This method works best when you get given a table above as this allows you to quickly eliminate options. However, it is still a viable method even if you don't get this information in the question. You might get given the equation with some numerical values present, so could apply the same method to that scenario.

Chemistry Calculations
Equations you **MUST** know for Section 2:

- Atomic Mass = Mass/Moles

- Amount (mol) = Concentration (mol/dm³) x Volume (dm³)

Avogadro's Constant:
One mole of anything contains 6×10^{23} of it e.g. 5 Moles of water contain $5 \times 6 \times 10^{23}$ number of water molecules.

Abundances:
The average atomic mass takes the abundances of all isotopes into account. Thus:
A_r = (Abundance of Isotope 1) x (Mass of Isotope 1) + (Abundance of Isotope 2) x (Mass of Isotope 2) +…

Converting between volumes:
It is useful to remember that $1 \ dm^3 = 1000 \ cm^3 = 1$ litre

CHEMISTRY QUESTIONS

Question 376:

Which of the following most accurately defines an isotope?

A. An isotope is an atom of an element that has the same number of protons in the nucleus but a different number of neutrons orbiting the nucleus.

B. An isotope is an atom of an element that has the same number of neutrons in the nucleus but a different number of protons orbiting the nucleus.

C. An isotope is any atom of an element that can be split to produce nuclear energy.

D. An isotope is an atom of an element that has the same number of protons in the nucleus but a different number of neutrons in the nucleus.

E. An isotope is an atom of an element that has the same number of protons in the nucleus but a different number of electrons orbiting it.

Question 377:

Which of the following is an example of a displacement reaction?

1. $Fe + SnSO4 \rightarrow FeSO_4 + Sn$
2. $Cl_2 + 2KBr \rightarrow Br_2 + 2KCl$
3. $H_2SO_4 + Mg \rightarrow MgSO_4 + H_2$
4. $NaHCO_3 + HCl \rightarrow NaCl + CO_2 + H_2O$

A. 1 only

B. 1 and 2 only

C. 2 and 3 only

D. 3 and 4 only

E. 1, 2, 3 and 4

Question 378:

What values of **a**, **b** and **c** are needed to balance the equation below?

$$aCa(OH)_2 + bH_3PO_4 \rightarrow Ca_3(PO_4)_2 + cH_2O$$

A. **a** = 3, **b** = 2, **c** = 6

B. **a** = 2, **b** = 2, **c** = 4

C. **a** = 3, **b** = 2, **c** = 1

D. **a** = 1, **b** = 2, **c** = 3

E. **a** = 4, **b** = 2, **c** = 6

Question 379:

What values of **s**, **t** and **u** are needed to balance the equation below?

$$sAgNO_3 + tK_3PO_4 \rightarrow 3Ag_3PO_4 + uKNO_3$$

A. **s** = 9, **t** = 3, **u** = 9

B. **s** = 6, **t** = 3, **u** = 9

C. **s** = 9, **t** = 3, **u** = 6

D. **s** = 9, **t** = 6, **u** = 9

E. **s** = 3, **t** = 3, **u** = 9

Question 380:

Which of the following statements are true with regard to displacement?

1. A less reactive halogen can displace a more reactive halogen.
2. Chlorine cannot displace bromine or iodine from an aqueous solution of its salts.
3. Bromine can displace iodine according to the reactivity series.
4. Fluorine can displace chlorine as it is higher up the group.
5. Lithium can displace francium as it is higher up the group.

A. 3 only   C. 1 and 2 only   E. 2, 3 and 5 only
B. 5 only   D. 3 and 4 only

Question 381:

What mass of magnesium oxide is produced when 75g of magnesium is burned in excess oxygen?
Relative Atomic Masses: Mg = 24, O = 16

A. 80g   B. 100g   C. 125g   D. 145g   E. 175g

Question 382:

Hydrogen can combine with hydroxide ions to produce water. Which process is involved in this?

A. Hydration   C. Reduction   E. Evaporation
B. Oxidation   D. Dehydration

Question 383:

Which of the following statements about ammonia are correct?

1. It has a formula of NH_3.
2. Nitrogen contributes 82% to its mass.
3. It can be broken down again into nitrogen and hydrogen.
4. It is covalently bonded.
5. It is used to make fertilisers.

A. 1 and 2 only   C. 3, 4 and 5 only   E. 1, 2, 3, 4 and 5
B. 1 and 4 only   D. 1, 2 and 5 only

Question 384:

What colour will a universal indicator change to in a solution of whole milk (neutral pH) and lipase?

A. From green to orange.   C. From purple to green.   E. From yellow to purple.
B. From red to green.   D. From purple to orange.

Question 385:

Vitamin C [$C_6H_8O_6$] can be artificially synthesised from glucose [$C_6H_{12}O_6$]. What type of reaction is this likely to be?

A. Dehydration

B. Hydration

C. Oxidation

D. Reduction

E. Displacement

Question 386:

Which of the following statements are true?

1. Cu^{64} will undergo oxidation faster than Cu^{65}.
2. Cu^{65} will undergo reduction faster than Cu^{64}.
3. Cu^{65} and Cu^{64} have the same number of electrons.

A. 1 only

B. 2 only

C. 3 only

D. 2 and 3 only

E. 1 and 3 only

F. 1, 2 and 3

Question 387:

6g of Mg^{24} is added to a solution containing 30g of dissolved sulguric acid (H_2SO_4). Which of the following statements are true?

Relative Atomic Masses: S = 32, Mg = 24, O = 16, H = 1

1. In this reaction, the magnesium is the limiting reagent
2. In this reaction, sulfuric acid is the limiting reagent
3. The mass of salt produced equals the original mass of sulfuric acid

A. 1 only

B. 2 only

C. 3 only

D. 1 and 2 only

E. 1 and 3 only

Question 388:

In which of the following mixtures will a displacement reaction occur?

1. $Cu + 2AgNO_3$
2. $Cu + Fe(NO_3)_2$
3. $Ca + 2H_2O$
4. $Fe + Ca(OH)_2$

A. 1 only

B. 2 only

C. 3 only

D. 4 only

E. 1 and 3 only

Question 389:

Which of the following statements is true about the following chain of metals?

$Na \rightarrow Ca \rightarrow Mg \rightarrow Al \rightarrow Zn$

Moving from left to right:

1. The reactivity of the metals increases.
2. The likelihood of corrosion of the metals increases.
3. More energy is required to separate these metals from their ores.
4. The metals lose electrons more readily to form positive ions.

A. 1 and 2 only   C. 2 and 3 only   E. None of the
B. 1 and 3 only   D. 1 and 4 only   above

Question 390:

In which of the following mixtures will a displacement reaction occur?

1. I_2 + 2KBr
2. Cl_2 + 2NaBr
3. Br_2 + 2KI

A. 1 only   C. 3 only   E. 2 and 3 only
B. 2 only   D. 1 and 2 only

Question 391:

Which of the following statements about Al and Cu are true?

1. Al is used to build aircraft because it is lightweight and resists corrosion.
2. Cu is used to build electrical wires because it is a good insulator.
3. Both Al and Cu are good conductors of heat.
4. Al is commonly alloyed with other metals to make coins.
5. Al is resistant to corrosion because of a thin layer of aluminium hydroxide on its surface.

A. 1 and 3 only   C. 1, 3 and 5 only   E. 2, 4 and 5 only
B. 1 and 4 only   D. 1, 3, 4, 5 only

Question 392:

21g of Li^7 reacts completely with excess water. Given that the molar gas volume is 24 dm³ under the conditions, what is the volume of hydrogen produced?

A. 12 dm³   C. 36 dm³   E. 72 dm³
B. 24 dm³   D. 48 dm³

Question 393:

Which of the following statements regarding bonding are true?

1. NaCl has stronger ionic bonds than $MgCl_2$.
2. Transition metals are able to lose varying numbers of electrons to form multiple stable positive ions.
3. All covalently bonded structures have lower melting points than ionically bonded compounds.
4. No covalently bonded structures conduct electricity.

A. 1 only   C. 3 only   E. 1 and 2 only

B. 2 only   D. 4 only

Question 394:

Consider the following two equations:

A. $C + O_2 \rightarrow CO_2$ ΔH = -394 kJ per mole

B. $CaCO_3 \rightarrow CaO + CO_2$ ΔH = + 178 kJ per mole

Which of the following statements are true?

1.   Reaction **A** is exothermic and Reaction **B** is endothermic.
2. CO_2 has less energy than C and O_2.
3.   CaO is more stable than $CaCO_3$.

A. 1 only   C. 3 only   E. 1 and 3

B. 2 only   D. 1 and 2

Question 395:

Which of the following are true of regarding the oxides formed by Na, Mg and Al?

1. All of the metals and their solid oxides conduct electricity.
2. MgO has stronger bonds than Na_2O.
3. Metals are extracted from their molten ores by fractional distillation.

A. 1 only   C. 3 only   E. 2 and 3 only

B. 2 only   D. 1 and 2 only

Question 396:

Which of the following pairs have the same electronic configuration?

1. Li^+ and Na^+
2. Mg^{2+} and Ne
3. Na^{2+} and Ne
4. O^{2+} and a Carbon atom

A. 1 only
B. 1 and 2 only
C. 1 and 3 only
D. 2 and 3 only
E. 2 and 4 only

Question 397:

In relation to the reactivity of elements in Groups 1 and 2, which of the following statements is correct?

1. Reactivity decreases as you go down Group 1.
2. Reactivity increases as you go down Group 2.
3. Group 1 metals are generally less reactive than Group 2 metals.

A. Only 1
B. Only 2
C. Only 3
D. 1 and 2
E. 2 and 3

Question 398:

What role do catalysts fulfil in an endothermic reaction?

A. They increase the temperature, causing the reaction to occur at a faster rate.
B. They decrease the temperature, causing the reaction to occur at a faster rate.
C. They reduce the energy of the reactants in order to trigger the reaction.
D. They reduce the activation energy of the reaction.
E. They increase the activation energy of the reaction.

Question 399:

Tritium H^3 is an isotope of hydrogen. Why is tritium commonly referred to as 'heavy hydrogen'?

A. Because H^3 contains 3 protons making it heavier than H^1 that contains 1 proton.
B. Because H^3 contains 3 neutrons making it heavier than H^1 that contains 1 neutron.
C. Because H^3 contains 1 neutron and 2 protons making it heavier than H^1 that contains 1 neutron and 1 proton.
D. Because H^3 contains 1 proton and 2 neutrons making it heavier than H^1 that contains 1 proton.
E. Because H^3 contains 3 electrons making it heavier than H^1 that contains 1 electron.

Question 400:

In relation to redox reactions, which of the following statements are correct?

1. Oxidation describes the loss of electrons.
2. Reduction increases the electron density of an ion, atom or molecule.
3. Halogens are powerful reducing agents.

A. Only 1
B. Only 2
C. Only 3
D. 1 and 2
E. 2 and 3

Question 401:

Which one of the following statements is correct?

A. At higher temperatures, gas molecules move at angles that cause them to collide with each other more frequently.
B. Gas molecules have lower energy after colliding with each other.
C. At higher temperatures, gas molecules attract each other resulting in more collisions.
D. The average kinetic energy of gas molecules is the same for all gases at the same temperature.
E. The momentum of gas molecules decreases as pressure increases.

Question 402:

Which of the following are exothermic reactions?

1. Burning magnesium in pure oxygen
2. The combustion of hydrogen
3. Aerobic respiration
4. Evaporation of water in the oceans
5. The reaction between a strong acid and a strong base

A. 1, 2 and 4
B. 1, 2 and 5
C. 1, 3 and 5
D. 2, 3 and 4
E. 1, 2, 3 and 5

Question 403:

Ethene reacts with oxygen to produce water and carbon dioxide. Which elements are oxidised/reduced?

A. Carbon is reduced and oxygen is oxidised.
B. Hydrogen is reduced and oxygen is oxidised.
C. Carbon is oxidised and hydrogen is reduced.
D. Hydrogen is oxidised and carbon is reduced.
E. Carbon is oxidised and oxygen is reduced.

Question 404:

In the reaction between zinc and copper (II) sulphate which elements act as oxidising + reducing agents?

A. Zinc is the reducing agent while sulfur is the oxidizing agent.
B. Zinc is the reducing agent while copper in $CuSO_4$ is the oxidizing agent.
C. Copper is the reducing agent while zinc is the oxidizing agent.
D. Oxygen is the reducing agent while copper in $CuSO_4$ is the oxidizing agent.
E. Sulfur is the reducing agent while oxygen is the oxidizing agent.

Question 405:

Which of the following statements is true?

A. Acids are compounds that act as proton acceptors in an aqueous solution.
B. Acids only exist in a liquid state.
C. Strong acids are partially ionized in a solution.
D. Weak acids generally have a pH of 6 to 7.
E. The reaction between a weak and strong acid produces water and salt.

Question 406:

An unknown element, Z, has 3 isotopes: Z^5, Z^6 and Z^8. Given that the atomic mass of Z is 7, and the relative abundance of Z^5 is 20%, which of the following statements are correct?

1. Z^5 and Z^6 are present in the same abundance.
2. Z^8 is the most abundant of the isotopes.
3. Z^8 is more abundant than Z^5 and Z^6 combined.

A. 1 only
B. 2 only
C. 3 only
D. 1, 2 and 3
E. 2 and 3 only

Question 407:

Which of following best describes the products when an acid reacts with a metal that is more reactive than hydrogen?

A. Salt and hydrogen
B. Salt and ammonia
C. Salt and water
D. A weak acid and a weak base
E. A strong acid and a strong base

Question 408:

Choose an option from the table below to balance the following equation:

a $FeSO_4$ + **b** $K_2Cr_2O_7$ + **c** H_2SO_4 → **d** $(Fe)_2(SO_4)_3$ + **e** $Cr_2(SO_4)_3$ + **f** K_2SO_4 + **g** H_2O

	a	b	c	d	e	f	g
A	6	1	8	3	1	1	7
B	6	1	7	3	1	1	7
C	2	1	6	2	1	1	6
D	12	1	14	4	1	1	14
E	4	1	12	4	1	1	12

Question 409:

Which of the following statements is correct?

A. Matter consists of atoms that have a net electrical charge.

B. Atoms and ions of the same element have different numbers of protons and electrons but the same number of neutrons.

C. Over 80% of an atom's mass comes from its protons.

D. Atoms of the same element that have different numbers of neutrons react at significantly different rates.

E. Protons in the nucleus of atoms repel each other as they are positively charged.

Question 410:

Which of the following statements is correct?

A. The noble gases are chemically inert and therefore useless to man.

B. All of the noble gases have a full outer electron shell.

C. The majority of noble gases are brightly coloured.

D. The boiling point of the noble gases decreases as you progress down the Group.

E. Neon is the most abundant noble gas.

Question 411:

In relation to alkenes, which of the following statements is correct?
1. They all contain double bonds.
2. They can all be reduced to alkanes.
3. The equation 'alkene + hydrogen → alkane' is an example of a hydration reaction.

A. Only 1   C. Only 3   E. 2 and 3

B. Only 2   D. 1 and 2

Question 412:

Chlorine is made up of two isotopes, Cl^{35} (atomic mass 34.969) and Cl^{37} (atomic mass 36.966). Given that the atomic mass of chlorine is 35.453, which of the following statements is correct?

A. Cl^{35} is about 3 times more abundant than Cl^{37}.

B. Cl^{35} is about 10 times more abundant than Cl^{37}.

C. Cl^{37} is about 3 times more abundant than Cl^{35}.

D. Cl^{37} is about 10 times more abundant than Cl^{35}.

E. Both isotopes are equally abundant.

Question 413:

Which of the following statements regarding transition metals is correct?

A. Transition metals form ions that have multiple colours.

B. Transition metals usually form covalent bonds.

C. Transition metals cannot be used as catalysts as they are too reactive.

D. Transition metals are poor conductors of electricity.

E. Transition metals are found in group 2 of the periodic table.

Question 414:

20 g of impure Na^{23} reacts completely with excess water to produce 8,000 cm^3 of hydrogen gas under standard conditions. What is the percentage purity of sodium?
[Under standard conditions 1 mole of gas occupies 24 dm^3]

A. 88.0%   B. 76.5%   C. 66.0%   D. 38.0%   E. 15.3%

Question 415:

An organic molecule contains 70.6% Carbon, 5.9% Hydrogen and 23.5% Oxygen. It has a molecular mass of 136. What is its chemical formula?

A. C_4H_4O B. C_5H_4O C. $C_8H_8O_2$ D. $C_{10}H_8O_2$ E. C_2H_2O

Question 416:

Choose an option from the table below to balance the following equation:

aS + **b**HNO$_3$ → **c**H$_2$SO$_4$ + **d**NO$_2$ + **e**H$_2$O

	a	b	c	d	e
A	3	5	3	5	1
B	1	6	1	6	2
C	6	14	6	14	2
D	2	4	2	4	4
E	2	3	2	3	2

Question 417:

Which of the following statements is true?

1. Ethane and ethene can both dissolve in organic solvents.
2. Ethane and ethene can both be hydrogenated in the presence of nickel.
3. Breaking C=C requires double the energy needed to break C-C.

A. 1 only   C. 3 only   E. 2 and 3 only
B. 2 only   D. 1 and 2 only

Question 418:

Diamond, graphite, methane and ammonia all contain covalent bonds. Which row in the table adequately describes the properties associated with each compound?

	Compound	Melting Point	Able to conduct electricity	Soluble in water
1.	Diamond	High	Yes	No
2.	Graphite	High	Yes	No
3.	CH$_{4 (g)}$	Low	No	No
4.	NH$_{3 (g)}$	Low	No	Yes

A. 1 and 2 only   C. 1 and 3 only   E. 2, 3 and 4
B. 2 and 3 only   D. 1 and 4 only

Question 419:

Which of the following statements about catalysts are true?

1. Catalysts reduce the energy required for a reaction to take place.
2. Catalysts are used up in reactions.

3. Catalysed reactions are almost always exothermic.

A. I only   B. 2 only   C. I and 2   D. 2 and 3   E. I, 2 and 3

Question 420:

What is the name of the molecule below?

A. But-1-ene   E. Pent-2-ene
B. But-2-ene   F. Pentane
C. Pent-3-ene   G. Pentanoic acid
D. Pent-1-ene

Question 421:

Which of the following statements is correct regarding Group I elements? [Excluding hydrogen]

A. The oxidation number of Group I elements usually decreases in most reactions.
B. Reactivity decreases as you progress down Group I.
C. Group I elements do not react with water.
D. All Group I elements react spontaneously with oxygen.
E. All of the above.
F. None of the above.

Question 422:

Which of the following statements about electrolysis are correct?

1. The cathode attracts negatively charged ions.
2. Atoms are reduced at the anode.
3. Electrolysis can be used to separate mixtures.

A. Only I
B. Only 2
C. 2 and 3
D. Only 3
E. None of the above

Question 423:

Which of the following is **NOT** an isomer of pentane?

A. $CH_3CH_2CH_2CH_2CH_3$ C. $CH_3(CH_2)_3CH_3$
B. $CH_3C(CH_3)CH_3CH_3$ D. $CH_3C(CH_3)_2CH_3$

Question 424:
Choose an option to balance the following equation:
$Cu + HNO_3 \rightarrow Cu(NO_3)_2 + NO + H_2O$

A. $8\ Cu + 3\ HNO_3 \rightarrow 8\ Cu(NO_3)_2 + 4\ NO + 2\ H_2O$
B. $3\ Cu + 8\ HNO_3 \rightarrow 2\ Cu(NO_3)_2 + 3\ NO + 4\ H_2O$
C. $5Cu + 7HNO_3 \rightarrow 5\ Cu(NO_3)_2 + 4\ NO + 8\ H_2O$
D. $6\ Cu + 10\ HNO_3 \rightarrow 6\ Cu(NO_3)_2 + 3\ NO + 7\ H_2O$
E. $3\ Cu + 8\ HNO_3 \rightarrow 3\ Cu(NO_3)_2 + 2\ NO + 4\ H_2O$

Question 425:
Which of the following statements regarding alkenes is correct?

A. Alkenes are an inorganic homologous series.
B. Alkenes always have three times as many hydrogen atoms as they do carbon atoms.
C. Bromine water changes from clear to brown in the presence of an alkene.
D. Alkenes are more reactive than alkanes because they are unsaturated.
E. Alkenes frequently take part in subtraction reactions.

Question 426:
Which one of the following statements is correct regarding Group 17?

A. All Group 17 elements are electrophilic and therefore form negatively charged ions.
B. The reaction between sodium and fluorine is less vigorous than sodium and iodine.
C. Some Group 17 elements are found naturally as unbonded atoms.
D. All of the above.
E. None of the above.

Question 427:
Why does the electrolysis of NaCl solution (brine) require the strict separation of the products of anode and cathode?

A. To prevent the preferential discharge of ions.
B. In order to prevent spontaneous combustion.
C. In order to prevent production of H_2.
D. In order to prevent the formation of HCl.
E. In order to avoid CO poisoning.

Question 428:

In relation to the electrolysis of brine (NaCl), which of the following statements are correct?

1. Electrolysis results in the production of hydrogen and chlorine gas.
2. Electrolysis results in the production of sodium hydroxide.
3. Hydrogen gas is released at the anode and chlorine gas is released at the cathode.

A. Only 1   C. Only 3   E. 1 and 3
B. Only 2   D. 1 and 2

Question 429:

Which of the following statements is correct?

A. Alkanes consist of multiple C-H bonds that are very weak.
B. An alkane with 14 hydrogen atoms is called heptane.
C. All alkanes consist purely of hydrogen and carbon atoms.
D. Alkanes burn in excess oxygen to produce carbon monoxide and water.
E. Bromine water is decolourised in the presence of an alkane.

Question 430:

Which of the following statements are correct?

1. All alcohols contain a hydroxyl functional group.
2. Alcohols are highly soluble in water.
3. Alcohols are sometimes used as biofuels.

A. Only 1   C. Only 3   E. 1, 2 and 3
B. Only 2   D. 1 and 2

Question 431:

Which row of the table below is correct?

	Non-Reducible Hydrocarbon			Reducible Hydrocarbon		
A	C_nH_{2n}	$Br_{2(aq)}$ remains brown	Saturated	C_nH_{2n+2}	Turns $Br_{2(aq)}$ colourless	Unsaturated
B	C_nH_{2n+2}	Turns $Br_{2(aq)}$ colourless	Unsaturated	C_nH_{2n}	$Br_{2(aq)}$ remains brown	Saturated
C	C_nH_{2n}	$Br_{2(aq)}$ remains brown	Unsaturated	C_nH_{2n+2}	Turns $Br_{2(aq)}$ colourless	Saturated
D	C_nH_{2n+2}	Turns $Br_{2(aq)}$ colourless	Saturated	C_nH_{2n}	$Br_{2(aq)}$ remains brown	Unsaturated
E	C_nH_{2n+2}	$Br_{2(aq)}$ remains brown	Saturated	C_nH_{2n}	Turns $Br_{2(aq)}$ colourless	Unsaturated

Question 432:

How many grams of magnesium chloride are formed when 10 grams of magnesium oxide are dissolved in excess hydrochloric acid? Relative atomic masses: $Mg = 24$, $O = 16$, $H = 1$, $Cl = 35.5$

A. 10.00   C. 20.00   E. 47.55

B. 14.95   D. 23.75

Question 433:

Pentadecane has the molecular formula $C_{15}H_{32}$. Which one of the following statements is true?

A. Pentadecane has a lower boiling point than pentane.

B. Pentadecane is more flammable than pentane.

C. Pentadecane is more volatile than pentane.

D. Pentadecane is more viscous than pentane.

E. All of the above.

Question 434:

The rate of reaction is normally dependent upon:

1. The temperature.
2. The concentration of reactants.
3. The concentration of the catalyst.
4. The surface area of the catalyst.

A. 1 and 2   C. 2, 3 and 4   E. 1, 2, 3 and 4

B. 2 and 3   D. 1, 3 and 4

Question 435:

The equation below shows the complete combustion of a sample of unknown hydrocarbon in excess oxygen.

$C_aH_b + O_2 \rightarrow cCO_2 + dH_2O$

The reaction yielded 176 grams of CO_2 and 108 grams of H_2O. What is the most likely formula of the unknown hydrocarbon? Relative atomic masses:

H = 1, C = 12, O = 16.

A. CH_4 B. CH_3 C. C_2H_6 D. C_3H_9 E. C_2H_4

Question 436:

What type of reaction must ethanol undergo in order to be converted to ethylene oxide (C_2H_4O)?

A. Oxidation   C. Dehydration   E. Redox

B. Reduction   D. Hydration

Question 437:

What values of a, b and c balance the equation below?

$$a\ Ba_3N_2 + 6H_2O \rightarrow b\ Ba(OH)_2 + c\ NH_3$$

	a	b	c
A	1	2	3
B	1	3	2
C	2	1	3
D	2	3	1
E	3	1	2

Question 438:

What values of a, b and c balance the equation below?

$$a\ FeS + 7O_2 \rightarrow b\ Fe_2O_3 + c\ SO_2$$

	a	b	c
A	3	2	2
B	2	4	1
C	3	1	5
D	4	1	3
E	4	2	4

Question 439:

Magnesium consists of 3 isotopes: Mg^{23}, Mg^{25}, and Mg^{26} which are found naturally in a ratio of 80:10:10. Calculate the relative atomic mass of magnesium.

A. 23.3
B. 23.4
C. 23.5
D. 23.6
E. 24.6

Question 440:

Consider the three reactions:

1. $Cl_2 + 2Br^- \rightarrow 2Cl^- + Br_2$
2. $Cu^{2+} + Mg \rightarrow Cu + Mg^{2+}$
3. $Fe_2O_3 + 3CO \rightarrow 2Fe + 3CO_2$

Which of the following statements are correct?

A. Cl_2 and Fe_2O_3 are reducing agents.
B. CO and Cu^{2+} are oxidising agents.
C. Br_2 is a stronger oxidising agent than Cl_2.
D. Mg is a stronger reducing agent than Cu.

Question 441:

Which row of the table below best describes the properties of NaCl?

| | Melting Point | Solubility in Water | Conducts electricity? | |
			As a solid	In solution
A	High	Yes	Yes	Yes
B	High	No	Yes	No
C	High	Yes	No	Yes
D	High	No	No	No
E	Low	Yes	Yes	Yes

Question 442:

80g of sodium hydroxide reacts with excess zinc nitrate to produce zinc hydroxide. Calculate the mass of zinc hydroxide produced. Relative atomic mass: N = 14, Zn = 65, O = 16, Na = 23.

A. 49g
B. 95g
C. 99g
D. 100g
E. 198g

Question 443:

Which of the following statements is correct?

A. The reaction between all Group I metals and water is exothermic.
B. Sodium reacts less vigorously with water than potassium does.
C. All Group I metals react with water to produce elemental hydrogen.
D. All Group I metals react with water to produce a metal hydroxide.
E. All of the above.

Question 444:

Which one of the following statements is correct?

A. NaCl can be separated using sieves.
B. CO_2 can be separated using electrolysis.
C. Dyes in a sample of ink can be separated using chromatography.
D. Oil and water can be separated using fractional distillation.
E. Methane and diesel can be separated using a separating funnel.

Question 445:

Which of the following statements about the reaction between caesium and fluoride are correct?

1. It is an exothermic reaction and therefore requires catalysts.
2. It results in the formation of a salt.
3. The addition of water will make the reaction safer.

A. Only 1   C. Only 3   E. 2 and 3
B. Only 2   D. 1 and 2

Question 446:

Which of the following statements is generally true about stable isotopes?

1. The nucleus contains an equal number of neutrons and protons.
2. The nuclear charge is equal and opposite to the peripheral charge due to the orbiting electrons.
3. They can all undergo radioactive decay into more stable isotopes.

A. Only 1   C. Only 3   E. 2 and 3
B. Only 2   D. 1 and 2

Question 447:

Why do most salts have very high melting points?

A. Their surface is able to radiate away a significant portion of the heat to their environment.
B. The ionic bonds holding them together are very strong.
C. The covalent bonds holding them together are very strong.
D. They tend to form large macromolecules as each salt molecule bonds with multiple other molecules.
E. All of the above.

Question 448:

A bottle of water contains 306ml of pure deionised water. How many protons are in the bottle from the water?
[Avogadro Constant = 6×10^{23} mol^{-1}]

A. 1×10^{22}　　　　B. 1×10^{23}　　　　C. 1×10^{24}　　　　D. 1×10^{25}　　　　E. 1×10^{26}

Question 449:

On analysis, an organic substance is found to contain 41.4% carbon, 55.2% oxygen and 3.45% hydrogen by mass. Which of the following could be the empirical formula of this substance?

A. $C_3O_3H_6$

B. $C_3O_3H_{12}$

C. $C_4O_2H_4$

D. $C_4O_4H_4$

E. More information needed

Question 450:

A is a Group 2 element and B is a Group 17 element. Which row best describes what happens when A reacts with B?

	B is	Formula
A	Reduced	AB
B	Reduced	A_2B
C	Reduced	AB_2
D	Oxidised	AB
E	Oxidised	A_2B

SECTION 2: PHYSICS

If you haven't done physics at AS level then you'll have to ensure that you are confident with commonly examined topics like Newtonian mechanics, electrical circuits and radioactive decay as you may not have covered these at GCSE level depending on the specification you did.

The first step to improving in this section of the BMAT is to memorise by rote all the equations listed on the next page, and build up an understanding of their relationship to the concepts listed on the BMAT specification.

The majority of the physics questions involve a fair bit of maths – this means you need to be comfortable with converting between units and also powers of 10. Manipulating numbers at speed without using a calculator is the key to success in this section, alongside memorising the equations listed. **Most BMAT physics questions require two-step calculations**. Consider the example:

A metal ball is released from the roof of a 20-metre building. Assuming air resistance is negligible; calculate the velocity at which the ball hits the ground. [$g = 10$ms^{-2}]

A. 5 ms^{-1} B. 10 ms^{-1} C. 15 ms^{-1} D. 20 ms^{-1} E. 25 ms^{-1}

When the ball hits the ground, all of its gravitational potential energy has been converted to kinetic energy.

Thus, $E_p = E_k$:
$$mg\Delta h = \frac{mv^2}{2}$$
Thus, $v = \sqrt{2gh} = \sqrt{2 \times 10 \times 20}$
$= \sqrt{400} = 20ms^{-1}$

Here, you were required to not only recall two equations but apply and rearrange them very quickly to get the answer - all in under 60 seconds. Thus, it is easy to understand why the physics questions are generally much harder to complete in the time given than the biology and some of the chemistry questions.

Note that if you were comfortable with basic Newtonian mechanics, you could have also solved this using a single suvat equation: $v^2 = u^2 + 2as$
$v = \sqrt{2 \times 10 \times 20} = 20ms^{-1}$

This is why you're **strongly advised to learn the 'suvat' equations** on the next page even if they're technically not on the syllabus for the BMAT.

SI Units

Remember that in order to get the correct answer you must always work in SI units i.e. do your calculations in terms of metres (not centimetres) and kilograms (not grams), etc.

> **Top tip!** Knowing SI units is extremely useful because they allow you to **'work out' equations** if you ever forget them e.g. the units for density are kg/m^3. Since kg is the SI unit for mass, and m^3 is represented by volume –the equation for density must be = Mass/Volume.
>
> This can also work the other way, for example we know that the unit for pressure is pascal (Pa). But based on the fact that Pressure = Force/Area, a pascal must be equivalent to N/m^2. Some physics questions will test your ability to manipulate units like this so it's important you are comfortable converting between them.

FORMULAE YOU MUST KNOW:

Equations of Motion:

- $s = ut + 0.5at^2$
- $v = u + at$
- $a = (v-u)/t$
- $v^2 = u^2 + 2as$

Equations relating to Force:

- Force = mass x acceleration
- Force = Momentum/Time
- Pressure = Force / Area
- Moment of a Force = Force x Distance
- Work done = Force x Displacement

For objects in equilibrium:

- Sum of clockwise moments = Sum of anti-clockwise moments
- Sum of all resultant forces = 0

Equations relating to Energy:

- Kinetic Energy = $0.5 \, mv^2$
- Δ in Gravitational Potential Energy = $mg\Delta h$
- Energy Efficiency = (Useful energy/ Total energy) x 100%

Equations relating to Power:

- Power = Work done / time
- Power = Energy transferred / time
- Power = Force x velocity

Electrical Equations:

- $Q = It$
- $V = IR$
- $P = IV = I^2R = V^2/R$
- V = Potential difference (V, Volts)
- R = Resistance (Ohms)
- P = Power (W, Watts)
- Q = Charge (C, Coulombs)
- t= Time (s, seconds)

Factor	Text	Symbol
10^{12}	Tera	T
10^{9}	Giga	G
10^{6}	Mega	M
10^{3}	Kilo	k
10^{2}	Hecto	h
10^{-1}	Deci	d
10^{-2}	Centi	c
10^{-3}	Milli	m
10^{-6}	Micro	μ
10^{-9}	Nano	n
10^{-12}	Pico	p

For Transformers: $\frac{V_p}{V_s} = \frac{n_p}{n_s}$ where:

- V: Potential difference
- n: Number of turns (on coil)
- p: Primary
- s: Secondary

Other:

- Weight = mass x g
- Density = Mass / Volume
- Momentum = Mass x Velocity
- $g = 9.81$ ms^{-2} (unless otherwise stated)

PHYSICS QUESTIONS

Question 451:

Which one of the following statements is **FALSE**?

A. Electromagnetic waves cause things to heat up.
B. X-rays and gamma rays can knock electrons out of their orbits.
C. Loud sounds can make objects vibrate.
D. Wave power can be used to generate electricity.
E. The amplitude of a wave determines its mass.

Question 452:

A spacecraft is analysing a newly discovered exoplanet. A rock of unknown mass falls on the planet from a height of 30 m. Given that $g = 5.4$ ms^{-2} on the planet, calculate the speed of the rock when it hits the ground and the time it took to fall.

	Speed (ms^{-1})	Time (s)
A	18	3.3
B	18	3.1
C	12	3.3
D	10	3.7
E	9	2.3

Question 453:

A canoe floating on the sea rises and falls 7 times in 49 seconds. The waves pass it at a speed of 5 ms^{-1}. How long are the waves?

A. 12 m   B. 22 m   C. 25 m   D. 35 m   E. 57 m

Question 454:

Miss Orrell lifts her 37.5 kg bike for a distance of 1.3 m in 5 s. The acceleration of free fall is 10 ms^{-2}. What is the average power that she generates?

A. 9.8 W
B. 12.9 W
C. 57.9 W
D. 79.5 W
E. 97.5W

Question 455:

A truck accelerates at 5.6 ms^{-2} from rest for 8 seconds. Calculate the final speed and the distance travelled in 8 seconds.

	Final Speed (ms^{-1})	Distance (m)
A	40.8	119.2
B	40.8	129.6
C	42.8	179.2
D	44.1	139.2
E	44.8	179.2

Question 456:

Which of the following statements is true when a skydiver jumps out of a plane?

A. The skydiver leaves the plane and will accelerate until the air resistance is greater than their weight.
B. The skydiver leaves the plane and will accelerate until the air resistance is less than their weight.
C. The skydiver leaves the plane and will accelerate until the air resistance equals their weight.
D. The skydiver leaves the plane and will accelerate until the air resistance equals their weight squared.
E. The skydiver will travel at a constant velocity after leaving the plane.

Question 457:

A 100 g apple falls on Isaac's head from a height of 20 m. Calculate the apple's momentum before the point of impact.
Take g = 10 ms^{-2}

A. 0.1 kgms^{-1} C. 1 kgms^{-1} E. 10 kgms^{-1}
B. 0.2 kgms^{-1} D. 2 kgms^{-1}

Question 458:

Which of the following characteristics do all electromagnetic waves all have in common?

1. They can travel through a vacuum.
2. They can be reflected.
3. They are the same length.
4. They have the same amount of energy.

A. 1, 2 and 3 only   C. 4 and 5 only   E. 1 and 2 only
B. 1, 2, 3 and 4 only   D. 3 and 4 only

Question 459:

A battery with an internal resistance of 0.8 Ω and e.m.f of 36 V is used to power a drill with resistance 1 Ω. What is the current in the circuit when the drill is connected to the power supply?

A. 5 A   B. 10 A   C. 15 A   D. 20 A   E. 25 A   F. 30 A

Question 460:

Officer Bailey throws a 20 g dart at a speed of 100 ms⁻¹. It strikes the dartboard and is brought to rest in 10 milliseconds. Calculate the average force exerted on the dart by the dartboard.

A. 0.2 N   C. 20 N   E. 2,000 N
B. 2 N   D. 200 N   F. 20,000 N

Question 461:

Professor Huang lifts a 50 kg bag through a distance of 0.7 m in 3 s. What average power does she generate to 3 significant figures? Take $g = 10$ms⁻²

A. 113 W   C. 115 W   E. 117 W
B. 114 W   D. 116 W

Question 462:

An electric scooter is travelling at a speed of 30 ms⁻¹ and is kept going at a constant speed against a 50 N frictional force by a driving force of 300 N in the direction of motion. Given that the engine runs at 200 V, calculate the current in the scooter.

A. 4.5 A   C. 450 A   E. 45,000 A
B. 45 A   D. 4,500 A

Question 463:

Which of the following statements about the physical definition of work are correct?

1. $Work\ done = \dfrac{Force}{distance}$
2. The unit of work is equivalent to kgms⁻².
3. Work is defined as a force causing displacement of the body upon which it acts.

A. Only 1   C. Only 3   E. 2 and 3
B. Only 2   D. 1 and 2

Question 464:

Which of the following statements about kinetic energy are correct?

1. It is defined as $E_k = \frac{mv^2}{2}$
2. The unit of kinetic energy is equivalent to Pa x m³.
3. Kinetic energy is equal to the amount of energy needed to decelerate the body in question from its current speed.

A. Only 1   C. 2 and 3   E. 1, 2 and 3
B. Only 2   D. 1 and 3

Question 465:

In relation to radiation, which of the following statements is **FALSE**?

A. Radiation is the emission of energy from a substance in the form of waves or particles.
B. Radiation can be either ionizing or non-ionizing.
C. Gamma radiation has very high energy.
D. Alpha radiation is of higher energy than beta radiation.
E. X-rays are an example of wave radiation.

Question 466:

In relation to the physical definition of half-life, which of the following statements are correct?

1. In radioactive decay, the half-life is independent of atom type and isotope.
2. Half-life is defined as the time required for exactly half of the entities to decay.
3. Half-life applies to situations of both exponential and non-exponential decay.

A. Only 1   C. Only 3   E. 2 and 3
B. Only 2   D. 1 and 2   F. 1 and 3

Question 467:

A radioactive element has a half life of 24 days. After 192 days, it has a count rate of 56. What was the original count rate?

A. 7,168
B. 14,280
C. 14,336
D. 28,672
E. 43,008

Question 468:

Which of the following statements concerning radioactive decay is / are true?

1. As a material decays, the rate of decay decreases.
2. All nuclei of the same element will have the same half-life, as it is an innate quality of an element.
3. Radioactive decay is a highly predictable process

A. Only 1   C. Only 3   E. 2 and 3
B. Only 2   D. 1 and 2

Question 469:

Two identical resistors (R_a and R_b) are connected in a series circuit. Which of the following statements are true?

1. The current through both resistors is the same.
2. The voltage through both resistors is the same.
3. The voltage across the two resistors is given by Ohm's Law.

A. Only 1   C. Only 3   E. 1, 2 and 3
B. Only 2   D. 1 and 2

Question 470:

The sun is 8 light-minutes away from the earth. Estimate the circumference of the earth's orbit around the sun. Assume that the earth is in a circular orbit around the sun. Speed of light = 3×10^8 ms^{-1}

A. 10^{24} m   C. 10^{18} m   E. 10^{12} m
B. 10^{21} m   D. 10^{15} m

Question 471:

Which of the following statements about calculating speed are true?

1. Speed is the same as velocity.
2. The internationally standardised unit for speed is ms^{-2}.
3. Velocity = distance/time.

A. Only 1   C. Only 2   E. None of
B. 1 and 2   D. Only 3   the above

Question 472:

Which of the following statements best defines Ohm's Law?

A. The current passing through an insulator between two points is indirectly proportional to the potential difference across the two points.
B. The current passing through an insulator between two points is directly proportional to the potential difference across the two points.
C. The current passing through a conductor between two points is inversely proportional to the potential difference across the two points.
D. The current passing through a conductor between two points is proportional to the square of the potential difference across the two points.
E. The current passing through a conductor between two points is directly proportional to the potential difference across the two points.

Question 473:

Which of the following statements regarding Newton's Second Law are correct?
1. For objects at rest, the resultant force must be 0 Newtons
2. Force = Mass x Acceleration
3. Force = Rate of change of Momentum

A. Only 1   C. 1 and 3   E. 1, 2 and 3
B. 2 and 3   D. 1 and 2

Question 474:

Which of the following equations concerning electrical circuits are correct?

1. $Charge = \dfrac{Voltage \; x \; time}{Resistance}$
2. $Charge = \dfrac{Power \; x \; time}{Voltage}$
3. $Charge = \dfrac{Current \; x \; time}{Resistance}$

A. Only 1   C. Only 3   E. 2 and 3
B. Only 2   D. 1 and 2

Question 475:

An elevator has a mass of 1,600 kg and is carrying passengers that have a combined mass of 200 kg. A constant frictional force of 4,000 N retards its motion upward. What force must the motor provide for the elevator to move with an upward acceleration of 1 ms^{-2}? Assume: $g = 10$ ms^{-2}

A. 1,190 N   B. 11,900 N   C. 18,000 N   D. 22,000 N   E. 23,800 N

Question 476:

A 1,000 kg car accelerates from rest at 5 ms^{-2} for 10 s. Then, a braking force is applied to bring it to rest within 20 seconds. What is the total distance travelled by the car?

A. 125 m
B. 250 m

C. 650 m
D. 750 m

E. More information needed

Question 477:

An electric heater is connected to 120 V mains by a copper wire that has a resistance of 8 ohms. What is the power of the heater?

A. 90W B. 180W C. 900W D. 1800W E. More information needed

Question 478:

In a particle accelerator, electrons are accelerated through a potential difference of 40 MV and emerge with an energy of 40MeV (1 MeV = 1.60 x 10^{-13} J). Each pulse contains 5,000 electrons. The current is zero between pulses. Assuming that the electrons have zero energy prior to being accelerated what is the power delivered by the electron beam?

A. 1 kW
B. 10 kW

C. 100 kW
D. 1,000 kW

E. More information needed

Question 479:

Which **one** of the following statements is **true**?

A. When an object is in equilibrium with its surroundings, there is no energy transferred to or from the object and so its temperature remains constant.
B. When an object is in equilibrium with its surroundings, it radiates and absorbs energy at the same rate and so its temperature remains constant.
C. Radiation is faster than convection but slower than conduction.
D. Radiation is faster than conduction but slower than convection.
E. None of the above.

Question 480:

A 6kg block is pulled from rest along a horizontal frictionless surface by a constant horizontal force of 12 N. Calculate the speed of the block after it has moved 300 cm.

A. $2\sqrt{3}\ ms^{-1}$
B. $4\sqrt{3}\ ms^{-1}$

C. $4\sqrt{3}\ ms^{-1}$
D. $12\ ms^{-1}$

E. $\sqrt{\frac{3}{2}}\ ms^{-1}$

Question 481:
A 100 V heater heats 1.5 litres of pure water from 10°C to 50°C in 50 minutes. Given that 1 kg of pure water requires 4,000 J to raise its temperature by 1°C, calculate the resistance of the heater.

A. 12.5 ohms
B. 25 ohms

C. 125 ohms
D. 250 ohms

E. 500 ohms

Question 482:
Which of the following statements are **true**?

1. The half life of a radioactive substance is equal to half the time taken for its nuclei to decay.
2. When a nucleus emits a beta particle, it is converted to a new element.
3. When a nucleus emits an alpha particle, one of its neutrons becomes a proton and an electron.

A. Only 1
B. 2 and 3

C. 2 only
D. 1 and 2

E. None of
    the above

Question 483:
Which of the following statements are **true**? Assume $g = 10$ ms^{-2}.

1. Gravitational potential energy is defined as $\Delta E_p = m \times g \times \Delta h$.
2. Gravitational potential energy is a measure of the work done against gravity.
3. A reservoir situated 1 km above ground level with 10^6 litres of water has a potential energy of 1 Giga Joule.

A. Only 1
B. Only 2

C. Only 3
D. 1 and 3

E. 1, 2 and 3

Question 484:
Which of the following statements are correct in relation to Newton's 3rd law?

1. For every action there is an equal and opposite reaction.
2. According to Newton's 3rd law, there are no isolated forces.
3. When a rifle recoils as a bullet is fired from it, according to Newton's third law of motion, the acceleration of the recoiling rifle is the same size as the acceleration of the bullet.

A. Only 1
B. Only 2

C. 2 and 3
D. 1 and 2

E. 1, 2 and 3

Question 485:

Which of the following statements are correct?

1. Positively charged objects have gained electrons.
2. Electrical charge in a circuit over a period of time can be calculated if the voltage and resistance are known.
3. Objects can be charged by friction.

A. Only 1	C. Only 3	E. 2 and 3	G. 1, 2 and 3
B. Only 2	D. 1 and 2	F. 1 and 3	

Question 486:

Which of the following statements is true?

A. The gravitational force between two objects is independent of their mass.
B. Each planet in the solar system exerts a gravitational force on the Earth.
C. Two objects dropped from the Eiffel tower will always land on the ground at the same time if they have the same mass.
D. All of the above.
E. None of the above.

Question 487:

Which of the following best defines an electrical conductor?

A. Conductors are usually made from metals, and they conduct electrical charge in multiple directions.
B. Conductors are usually made from non-metals, and they conduct electrical charge in multiple directions.
C. Conductors are usually made from metals, and they conduct electrical charge in one fixed direction.
D. Conductors are usually made from non-metals, and they conduct electrical charge in one fixed direction.
E. Conductors allow the passage of electrical charge with zero resistance because they contain freely mobile charged particles.

Question 488:

An 800 kg compact car delivers 20% of its power output to its wheels. If the car has a mileage of 30 miles/gallon and travels at a speed of 60 miles/hour, how much power is delivered to the wheels? 1 gallon of petrol contains 9×10^8 J.

A. 10 kW	B. 20 kW	C. 40 kW	D. 50 kW	E. 100 kW

Question 489:

Which of the following statements about beta radiation are true?

1. After a beta particle is emitted, the atomic mass number is unchanged.
2. Beta radiation can penetrate paper but not aluminium foil.
3. A beta particle is emitted from the nucleus of the atom when an electron transforms to a neutron.

A. 1 only

B. 2 only

C. 1 and 3

D. 1 and 2

E. 2 and 3

F. 1, 2 and 3

Question 490:

A car with a weight of 15,000 N is travelling at a speed of 15 ms^{-1} when it crashes into a wall and is brought to rest in 10 milliseconds. Calculate the average braking force exerted on the car by the wall. Take g = 10 ms^{-2}

A. $1.25 \times 10^4 N$

B. $1.25 \times 10^5 N$

C. $1.25 \times 10^6 N$

D. $2.25 \times 10^4 N$

E. $2.25 \times 10^6 N$

Question 491:

Which of the following statements are correct?

1. Electrical insulators are usually metals e.g. copper.
2. The flow of charge through electrical insulators is extremely low.
3. Electrical insulators can be charged by rubbing them together.

A. Only 1

B. Only 2

C. Only 3

D. 1 and 2

E. 2 and 3

The following information is needed for Questions 492 and 493:

This graph represents a car's movement. At t=0 the car's displacement was 0 m.

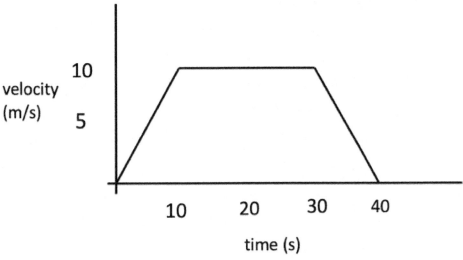

Question 492:
Which of the following statements are **NOT true**?

1. The car is reversing after t = 30.
2. The car moves with constant acceleration from t = 0 to t = 10.
3. The car moves with constant speed from t = 10 to t = 30.

A. I only   C. 3 only   E. I and 2
B. 2 only   D. I and 3

Question 493:
Calculate the distance travelled by the car.

A. 200 m   C. 350 m   E. More information needed
B. 300 m   D. 400 m

Question 494:
A 1,000 kg rocket is launched during a thunderstorm and reaches a constant velocity 30 seconds after launch. Suddenly, a strong gust of wind acts on the rocket for 5 seconds with a force of 10,000 N in the direction of movement.
What is the resulting change in velocity?

A. 0.5 ms^{-1} C. 50 ms^{-1} E. More information needed
B. 5 ms^{-1} D. 500 ms^{-1}

Question 495:
A 0.5 tonne crane lifts a 0.01 tonne wardrobe by 100 cm in 5,000 milliseconds.
Calculate the average power generated by the crane. Take $g = 10$ ms^{-2}.

A. 0.2 W
B. 2 W

C. 5 W
D. 20 W

E. More information needed

Question 496:
A 20 V battery is connected to a circuit consisting of a 1 Ω and 2 Ω resistor in parallel. Calculate the overall current of the circuit.

A. 6.67 A
B. 8 A

C. 12 A
D. 20 A

E. 30 A

Question 497:
Which **one** of the following statements is correct?

A. The speed of light changes when it enters water.
B. The speed of light changes when it leaves water.
C. The direction of light changes when it enters water.
D. The direction of light changes when it leaves water.
E. All of the above.
F. None of the above.

Question 498:
In a parallel circuit, a 60 V battery is connected to two branches. Branch A contains 6 identical 5 Ω resistors and branch B contains 2 identical 10 Ω resistors.

Calculate the current in branches A and B.

	I_A (A)	I_B (A)
A	0	6
B	6	0
C	2	3
D	3	2
E	1	5

Question 499:
Calculate the voltage of an electrical circuit that has a power output of 50,000,000,000 nW and a current of 0.000000004 GA.

A. 0.0125 GV
B. 0.0125 MV

C. 0.0125 kV
D. 0.0125 nV

E. 0.0125 mV

Question 500:
Which of the following statements about radioactive decay is correct?

A. Radioactive decay is highly predictable.
B. An unstable element will continue to decay until it reaches a stable nuclear configuration.
C. All forms of radioactive decay release gamma rays.
D. All forms of radioactive decay release X-rays.
E. None of the above.

Question 501:
A circuit contains three identical resistors of unknown resistance connected in series with a 15 V battery. The power output of the circuit is 60 W.
Calculate the overall resistance of the circuit when two further identical resistors are added to it.

A. 0.125 Ω
B. 1.25 Ω

C. 3.75 Ω
D. 6.25 Ω

E. 18.75 Ω

Question 502:
The engine in a 5,000 kg tractor uses 1 litre of fuel to move 0.1 km. 1 ml of the fuel contains 20 kJ of energy.
Calculate the engine's efficiency. Take $g = 10$ ms^{-2}

A. 2.5 %
B. 25 %

C. 38 %
D. 50 %

E. More information needed.

Question 503:
Which of the following statements are correct?

1. Electromagnetic induction occurs when a wire moves relative to a magnet.
2. Electromagnetic induction occurs when a magnetic field changes.
3. An electrical current is generated when a coil rotates in a magnetic field.

A. Only 1
B. 1 and 3

C. 2 and 3
D. 1 and 2

E. 1, 2 and 3

Question 504:
Which of the following statements are correct regarding parallel circuits?

1. The current flowing through a branch is dependent on the resistance of that branch.
2. The total current flowing into the branches is equal to the total current flowing out of the branches.
3. An ammeter will always give the same reading regardless of its location in the circuit.

A. Only 1
B. Only 2
C. 2 and 3
D. 1 and 2
E. All of the above

Question 505:
Which of the following statements regarding series circuits are true?

1. The overall resistance of a circuit is given by the sum of all resistors in the circuit.
2. Electrical current moves from the positive terminal to the negative terminal.
3. Electrons move from the positive terminal to the negative terminal.

A. Only 1
B. Only 2
C. Only 3
D. 1 and 2
E. 1 and 3

Question 506:
The graphs below show current vs. voltage plots for 4 different electrical components.

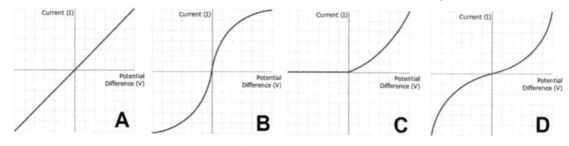

Which of the following graphs represents a resistor at constant temperature, and which represents a filament lamp?

	Fixed Resistor	Filament Lamp
A	A	B
B	A	C
C	D	D
D	C	A
E	C	C

Question 507:

Which of the following statements are true about vectors?

A. Vectors can be added or subtracted.
B. All vector quantities have a defined magnitude.
C. All vector quantities have a defined direction.
D. All of the above.
E. None of the above.

Question 508:

The acceleration due to gravity on the Earth is six times greater than that on the moon. Dr Tyson records the weight of a rock as 250 N on the moon.

Calculate the density of the rock given that it has a volume of 250 cm³. Take g_{Earth} = 10 ms⁻²

A. 0.2 kg/cm³   C. 0.6 kg/cm³   E. More information needed.
B. 0.5 kg/cm³   D. 0.7 kg/cm³

Question 509:

A radioactive element X_{78}^{225} undergoes alpha decay. What is the atomic mass and atomic number after 5 alpha particles have been released?

	Mass Number	Atomic Number
A	200	56
B	200	58
C	215	64
D	205	68
E	215	58

Question 510:

A 20 A current passes through a circuit with resistance of 10 Ω. The circuit is connected to a transformer that contains a primary coil with 5 turns and a secondary coil with 10 turns. Calculate the potential difference exiting the transformer.

A. 100 V   C. 400 V   E. 2,000 V
B. 200 V   D. 500 V

Question 511:

A metal ball of unknown mass is dropped from an altitude of 1 km and reaches terminal velocity 300 m before it hits the ground. Given that resistive forces do a total of 10 kJ of work for the last 100 m before the ball hits the ground, calculate the mass of the ball. Take $g = 10ms^{-2}$.

A. 1 kg
B. 2 kg

C. 5 kg
D. 10 kg

E. More information needed.

Question 512:

Which of the following statements about the electromagnetic spectrum is correct?

A. The wavelength of ultraviolet waves is shorter than that of x-rays.
B. For waves in the electromagnetic spectrum, wavelength is directly proportional to frequency.
C. Waves in the electromagnetic spectrum travel at the speed of sound.
D. Humans are able to visualise the majority of the electromagnetic spectrum.
E. None of the above.

Question 513:

In relation to the Doppler effect, which of the following statements are true?
1. If an object emitting a wave moves towards the sensor, the wavelength increases and frequency decreases.
2. An object that originally emitted a wave of a wavelength of 20 mm followed by a second reading delivering a wavelength of 15 mm is moving towards the sensor.
3. The faster the object is moving away from the sensor, the greater the increase in frequency.

A. Only 1
B. Only 2

C. Only 3
D. 1 and 2

E. 2 and 3

Question 514:

A 5 g bullet travels at 1 km/s and hits a brick wall. It penetrates 50 cm before being brought to rest 100 ms after impact. Calculate the average braking force exerted by the wall on the bullet.

A. 50 N
B. 500 N

C. 5,000 N
D. 50,000 N

E. More information needed.

Question 515:

Polonium (Po) is a highly radioactive element that has no known stable isotope. Po^{210} undergoes radioactive decay to Pb^{206} and Y. Calculate the number of protons in 10 moles of Y. [Avogadro's Constant $= 6 \times 10^{23}$ mol^{-1}]

A. 0

B. 1.2×10^{24}

C. 1.2×10^{25}

D. 2.4×10^{24}

E. 2.4×10^{25}

Question 516:

Dr Sale measures the background radiation in a nuclear wasteland to be 1,000 Bq. He then detects a spike of 16,000 Bq from a nuclear rod made up of an unknown material. 300 days later, he visits and can no longer detect a reading higher than 1,000 Bq from the rod, even though it hasn't been disturbed. What is the longest possible half-life of the nuclear rod?

A. 25 days

B. 50 days

C. 75 days

D. 100 days

E. More information needed

Question 517:

A radioactive element Y_{89}^{200} undergoes several stages of beta (β^-) and gamma decay. What are the numbers of protons and neutrons in the element after the emission of 5 beta particles and 2 gamma waves?

	Protons	Neutrons
A	79	101
B	84	116
C	89	111
D	94	111
E	94	106

Question 518:

Most symphony orchestras tune to 'standard pitch' (frequency = 440 Hz). When they are tuning, sound directly from the orchestra reaches audience members that are 500 m away in 1.5 seconds.
Estimate the wavelength of 'standard pitch'.

A. 0.05 m

B. 0.5 m

C. 0.75 m

D. 1.5 m

E. More information needed

Question 519:

A 1 kg cylindrical artillery shell with a radius of 50 mm is fired at a speed of 200 ms^{-1}. It strikes an armour-plated wall and is brought to rest in 500 μs.

Calculate the average pressure exerted on the artillery shell by the wall at the time of impact.

A. 5×10^6 Pa   C. 5×10^8 Pa   E. More information needed

B. 5×10^7 Pa   D. 5×10^9 Pa

Question 520:

A 1,000 W display fountain launches 120 litres of water straight up every minute. Given that the fountain is 10% efficient, calculate the maximum possible height that the stream of water could reach.
Assume that there is negligible air resistance and $g = 10$ ms^{-2}.

A. 1 m   C. 10 m   E. 50m

B. 5 m   D. 20 m

Question 521

In relation to transformers, which of the following is true?

1. Step up transformers increase the voltage leaving the transformer.
2. In step down transformers, the number of turns in the primary coil is smaller than in the secondary coil.
3. For transformers that are 100% efficient: $I_p V_p = I_s V_s$

A. Only 1   C. Only 3   E. 1 and 3

B. Only 2   D. 2 and 3

Question 522:

The half-life of Carbon-14 is 5,730 years. A bone is found that contains 6.25% of the amount of C^{14} that would be found in a modern-day bone. How old is the bone?

A. 11,460 years   C. 22,920 years   E. 34,380 years

B. 17,190 years   D. 28,650 years

Question 523:

A wave has a velocity of 2,000 mm/s and a wavelength of 250 cm. What is its frequency in MHz?

A. 8×10^{-3} MHz   C. 8×10^{-5} MHz   E. 8×10^{-7} MHz

B. 8×10^{-4} MHz   D. 8×10^{-6} MHz

Question 524:

A radioactive element has a half-life of 25 days. After 350 days it has a count rate of 50. What was its original count rate?

A. 102,400   C. 204,800   E. 819,200
B. 162,240   D. 409,600

Question 525:

Which of the following units is **NOT** equivalent to a Volt (V)?

A. $JA^{-1}s^{-1}$ C. $Nms^{-1}A^{-1}$ E. JC^{-1}
B. WA^{-1} D. NmC

SECTION 2: MATHS

BMAT maths questions are designed to be time-consuming, and a lack of proper exam technique can really trip up some candidates. During your question practice, you will have the opportunity to assess how quickly you are able to complete the calculations. If you find yourself consistently not finishing, it might be worth leaving the maths (and probably physics) questions until the very end of the paper, so you are able to complete as many of the other questions as possible under timed conditions. Good students sometimes have a habit of making easy questions difficult; remember that the BMAT only tests GCSE level knowledge so you are not expected to know or use calculus or trigonometry in any part of the exam.

Formulae you **MUST** know:

2D Shapes			3D Shapes		
Area				**Surface Area**	**Volume**
Circle	πr^2		**Cuboid**	Sum of all 6 faces	Length x width x height
Parallelogram	Base x Vertical height		**Cylinder**	$2\pi r^2 + 2\pi rl$	πr^2 x l
Trapezium	0.5 x h x (a+b)		**Cone**	$\pi r^2 + \pi rl$	πr^2 x (h/3)
Triangle	0.5 x base x height		**Sphere**	$4\pi r^2$	$(4/3)\pi r^3$

Even good students who are studying maths at A2 level can struggle with certain BMAT maths topics because they're usually glossed over at school. These include:

Quadratic Formula

The solutions for a quadratic equation in the form $ax^2 + bx + c = 0$ are given by: $x = \frac{-b \pm \sqrt{b^2 - 4ac}}{2a}$

Remember that you can also use the discriminant to quickly see if a quadratic equation has any solutions:

$$If\ b^2 - 4ac < 0: No\ solutions$$
$$If\ b^2 - 4ac = 0: One\ solution$$
$$If\ b^2 - 4ac > 2: Two\ solutions$$

Completing the Square

If a quadratic equation cannot be factorised easily and is in the format $ax^2 + bx + c = 0$ then you can

rearrange it into the form $a\left(x + \dfrac{b}{2a}\right)^2 + \left[c - \dfrac{b^2}{4a}\right] = 0$

This looks more complicated than it is – remember that in the BMAT, you're extremely unlikely to get quadratic equations where $a > 1$ and an equation that doesn't have any easy factors. This gives you an

easier equation: $\left(x + \dfrac{b}{2}\right)^2 + \left[c - \dfrac{b^2}{4}\right] = 0$ and is best understood with an example.

Consider: $x^2 + 6x + 10 = 0$

This equation cannot be factorised easily but note that: $x^2 + 6x - 10 = (x + 3)^2 - 19 = 0$

Therefore, $x = -3 \pm \sqrt{19}$. Completing the square is an important skill – make sure you're comfortable with it.

Difference between 2 Squares

If you are asked to simplify expressions and find that there are no common factors but find that the expressions involve square numbers – you might be able to factorise by using the 'difference between two squares'.

For example, $x^2 - 25$ can also be expressed as $(x + 5)(x - 5)$.

MATHS QUESTIONS

Question 526:
Robert has a box of building blocks. The box contains 8 yellow blocks and 12 red blocks. He picks three blocks from the box and stacks them up high. Calculate the probability that he stacks two red building blocks and one yellow building block, in **any** order.

A. $\dfrac{8}{20}$ B. $\dfrac{44}{95}$ C. $\dfrac{11}{18}$ D. $\dfrac{8}{19}$ E. $\dfrac{12}{20}$

Question 527:
Solve $\dfrac{3x+5}{5} + \dfrac{2x-2}{3} = 18$

A. 12.11   B. 13.21   C. 13.95   D. 15.2   E. 19

Question 528:
Solve $3x^2 + 11x - 20 = 0$

A. 0.75 and $-\dfrac{4}{3}$ C. -5 and $\dfrac{4}{3}$ E. 12 only

B. -0.75 and $\dfrac{4}{3}$ D. 5 and $\dfrac{4}{3}$

Question 529:
Express $\dfrac{5}{x+2} + \dfrac{3}{x-4}$ as a single fraction.

A. $\dfrac{15x-120}{(x+2)(x-4)}$ C. $\dfrac{8x-14}{(x+2)(x-4)}$ E. 24

B. $\dfrac{8x-26}{(x+2)(x-4)}$ D. $\dfrac{15}{8x}$

Question 530:
The value of p is directly proportional to the cube root of q. When p = 12, q = 27. Find the value of q when p = 24.

A. 32   B. 64   C. 124   D. 128   E. 216

Question 531:

Which of the following is equivalent to $(\sqrt{7} - 2)^4$?

A. $233 - 88\sqrt{7}$
B. $233 + 88\sqrt{7}$
C. $49 - 16$
D. $121 - 44\sqrt{7}$
E. $49 - 2$

Question 532:

Calculate: $\dfrac{2.302 \; x \; 10^5 + 2.302 \;\; 10^2}{1.151 \; x \; 10^{10}}$

A. 0.0000202   C. 0.00002002   E. 0.000002002
B. 0.00020002   D. 0.00000002

Question 533:

Given that $y^2 + \mathbf{a}y + \mathbf{b} = (y + 2)^2 - 5$, find the values of **a** and **b**.

	a	b
A	-1	4
B	1	9
C	-1	-9
D	-9	1
E	4	-1

Question 534:

Express $\dfrac{4}{5} + \dfrac{m-2n}{m+4n}$ as a single fraction in its simplest form:

A. $\dfrac{6m+6n}{5(m+4n)}$ C. $\dfrac{20m+6n}{5(m+4n)}$ E. $\dfrac{3(3m+2n)}{5(m+4n)}$

B. $\dfrac{9m+26n}{5(m+4n)}$ D. $\dfrac{3m+9n}{5(m+4n)}$

Question 535:
A is inversely proportional to the square root of B. When A = 4, B = 25.
Calculate the value of A when B = 16.

A. 0.8   B. 4   C. 5   D. 6   E. 10

Question 536:
S, T, U and V are points on the circumference of a circle, and O is the centre
of the circle.

Given that angle SVU = 89°, calculate the size of the smaller angle SOU.

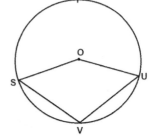

A. 89°   C. 102°   E. 182°
B. 91°   D. 178°

Question 537:
Open cylinder A has a surface area of 8π cm² and a volume of 2π cm³. Open cylinder B is an enlargement
of A and has a surface area of 32π cm². Calculate the volume of cylinder B.

A. 2π cm³   C. 10π cm³   E. 16π cm³
B. 8π cm³   D. 14π cm³

Question 538:
Express $\dfrac{8}{x(3-x)} - \dfrac{6}{x}$ in its simplest form.

A. $\dfrac{3x-10}{x(3-x)}$ C. $\dfrac{6x-10}{x(3-2x)}$ E. $\dfrac{6x-10}{x(3-x)}$

B. $\dfrac{3x+10}{x(3-x)}$ D. $\dfrac{6x-10}{x(3+2x)}$

Question 539:
A bag contains 10 balls. 9 of the balls are white and 1 is black. What is the probability that the black ball
is drawn in the tenth and final draw if the drawn balls are not replaced?

A. 0   B. $\dfrac{1}{10}$ C. $\dfrac{1}{100}$ D. $\dfrac{1}{10^{10}}$ E. $\dfrac{1}{362,880}$

Question 540:
Gambit has an ordinary deck of 52 cards. What is the probability of Gambit drawing 2 Kings (without
replacement)?

A. 0   B. $\dfrac{1}{169}$ C. $\dfrac{1}{221}$ D. $\dfrac{4}{663}$ E. None of the above

Question 541:

I have two identical unfair dice. The probability of rolling a 6 is twice as high as the probability of rolling any other number.

What is the probability that when I roll both dice the total will be 12?

A. 0   B. $\frac{4}{49}$ C. $\frac{1}{9}$ D. $\frac{2}{7}$ E. None of the above

Question 542:

A roulette wheel consists of 36 numbered spots and 1 zero spot (i.e. 37 spots in total).

What is the probability that the ball will stop in a spot either divisible by 3 or 2?

A. 0   B. $\frac{25}{37}$ C. $\frac{25}{36}$ D. $\frac{18}{37}$ E. $\frac{24}{37}$

Question 543:

I have a fair coin that I flip 4 times. What is the probability I get 2 heads and 2 tails?

A. $\frac{1}{16}$ B. $\frac{3}{16}$ C. $\frac{3}{8}$ D. $\frac{9}{16}$ E. None of the above

Question 544:

Shivun rolls two fair dice. What is the probability that he gets a total of 5, 6 or 7?

A. $\frac{9}{36}$ B. $\frac{7}{12}$ C. $\frac{1}{6}$ D. $\frac{5}{12}$ E. None of the above

Question 545:

Dr Savary has a bag that contains x red balls, y blue balls and z green balls (and no others). He pulls out a ball, replaces it, and then pulls out another. What is the probability that he picks one red ball and one green ball?

A. $\frac{2(x+y)}{x+y+z}$ C. $\frac{2xz}{(x+y+z)^2}$ E. $\frac{4xz}{(x+y+z)^4}$

B. $\frac{xz}{(x+y+z)^2}$ D. $\frac{(x+z)}{(x+y+z)^2}$

Question 546:

Mr Kilbane has a bag that contains x red balls, y blue balls and z green balls (and no others). He pulls out a ball, does **NOT** replace it, and then pulls out another. What is the probability that he picks one red ball and one blue ball?

A. $\dfrac{2xy}{(x+y+z)^2}$

B. $\dfrac{2xy}{(x+y+z)(x+y+z-1)}$

C. $\dfrac{2xy}{(x+y+z)^2}$

D. $\dfrac{xy}{(x+y+z)(x+y+z-1)}$

E. $\dfrac{4xy}{(x+y+z-1)^2}$

Question 547:

There are two tennis players. The first player wins the point with probability p, and the second player wins the point with probability 1-p. The rules of tennis say that the first player to score four points wins the game, unless the score is 4-3. At this point the first player to be two points ahead wins the game.

What is the probability that the first player wins in exactly 5 rounds?

A. 4p⁴(1-p)

B. p⁴(1-p)

C. 4p(1-p)

D. 4p(1-p)⁴

E. 4p⁵(1-p)

Question 548:

The equation below gives y in terms of x.

$$y = 2(\frac{x}{4} - 7)^2 - 5$$

Rearrange the equation to give an expression for x in terms of y.

A. $x = 28 \pm 4\sqrt{\dfrac{y+5}{2}}$

B. $x = 7 \pm 4\sqrt{\dfrac{y+5}{2}}$

C. $x = 28 \pm 4\sqrt{\dfrac{y-5}{2}}$

D. $x = 7 \pm 4\sqrt{\dfrac{y-5}{2}}$

E. $x = 28 \pm 4\sqrt{\dfrac{y\pm5}{2}}$

Question 549:

The volume of a sphere is $V = \frac{4}{3}\pi r^3$, and the surface area of a sphere is $S = 4\pi r^2$. Express S in terms of V

A. $S = (4\pi)^{2/3}(3V)^{2/3}$

B. $S = (8\pi)^{1/3}(3V)^{2/3}$

C. $S = (4\pi)^{1/3}(9V)^{2/3}$

D. $S = (4\pi)^{1/3}(3V)^{2/3}$

E. $S = (16\pi)^{1/3}(9V)^{2/3}$

Question 550:

Express the volume of a cube, V, in terms of its surface area, S.

A. $V = (S/6)^{3/2}$

B. $V = S^{3/2}$

C. $V = (6/S)^{3/2}$

D. $V = (S/6)^{1/2}$

E. $V = (S/36)^{1/2}$

Question 551:

Solve the equations $4 + 3y = 7$ and $2x + 8y = 12$

A. $(x, y) = \left(\frac{17}{13}, \frac{10}{13}\right)$

B. $(x, y) = \left(\frac{10}{13}, \frac{17}{13}\right)$

C. $(x, y) = (1, 2)$

D. $(x, y) = (2, 1)$

E. $(x, y) = (6, 3)$

Question 552:

Rearrange $\frac{(7x+10)}{(9x + 5)} = 3y^2 + 2$, to make x the subject.

A. $x = \frac{15\,y^2}{7 - 9(3y^2+2)}$

B. $x = \frac{15\,y^2}{7 + 9(3y^2+2)}$

C. $x = -\frac{15\,y^2}{7 - 9(3y^2+2)}$

D. $x = -\frac{15\,y^2}{7 + 9(3y^2+2)}$

E. $x = -\frac{5\,y^2}{7 + 9(3y^2+2)}$

F. $x = \frac{5\,y^2}{7 + 9(3y^2+2)}$

Question 553:

Simplify $3x\left(\frac{3x^7}{x^{\frac{1}{3}}}\right)^3$

A. $27\,x^{20}$

B. $87\,x^{20}$

C. $9\,x^{21}$

D. $27\,x^{21}$

E. $81\,x^{21}$

Question 554:

Simplify $2x[(2x)^7]^{\frac{1}{14}}$

A. $2x\sqrt{2\,x^4}$

B. $2x\sqrt{2x^3}$

C. $2\sqrt{2\,x^4}$

D. $2\sqrt{2x^3}$

E. $8\,x^3$

Question 555:
What is the circumference of a circle with an area of 10π?

A. $2\pi\sqrt{10}$

B. $\pi\sqrt{10}$

C. 10π

D. 20π

E. $\sqrt{10}$

Question 556:
If $a.b = (ab) + (a+b)$, then calculate the value of $(3.4).5$

A. 19

B. 54

C. 100

D. 119

E. 132

Question 557:
If $a.b = \dfrac{a^b}{a}$, calculate $(2.3).2$

A. $\dfrac{16}{3}$

B. 1

C. 2

D. 4

E. 8

Question 558:
Solve $x^2 + 3x - 5 = 0$

A. $x = -\dfrac{3}{2} \pm \dfrac{\sqrt{11}}{2}$

B. $x = \dfrac{3}{2} \pm \dfrac{\sqrt{11}}{2}$

C. $x = -\dfrac{3}{2} \pm \dfrac{\sqrt{11}}{4}$

D. $x = \dfrac{3}{2} \pm \dfrac{\sqrt{29}}{2}$

E. $x = -\dfrac{3}{2} \pm \dfrac{\sqrt{29}}{2}$

Question 559:
How many times do the curves $y = x^3$ and $y = x^2 + 4x + 14$ intersect?

A. 0

B. 1

C. 2

D. 3

E. 4

Question 560:
Which of the following graphs **do not** intersect?

1. $y = x$

2. $y = x^2$

3. $y = 1-x^2$

4. $y = 2$

A. 1 and 2

B. 2 and 3

C. 3 and 4

D. 1 and 3

E. 1 and 4

Question 561:
Calculate the product of 897,653 and 0.009764.

A. 87646.8

B. 8764.68

C. 876.468

D. 87.6468

E. 8.76468

Question 562:

Solve for x: $\frac{7x+3}{10} + \frac{3x+1}{7} = 14$

A. $x = \frac{929}{51}$
B. $x = \frac{949}{47}$
C. $x = \frac{949}{79}$
D. $x = \frac{980}{79}$

Question 563:

What is the area of an equilateral triangle with side length x.

A. $\frac{x^2\sqrt{3}}{4}$
B. $\frac{x\sqrt{3}}{4}$
C. $\frac{x^2}{2}$
D. $\frac{x}{2}$
E. x^2

Question 564:

Simplify $3 - \frac{7x(25x^2 - 1)}{49x^2(5x+1)}$

A. $3 - \frac{5x-1}{7x}$
C. $3 + \frac{5x-1}{7x}$
E. $3 - \frac{5x^2}{49}$

B. $3 - \frac{5x+1}{7x}$
D. $3 + \frac{5x+1}{7x}$

Question 565:

Solve the equation $x^2 - 10x - 100 = 0$

A. $-5 \pm 5\sqrt{5}$
C. $5 \pm 5\sqrt{5}$
E. $5 \pm 5\sqrt{125}$

B. $-5 \pm \sqrt{5}$
D. $5 \pm \sqrt{5}$
F. $-5 \pm \sqrt{125}$

Question 566:

Rearrange $x^2 - 4x + 7 = y^3 + 2$ to make x the subject.

A. $x = 2 \pm \sqrt{y^3 + 1}$
C. $x = -2 \pm \sqrt{y^3 - 1}$
E. x cannot be made the subject of this equation.

B. $x = 2 \pm \sqrt{y^3 - 1}$
D. $x = -2 \pm \sqrt{y^3 + 1}$

Question 567:

Rearrange $3x + 2 = \sqrt{7x^2 + 2x + y}$ to make y the subject.

A. $y = 4x^2 + 8x + 2$
C. $y = 2x^2 + 10x + 2$
E. $y = x^2 + 10x + 2$

B. $y = 4x^2 + 8x + 4$
D. $y = 2x^2 + 10x + 4$
F. $y = x^2 + 10x + 4$

Question 568:

Jane has a bag containing 12 sweets. She has 4 bonbons, 6 gobstoppers and 2 toffees.

Calculate the probability that she picks out 2 bonbons and 1 toffee in any order when she removes 3 sweets from the bag.

A. $\dfrac{24}{1728}$ B. $\dfrac{24}{1320}$ C. $\dfrac{72}{1320}$ D. $\dfrac{72}{1728}$ E. $\dfrac{9}{12}$

Question 569:

The aspect ratio of my television screen is 4:3 and the diagonal is 50 inches. What is the area of my television screen?

A. 1,200 inches C. 120 inches² E. More information
B. 1,000 inches² D. 100 inches² needed.

Question 570:

Rearrange the equation $\sqrt{1 + 3x^{-2}} = y^5 + 1$ to make x the subject.

A. $x = \dfrac{(y^{10} + 2y^5)}{3}$ C. $x = \sqrt{\dfrac{3}{y^{10} + 2y^5}}$ E. $x = \sqrt{\dfrac{y^{10} + 2y^5 + 2}{3}}$

B. $x = \dfrac{3}{(y^{10} + 2y^5)}$ D. $x = \sqrt{\dfrac{y^{10} + 2y^5}{3}}$

Question 571:

Solve $3x - 5y = 10 \ and \ 2x + 2y = 13$.

A. $(x, y) = (\dfrac{19}{16}, \dfrac{85}{16})$ C. $(x, y) = (\dfrac{85}{16}, \dfrac{19}{16})$ E. No solutions possible.

B. $(x, y) = (\dfrac{85}{16}, -\dfrac{19}{16})$ D. $(x, y) = (-\dfrac{85}{16}, -\dfrac{19}{16})$

Question 572:

The two inequalities $x + y \leq 3 \ and \ x^3 - y^2 < 3$ define a region on a plane. Which of the following points lies inside the region?

A. (2, 1) C. (1, 2) E. (1, 2.5)
B. (2.5, 1) D. (3, 5) F. None of the above.

Question 573:
How many times do $y = x + 4$ and $y = 4x^2 + 5x + 5$ intersect?

A. 0   B. I   C. 2   D. 3   E. 4

Question 574:
How many times do $y = {}^{3}$ and $y = x$ intersect?

A. 0   B. I   C. 2   D. 3   E. 4

Question 575:
A cube has unit length sides. What is the length of a line joining a vertex to the midpoint of the opposite side?

A. $\sqrt{2}$ C. $\sqrt{3}$ E. $\frac{\sqrt{5}}{2}$

B. $\sqrt{\frac{3}{2}}$ D. $\sqrt{5}$

Question 576:
Simplify the following:

$$5x\left(2x^{-\frac{1}{3}}\right)^3$$

A. $10x^{\frac{26}{27}}$

B. $40x^{\frac{26}{27}}$

C. $40x$

D. 40

E. 10

Question 577:
Fully factorise: $3a^3 - 30a^2 + 75a$

A. $3a(a - 3)^3$ C. $3a(a^2 - 10a + 25)$ E. $3a(a + 5)^2$

B. $a(3a - 5)^2$ D. $3a(a - 5)^2$

Question 578:

Solve for x and y:

$$4x + 3y = 48$$
$$3x + 2y = 34$$

	x	y
A	8	6
B	6	8
C	3	4
D	4	3
E	30	12
F	12	30
G	No solutions possible	

Question 579:

Evaluate: $\dfrac{-\left(5^2 - 4 \times 7\right)^2}{-6^2 + 2 \times 7}$

A. $-\dfrac{3}{50}$ B. $\dfrac{11}{22}$ C. $-\dfrac{3}{22}$ D. $\dfrac{9}{50}$ E. $\dfrac{9}{22}$

Question 580:

All license plates are 6 characters long. The first 3 characters are letters, and the next 3 characters are numbers. If all letters and numbers can be used, how many unique license plates are possible?

A. 676,000   C. 67,600,000   E. 17,576,000
B. 6,760,000   D. 1,757,600   F. 175,760,000

Question 581:

How many solutions are there for the following equation: $2(2(x^2 - 3x)) = -9$

A. 0   B. 1   C. 2   D. 3   E. Infinite solutions.

Question 582:

Evaluate: $\left(x^{\frac{1}{2}} y^{-3}\right)^{\frac{1}{2}}$

A. $\dfrac{x^{\frac{1}{2}}}{y}$ B. $\dfrac{x}{y^{\frac{3}{2}}}$ C. $\dfrac{x^{\frac{1}{4}}}{y^{\frac{3}{2}}}$ D. $\dfrac{y^{\frac{1}{4}}}{x^{\frac{3}{2}}}$

Question 583:
Given that:
$$4^x \times 16^y = 2^z$$
Express z in terms of x and y.

A. $z = 2x + 4y$
B. $z = x + y + 6$
C. $z = \frac{x}{2} \times \frac{y}{4}$
D. $z = \frac{2}{x} + \frac{4}{y}$
E. $z = 2x \times 4y$

Question 584:

Evaluate: $5\left[5(6^2 - 5 \times 3) + 400^{\frac{1}{2}}\right]^{1/3} + 7$

A. 0   B. 25   C. 32   D. 49   E. 56   F. 200

Question 585:
What is the area of a regular hexagon with side length 1 unit?

A. $3\sqrt{3}$ C. $\sqrt{3}$ E. 6

B. $\frac{3\sqrt{3}}{2}$ D. $\frac{\sqrt{3}}{2}$ F. More information needed

Question 586:
Dexter moves into a new rectangular room that is 19 metres longer than it is wide, and its total area is 780 square metres. What are the dimensions of this room?

A. Width = 20 m; Length = -39 m
B. Width = 20 m; Length = 39 m
C. Width = 39 m; Length = 20 m
D. Width = -39 m; Length = 20 m
E. Width = -20 m; Length = 39 m

Question 587:
Tom uses 34 meters of fencing to enclose his rectangular plot. He measured the diagonals to 13 metres long. What is the length and width of the plot?

A. 3 m by 4 m
B. 5 m by 12 m
C. 6 m by 12 m
D. 8 m by 15 m
E. 9 m by 15 m
F. 10 m by 10 m

Question 588:

Solve $\frac{3x-5}{2} + \frac{x+5}{4} = x + 1$

A. 1   C. 3   E. 4.5
B. 1.5   D. 3.5   F. None of the above

Question 589:

Calculate: $\frac{5.226 \times 10^6 + 5.226 \times 10^5}{1.742 \times 10^{10}}$

A. 0.033   C. 0.00033   E. 0.0000033
B. 0.0033   D. 0.000033

Question 590:

Calculate the area of the triangle shown to the right:

A. $3 + \sqrt{2}$
B. $\frac{2 + 2\sqrt{2}}{2}$
C. $2 + 5\sqrt{2}$
D. $3 - \sqrt{2}$
E. 3

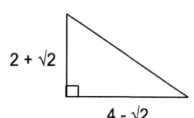

$2 + \sqrt{2}$

$4 - \sqrt{2}$

Question 591:

Rearrange $\sqrt{\frac{4}{x} + 9} = y - 2$ to make x the subject.

A. $x = \frac{11}{(y-2)^2}$

B. $x = \frac{9}{(y-2)^2}$

C. $x = \frac{4}{(y+1)(y-5)}$

D. $x = \frac{4}{(y-1)(y+5)}$

E. $x = \frac{4}{(y+1)(y+5)}$

Question 592:
When 5 is subtracted from 5x the result is half the sum of 2 and 6x. What is the value of x?

A. 0   B. 1   C. 2   D. 3   E. 4

Question 593:
Estimate $\dfrac{54.98 + 2.25^2}{\sqrt{905}}$

A. 0   B. 1   C. 2   D. 3   E. 4

Question 594:
At a restaurant called Pizza Parlour, you can order single, double or triple cheese in the crust. You also have the option to include ham, olives, pepperoni, bell pepper, meatballs, tomato slices, and pineapple. How many different types of pizza are available at Pizza Parlour?

A. 10   C. 192   E. 768
B. 96   D. 384

Question 595:
Solve the simultaneous equations $x^2 + y^2 = 1$ and $x + y = \sqrt{2}$, for x, y > 0

A. $(x, y) = (\frac{\sqrt{2}}{2}, \frac{\sqrt{2}}{2})$
B. $(x, y) = (\frac{1}{2}, \frac{\sqrt{3}}{2})$
C. $(x, y) = (\sqrt{2} - 1, 1)$
D. $(x, y) = (\sqrt{2}, \frac{1}{2})$
E. $(x, y) = (2, 2)$

Question 596:
Which of the following statements is **FALSE**?

A. Congruent objects always have the same dimensions and shape.
B. Congruent objects can be mirror images of each other.
C. Congruent objects do not always have the same angles.
D. Congruent objects can be rotations of each other.
E. Two triangles are congruent if they have two sides and one angle of the same magnitude.

Question 597:

Solve the inequality $x^2 \geq 6 - x$

A. $x \leq -3$ and $x \leq 2$

B. $x \leq -3$ and $x \geq 2$

C. $x \geq -3$ and $x \leq 2$

D. $x \geq 2$ only

E. $x \geq -3$ only

Question 598:

The hypotenuse of an isosceles right-angled triangle is x cm. What is the area of the triangle in terms of x?

A. $\frac{\sqrt{x}}{2}$

B. $\frac{x^2}{4}$

C. $\frac{x}{4}$

D. $\frac{3x^2}{4}$

E. $\frac{x^2}{10}$

Question 599:

Mr Heard derives a formula: $Q = \frac{(X+Y)^2 A}{3B}$. He doubles the values of X and Y, halves the value of A and triples the value of B. What happens to the value of Q?

A. Decreases by $\frac{1}{3}$

B. Increases by $\frac{1}{3}$

C. Decreases by $\frac{2}{3}$

D. Increases by $\frac{2}{3}$

E. Increases by $\frac{4}{3}$

Question 600:

Consider the graphs $y = x^2 - 2x + 3$, and $y = x^2 - 6x - 10$. Which of the following is true?

A. Both equations intersect the x-axis.

B. Neither equation intersects the x-axis.

C. The first equation does not intersect the x-axis; the second equation intersects the x-axis.

D. The first equation intersects the x-axis; the second equation does not intersect the x-axis.

E. More information is required to determine if the equations intersect the x-axis.

SECTION 3

The Basics

In BMAT Section 3, you have to write a short essay in response to one of three questions. The essay must not exceed one side of A4 and is a test of your ability to communicate clearly and concisely. The essay questions can span a wide variety of topics and thus demand different levels of comprehension and knowledge, but it is important to realise that one of the major skills being tested is actually your ability to construct a logical and coherent argument- and to convey it to the lay-reader.

Section 3 of the BMAT is frequently neglected by lots of students, who choose to spend their time on sections 1 & 2 instead due to the large amount of content on the specification. However, it's important to put in the work for Section 3 as it's testing skills directly relevant to medicine, and your essay will frequently be discussed at interview. Happily, our experience shows that Section 3 has the highest returns per hour of work out of all three sections, so it is well worth putting time into.

The aim of Section 3 is not to write as much as you can to fill the space available. Rather, the examiner is looking for you to make interesting and well-supported points to construct tight arguments, and for you to tie everything neatly together for a strong conclusion. Make sure you're writing critically and concisely; rambling costs you precious space and time. **Irrelevant material can actually lower your score.** You only get one side of A4 for your BMAT essay, so make it count!

Essay Structure

Most BMAT essay questions require you to address 3 main areas:

1) Explain what a quote or a statement means.
2) Argue for or against the statement.
3) Ask you "to what extent" you agree with the statement.

Part 1 should be the smallest portion of the essay (no more than 4 lines) and be used to provide a smooth introduction into the rather more demanding "argue for/against" part of the question. This is the main body of the essay and requires you to demonstrate a firm grasp of the concept being discussed and the ability to strengthen and support the argument with a wide variety of examples from multiple fields. This section should present a balanced response to the essay question, exploring **at least two distinct ideas**. Supporting evidence should be provided throughout the essay, with examples referred to when possible.

The third and final part effectively asks for your personal opinion and is a chance for you to shine - be brave and make an **innovative yet firmly grounded conclusion** for an exquisite mark. The conclusion should bring together all sides of the argument, in order to reach a clear and concise answer to the question. In short, this final section of the essay is your opportunity to demonstrate a unique perspective. It is important not to get too carried away here; ensure that your conclusion is an obvious or logical next step from the overall structure of your essay, which reflects careful planning and preparation.

Paragraphs

Paragraphs are an important formatting tool which show that you have thought through your arguments and are able to structure your ideas clearly. A new paragraph should be used every time a new idea is introduced. There is no single correct way to arrange paragraphs, but it's important that each paragraph flows smoothly from the last using connecting words and phrases. Examples of useful phrases to connect adjacent paragraphs might be "as a result of this", "in addition to", "in contrast to" etc. A slick, interconnected essay shows that you have the ability to communicate and organise your ideas effectively.

Given that you only have a limit of one A4 page to write in, **you shouldn't have more than 5 paragraphs. To save space,** use indents to inidicate individual paragraphs – don't leave empty lines! In general, 2 of these 5 paragraphs will be taken up by the introduction and conclusion respectively, leaving you with 3 larger paragraphs in which to make the main arguments of your essay.

Remember- the emphasis should remain on the quality and not quantity of writing. An essay with fewer paragraphs, but with well-developed ideas, is much more effective than a number of short, unsubstantial paragraphs that fail to fully address the question at hand, or an essay that isn't sufficiently focused.

Approaching the Essay

Section 3 can be broken down into 3 components; selecting your essay title, planning and writing it.

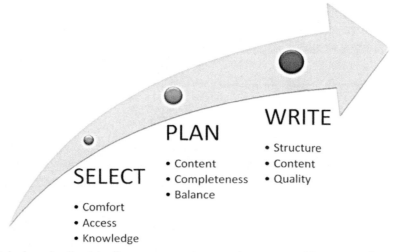

WRITE
- Structure
- Content
- Quality

PLAN
- Content
- Completeness
- Balance

SELECT
- Comfort
- Access
- Knowledge

Most students think that the "writing" component is most important. This is simply not true. The vast **majority of problems are caused by a lack of planning and a poor choice of essay title** - usually because students just want to get writing as quickly as possible since they are worried about finishing on time. Thirty minutes is long enough to be able to plan your essay well and *still* have time to write it so don't feel pressured to immediately start writing. Investing time in the planning process will make the writing much more efficient.

Step 1: Selecting

Making the right choice of essay titles is crucial to your success in Section 3. It is imperative that you are comfortable with the topic and that you fully understand the question being asked - it sounds silly but about 25% of essays that we mark score poorly because they don't actually answer the question!

Take two minutes to read all the questions at the start of Section 3. While one essay might initially stand out as being the easiest, if you haven't invested time to carefully think through it you might quickly find yourself running out of ideas. Likewise, a seemingly difficult essay might actually offer you a good opportunity to make interesting points and feel more accessible after spending a few moments working through the facets of the question.

Use this time to carefully select which question you will answer by gauging how comfortable you are with it given your background knowledge and suitable examples. Remember that Section 3 is not primarily a test of knowledge but rather a test of how well you are able to argue, so you will need to use relevant examples to support your points.

It's surprisingly easy to change a question into something similar, but with a different meaning. Thus, you may end up answering a completely different essay title by misconstruing the original question. Once you've decided which question you're going to do, read it very carefully a few times to make sure you fully understand what is being raised by the question, and ensure you can answer all aspects of the question. Keep reading it as you answer to ensure you stay on track!

Step 2: Planning
Why should I plan my essay?
There are multiple reasons you should plan your essay for the first 5-10 minutes of section 3:
- Section 3 is a test of communication – your writing will be clearer and more effective if you plan
- You don't have much space to write – make the most of it by writing a very well organised essay.
- Planning enables you to organise your thoughts and change the order before you start writing
- You run the risk of missing the point of the essay or only answering part of it if you don't plan adequately
- Section 3 is time pressured – you'll write the essay faster if you have a clear plan

How much time should I spend planning?
As a rough guide, it is **worth spending about 5-10 minutes to plan** and the remaining time to write the essay. However, this is not a strict rule, and you are advised to tailor your time management to suit your individual style. The planning process should enable you to write the essay more efficiently, as you will not have to spend time thinking up your next argument or example once you are writing it. However, it is important to strike a balance between developing a robust plan and leaving adequate time to write the essay in full. Completing past papers under timed conditions will enable you to find a ratio of time spent planning vs writing that works for you.

How should I go about the planning process?

There are a variety of methods that can be employed in order to plan essays (e.g. bullet-points, mind-maps etc). If you don't already know what works best for you, it's a good idea to experiment with different methods using past BMAT questions as examples.

Generally, the first step is to gather ideas relevant to the question, which will form the basic arguments around which the essay can be composed. You can then begin to structure your essay, including the way that points will be linked. At this stage it is worth considering the balance of your argument, and confirming that you have included arguments to support both sides of the debate. Once this general structure has been established, you must plan the examples or real world information you can use to help to support your arguments. Finally, you should assess the plan as a whole, and establish what your conclusion will be based on your arguments.

Step 3: Writing
Introduction

Why are introductions important?

An introduction provides the examiner with their first opportunity to evaluate your work. The introduction is where first impressions are formed of both your writing style and the strength of the arguments you present throughout your essay. A well-constructed introduction shows that you have really thought about the question in a logical way, and gives an indication of the arguments that will follow. It should leave the reader in no doubt as to what to expect from your essay.

What should an introduction do?

A good introduction should **briefly explain the statement or quote** in the essay question and give any relevant background information in a concise manner. However, don't fall into the trap of just repeating the statement in a different way. The introduction is the first opportunity to suggest an answer to the question posed, and the main body of your essay is effectively your chance to justify your answer.

Main Body
How do I go about making a convincing point?

Each idea that you propose should be supported and justified, in order to build a convincing overall argument. A point can be solidified through a basic Point → Evidence → Evaluation structure. This is a useful framework to ensure that each sentence within a paragraph builds upon the last, and that all the ideas presented are well developed.

How do I achieve a logical flow between ideas?

One of the most effective ways to display a good understanding of the question is to keep a logical flow throughout your essay. This means linking points effectively between paragraphs, and creating a congruent train of thought for the examiner as the argument develops. A good way to generate this flow of ideas is to provide ongoing comparisons of arguments, and discussing whether points support or dispute one another as you progress through the essay.

Should I use examples?

In short – yes! Examples can help boost the validity of arguments, and can help display high quality writing skills. Examples can add a lot of weight to your argument and make an essay much more relevant to the reader by demonstrating real-world representations of your points. When using examples, you should ensure that they are relevant and supportive of the point being made, in order to bolster your overall argument.

Some questions will provide more opportunities to include examples than others so don't worry if you aren't able to use as many examples as you would have liked. There is no set rule about how many examples should be included!

> *Top tip!* Remember that there is no single correct answer to these questions and you're not expected to be able to fit everything onto one page. Instead it's better to pick a few key points to produce a focused essay.

Conclusion

The conclusion provides an opportunity to emphasise the **overall position of your essay**, and package it into a neat summary for readers to take away. The conclusion serves two key purposes: firstly, it should summarise what has been discussed during the main body of the essay and secondly it should give a definitive answer to the question.

Some students use the conclusion to **introduce a new idea that hasn't been discussed**. This can be an interesting addition to an essay, and can help make you stand out. However, adopting this approach takes some considerable skill and our advice to the majority of students would be to present a well-organised, 'standard' conclusion since it is likely to be more effective than an adventurous but poorly executed one. It's worth bearing in mind throughout the whole process that the BMAT essay is a test of clear communication rather than literary style, so it's important to ensure you are doing the basics well rather than taking too many stylistic risks.

Medical Ethics

Usually there is a medical ethics question in the BMAT section 3, so it's well worth knowing the basics. There are huge ethics textbooks available, and indeed it's possible to study medical ethics at PhD level, however for the purposes of the BMAT you really only need to be familiar with the basic principles.

These ethical principles can be applied to all cases regardless of the social/ethnic background the healthcare professional or patient is from. In addition to being helpful in the BMAT, you'll need to know them for the interview stage of the application process so they're well worth learning now. The principles are:

Beneficence: The wellbeing of the patient should be the doctor's first priority. In medicine this means that one must act in the patient's best interests to ensure the best outcome is achieved for them i.e. 'Do Good'.

Non-Maleficence: This is the principle of avoiding harm to the patient (i.e. do no harm). There can be a danger that in a willingness to treat, doctors can sometimes cause more harm to the patient than good. This can especially be the case with major interventions, such as chemotherapy or surgery. Where a course of action has both potential harms and potential benefits, non-maleficence must be balanced against beneficence.

Autonomy: The patient has the right to decide about the treatment they receive. This principle requires the doctor to be a good communicator, so that the patient has sufficient information to make a meaningful decision. 'Informed consent' is thus a vital precursor to any treatment. A doctor must respect a patient's refusal of treatment even if they think it is not the correct choice. Note that patients cannot <u>demand</u> treatment – only refuse it, e.g. an alcoholic patient can refuse rehabilitation but cannot demand a liver transplant.

There are many situations where the application of autonomy can be quite complex, for example:
- **Treating children**: consent is required from the parents, although the autonomy of the child is taken into account increasingly as they get older.
- **Treating adults without the capacity** to make important decisions. The first challenge with this is in assessing whether or not a patient has the capacity to make the decisions. Just because a patient has a mental illness does not necessarily mean that they lack the capacity to make decisions about their health care. Where patients do lack capacity, the power to make decisions is transferred to the next of kin (or Legal Power of Attorney, if one has been set up).

Justice: This principle deals with the fair distribution and allocation of healthcare resources for the population.

Consent: This is an extension of the principle of autonomy - patients must agree to a procedure or intervention. For consent to be valid, it must be **voluntary informed consent.** This means that the patient must have sufficient mental capacity to make the decision and must be presented with all the relevant information (benefits, side effects and the likely complications) in a way they can understand.

Confidentiality: Patients expect that the information they reveal to doctors will be kept private- this is a key component in maintaining the trust between patients and doctors. You must ensure that patient details are kept confidential. Confidentiality can be broken if you suspect that a patient is a risk to themselves or to others e.g. where a doctor reasonably believes a patient might be involved in terrorism or has plans to complete suicide.

When answering a question on medical ethics, you need to ensure that you show an appreciation for two (or more) approaches to the ethical problem. Where appropriate, you should outline each point of view and how it pertains to the main principles of medical ethics before coming to a reasoned judgement to give a conclusion to the essay.

Common Mistakes
Ignoring the other side of the argument
Although you're normally required to support one side of the debate, it's important to **consider arguments against your judgement** in order to get higher marks. A good way to do this is to propose an argument that might be used against you, and then to argue why it doesn't hold true. You may use the format: *"some may say that…but this doesn't seem to be important because…"* in order to dispel opposition arguments, whilst still displaying that you have considered them. For example, *"some say that fox hunting shouldn't be banned because it is a tradition. However, witch hunting was also once a tradition – we must move with the times"*.

Answering the topic/Answering only part of the question
One of the most common mistakes is to only answer a part of the question whilst ignoring the rest of it as it feels inaccessible. According to the official mark scheme, **in order to get a score of 3 or more, you must write "…*an answer that addresses ALL aspects of the question*".** This should be your minimum standard- anything else that you write should then point you towards achieving 4/5.

Long Introductions
A key mistake to avoid is an excessively long, rambling introduction. Not only will it take up valuable space on your answer sheet, but it will also signal to the examiner that your answer is not clearly focused. Although background information about the topic can be useful, it is normally not necessary. Instead, the **emphasis should be placed on responding to the question**. Some students also just **rephrase the question** rather than actually explaining it. The examiner knows what the question is, and repeating it in the introduction is simply a waste of space in an essay where you are limited to just one A4 side.

Not including a Conclusion
An essay that lacks a conclusion is incomplete. This is easily avoided by ensuring you plan your essay carefully and stick to strict time-keeping while writing to allow yourself a few minutes at the end to construct a conclusion. Practising under timed conditions is therefore essential to avoid this costly mistake – an essay without a conclusion demonstrates to the examiner that you have not fully considered the question, and worse, that your organisational and planning skills are lacking. **The conclusion should be a distinct paragraph** in its own right and not just a couple of rushed lines at the end of the essay – you should invest time planning your conclusion at the beginning.

Sitting on the Fence
Students sometimes don't reach a clear conclusion. Although this is generally acceptable in a longer piece of writing, the BMAT requires you to **give a decisive answer to the question** and clearly explain how you've reached this judgement. Essays that do not come to a clear conclusion generally have a smaller impact and score lower. Remember, it is a test of communicating complex ideas clearly, so the reader must be provided with an obvious take-home message. When planning, keep this in mind, and ensure you are clear about the message before you write.

Exceeding the one page limit
The page limit is there for a reason – don't exceed it under any circumstances as any material over the limit won't be marked and it will appear that you haven't read the instructions.

Not using all the available space

Remember that you only have one A4 side to write on so ensure you make the maximum use of the space available to you. Don't leave lines to show paragraphs – instead, you should use indents. Similarly, you should also use the top-most line in the response sheet and avoid crossing entire sentences out unless in a dire emergency! Investing time in planning your essay and practising under timed conditions will make it easier to maximise your use of the space.

Marking your Essays

Practising section 3 can be tricky because most students don't know how to mark their essay. However, if you have a willing friend/family member, it is possible to get useful feedback and mark your work. You can use the diagram below to get an idea of your score. Keep in mind that this is just a very rough guide – examiners will look at several other factors when deciding on your overall score.

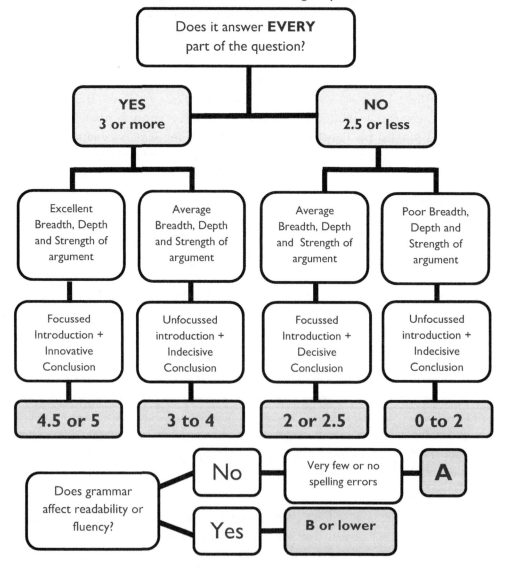

ANNOTATED ESSAYS

Example Essay 1:

"A doctor should never disclose medical information about his patients"
What does this statement mean? Argue to the contrary using examples to strengthen your response. To what extent do you agree with this statement?

The statement suggests that one of a doctor's most vital qualities is maintaining confidentiality of a patient's medical record. This involves all doctors with various specialities in different work places such as the clinics.

Disclosing medical informations regarding to their patients by doctors is considered as an unacceptable act within the medical society. For Example, by informing unrelated people about the patient might result in the individual's most embarrased health situation to be exposed. For example, suffering pain from their private parts and this may disgust other. This situation would inevitably upset the patient as their health privacy has been breached by others without consent leading to a sence of distrust towards doctors.

However, disclosing such matters to certain suitable people such as family and relatives may be crucial. For example, if the patient is the head of the family or the guardian to the children. As these individuals are in charge of leading and taking are of the family, they need to be able to perform mundane task (such as providing good support to the children) at their optimum. Also, to ensure that the members realise that the patient should not over exert him or herself despite their health conditions. Also, a sudden collapse will reduce shock when the family is to rush the patient back to the hospital knowing that the illness is related to the situation.

Overall, a patient's confidentially should not be disclosed without consent or any importance by all means. This is to respect their health privacy and to avoid any inconvenience within the medical society.

Examiner's Comments:
Introduction: The student appears to have an understanding of the topic but frequently makes statements that don't add much to the argument e.g. the second sentence of the first paragraph. The introduction would be better used to set up the counter arguments that will form the bulk of the main body.

Main Body: The first paragraph actually supports the statement and is therefore not actually answering the question (argue to the contrary). The example doesn't really add much either. The key issue that needed to be discussed was a doctor's duty to the patient – not about "disgusting others". The last sentence of the first paragraph is good and starts to address the question but it comes far too late. By that point the examiner will have already formed an opinion on the quality of writing and content of the essay, which could be hard to reverse.

The second paragraph is better but misses key points of the essay that needed discussion i.e. when can patient confidentiality be broken? Examples would include suspected terrorism, notifiable diseases, criminal activity, suicides etc, and you would be expected to be able to include and expand on at least a few of these points.

Conclusion: The conclusion doesn't really address the counter-arguments for breaking confidentiality. It gives a clear position but it is unclear how this has been arrived at and doesn't link well to the main essay. It also contains confusing terminology e.g. one discloses confidential information, not 'confidentiality'. The sentence concerning "inconvenience within the medical community" is also somewhat ambiguous.

Language: The poor grammar hinders the points that the student is trying to make throughout the essay e.g. "suffering pain from their private parts and this may disgust other". There are also frequent spelling mistakes like "embarrased" and "sence" that reduce the fluency significantly. Writing clearly is crucial in BMAT essays.

Score: D2

Example Essay 2

"A doctor should never disclose medical information about his patients"
What does this statement mean? Argue to the contrary using examples to strengthen your response. To what extent do you agree with this statement?

This statement is one of the duties set out by the General Medical Council for doctors to comply with, which is to respect patient's autonomy. It means that a doctor cannot share patient's medical information with other parties unless the patients, themselves, have granted permission to do so.

The ethical principle of respecting patients' autonomy cannot be applied in all cases, as some cases require doctors to disclose medical information about patients. First, when it involves a criminal act that has been comitted by a patient, a doctor has to report to an appropriate authority, such as the police. This is because the patient may potentially cause even more harm to others and as doctors, they have to prevent that from happening. An example of a case would be if a patient has a gunshot wound and he told his doctor that he had killed someone in the fight.

Next, another incident when a doctor has no choice but to disclose patients' medical information is when it may affect the health of society and could potentially cause an epidemic. Such patients might have an infectious disease and do not wish to let other people know about it. For instance, there has been many cases in West Africa where people who have Ebola are afraid to let their neighbours or friends find out because they do not want to be stigmatised and ostrilised from the society. However, these patients could spread the disease and so a doctor must not withold the information. Last of all, if a patient is underage, then he/ she is still not competent enough to make her own decision. Therefore, any medical information must be shared with his/ her legal guardian.

Respecting patient's autonomy by never disclosing their information is also important because patients have the right to chose who gets to know about it. It is his own body. He is the only person who knows the consequences of sharing this sensitive information. In conclusion, I believe that never disclosing patients' medical information cannot be complied in every incident. Respecting their autonomy is important but we have to treat each case separately.

Examiner's Comments:

Introduction: The introduction is well written but could be improved by making it explicitly clear that confidentiality can be breached in certain circumstances. This would then set up the main body nicely as the student would then be able to go straight into giving examples.

Main Body: The first sentence of the second paragraph is well written but should have gone in the introduction. There is a good breadth of argument with the important points being covered like a doctor's duty to prevent harm to others, 'public good' and the issue of 'capacity'. However, there is unnecessary padding that doesn't add much e.g. there was no need to expand on your example of infectious diseases. The extra space from avoiding this would have allowed the student to write about the fact that although confidentiality must sometimes be breached, a doctor has certain professional duties. For example, informing the patient both before and after and explaining why they have disclosed what they have to try and mitigate any damage to the doctor-patient relationship. In this way, public trust in medical professionals would be maintained.

Conclusion: The conclusion concisely summarises the arguments put forth in the main body and offers a nice resolution by saying that each case is different.

Language: Whilst it is clear that the student understands the question, there is some confusion as to the difference between "autonomy" and "confidentiality" (It's important to know the basics of medical ethics as they'll be helpful for the interview stage as well). Furthermore, there are minor spelling & grammar mistakes like "comitted" and "withold". In the conclusion, they assume that patients are male – "it is his own body" vs. "it is their own body".

Score: B3.5

Example Essay 3

"A doctor should never disclose medical information about his patients"

What does this statement mean? Argue to the contrary using examples to strengthen your response. To what extent do you agree with this statement?

Confidentiality is a basic patient right. The patient provides information to the doctor not to be unnecessarily shared with others without their knowledge or permission. On this basis, the statement argues that a doctor should never reveal medical data, such as results from tests or prescriptions given.

However, it can be argued that there are many circumstances whereby it is necesary to breach patient confidentiality and disclose medical information. More specifically, if the patient poses a threat to the public health, their medical situation should be disclosed immediately so that actions can be taken to prevent the spread. For instance, under the Public Health Act 1988, if a patient is suspected of communicable diseases such as tuberculosis, the doctor is required to inform the local health authorities immediately so that they can make precautions to protect the other citizens. In addition, a doctor should also disclose medical information if the patient has broken the law. For instance, the doctor should reveal medical data to the detectives and other relevant professionals if they request for it, to enable them to come to a conclusion of the case more quickly and accurately.

However, I agree with this statement to a large extent. After all, the patient should have the right over what happens to his medical documents and information. Revealing information about the patient unnecessarily will take this basic right away, and it is extremely unfair for the patient. Furthermore, this unprofessional decision may undermine the confidence between the patient and doctor. The patient may be less willing to reveal vital personal information to the doctor in the future, in fear that he might release this information as well. This would be extremely detrimental to the diagnoses and treatment for the patient or the doctor might not be able to gain sufficient information to make a more informed decision.

In conclusion, a doctor should never disclose medical information about his patients unless there are other external circumstances that oblige the doctor to do so. Breaking this confidentiality will cause the patient-doctor relationship to collapse, compromising the trust between them. However, in some cases, the decision to disclose is not that clear-cut; if a patient had sexually-transmitted infections, should the doctor disclose this information to his spouse? Such situations have to be decided on a case-by-case basis.

Examiner's Comments:

Introduction: This is a bold introduction that catches one's attention and gets straight to the point. The student however does make a rather generalised statement - "confidentiality is a basic patient right". A pedantic examiner could easily challenge this so it's important to be careful with wording statements like this to avoid tripping up.

Main Body: There is a good level of breadth and depth of argument here. However, there are again some generalised statements that are incorrect e.g. doctors don't need to break confidentiality if a patient has broken ANY law – just serious ones e.g. committed or intend to commit murder/terrorism etc. There is however, excellent discussion of the consequences of breaking confidentiality and a good level of detail (e.g. Public Health Act 1988).

Conclusion: this is an excellent conclusion that not only summarises the main arguments from both sides but also builds upon these to offer a solution as to when to break confidentiality by treating it on a case-by-case basis.

Language: There is only one spelling mistake (*"necessry"*) and although the somewhat general phrases stop this from scoring a perfect A5, it is still an excellent essay that displays good insight. As a general point, it's important to spend time making sure that you write exactly what you mean, and ensure you don't overcomplicate sentences or make over-ambitious arguments such that the overall meaning becomes unclear.

Score: A4.5

Example Essay 4

"Medicine is a science; not an art."
Explain what this statement means. Argue to the contrary that medicine is in fact an art using examples to illustrate your answer. To what extent, if any, is medicine a science?

I will explain the following statement "Medicine is a science, not an art" and argue that medicine is in fact an art with supports and to what extent is medicine a science.

I will talk about medicine and art in my opinion why, I think medicine is in fact an art and argue with the above statement. I think medicine is an art because of the human body its like a piece of artwork with creations and it is fascinating even more than a artwork of Picccasso or Van Gough's painting. It's like the artries and vessels ressemble the brushstroke of a painting and the heart is the meaning of the painting. To study about that and become a Doctor is like studying art to become and artist. Both Medicine and Art depend on passion if you don't have the passion you will not enjoy saving lives and will not create beautiful paintings. Medicine and Art are the most fascinating majors. They are completely different but at the sane time completely the same. Emotionally the same.

Now I will talk about the what extent is medicine a science. It depends on what course you are choosing in medicine For e.g. Biomedical it all counts as science it includes chemistry and maths which makes you a Biochemist or Phisyology, these are some interesting courses. Medicine is science because it depends on knowledge the years to study to become a doctor and save lives which is not an easy opprotunity to get.

In my essay I wrote about why I think that medicine is in face not all about sciene but about art too and argued with the above essay. I think Medicine is as fascinating and breathtaking as Art is with all the colours and bueatiful creations made my artists and saved by doctors.

Examiner's Comments:

Introduction: Although it is sometimes useful to outline what you are going to discuss when writing academic essays – you simply do not have enough space to do this in the BMAT. The introduction should be a clear and concise explanation of the statement, which isn't really done in this essay. Repeating the statement in the essay should also be avoided as it simply wastes valuable space and time.

Main Body: The student doesn't really have a good grasp of what the question is asking and as a result, the argument is off topic and slightly incoherent. The question requires the student to discuss things like medicine is a science because doctors put into practice medical principles that have an empirical factual basis. Whilst the third paragraph does this to a certain extent, the message is diluted somewhat because of a lack of focus. Medicine is also an 'art' because doctors also need to be able to communicate well with patients, to interpret clinical signs etc. The student seems have interpreted the question to literally mean "medicine is art" vs. "medicine is an art".

Conclusion: When writing a conclusion it is good practice to just make your point, as opposed to telling the reader you are making it. Thus, meta-writing is again not necessary in the conclusion. The final sentence, although interesting, is again off-topic.

Language: There are frequent spelling mistakes e.g. "sane time", "ressemble" and "*bueatiful*". The phrasing and grammatical errors are more serious and significantly affect the essay's fluency e.g. "talk about the what extent".

Score: D1

Example Essay 5

"Medicine is a science; not an art."

Explain what this statement means. Argue to the contrary that medicine is in fact an art using examples to illustrate your answer. To what extent, if any, is medicine a science?

Medicine. Arguably one of the most advancing fields in today's society. As a result, many of us have often thought about what medicine actually encompasses. The statement, "medicine is a science; not an art." is one that is constantly the subject of debate today. It questions to what extent that medicine may be considered an art, and to what extent a science. I feel that the statement does not suggest that medicine is an art form but instead, is well and truly a science, and is well within the definition of one.

An art form is usually something which is viewed as being expressive and emotional, as well as also being delicate. One can argue that the notion of caring for patients can be viewed as the non-scientific aspect of medicine, and therefore could be considered art as it requires one to be emotional and expressive. In order for a patient's rehabilitation process to be complete and succesful, the care directed at the patient should be tailored for them as the recovery of the psychological side of the human body is just as important as the physiological aspect. In order to be truly effective, the doctor should be able to be empathetic and try and understand the patient's pain when comforting them. This aspect of medicine does not involve the science of the human body or the knowledge of the intricate metabolic reactants which allow the human body to function so effectively.

However, another side of the debate could be that medicine is very much a science. This is due to the fact that medicine involves the analysis of a human body, for example, when diagnosing a patient or maybe understanding the effects a drug could bring to specific situations in the human body, something that is viewed with the utmost importance when administering a drug. The fact that in order to succesfully become a medical practitioner, one has to understand the physiology of the human body and have an immense and thorough knowledge of the anatomy of the human body, is the reason that medicine is associated with mainly being a science. This stems back to the days when we used to learn biology and chemistry back in school, and because medicine is largely based on those two core subjects, medicine, as a result, is widely regarded as solely being a science.

Examiner's Comments:

Introduction: The opening is catchy although unnecessarily long-winded. Thus, whilst the first sentence grabs the attention, the rest of it does little to keep it the next few sentences are very wordy and don't actually say very much. This is a prime example of just rephrasing the statement rather than explaining it, which wastes time and space.

Main Body: There are a good range of examples in the second paragraph - especially those about the rehabilitation process and empathy. Whilst the pro-science arguments are also well made, they are a bit one-dimensional. It would be better to discuss how trials are done to ensure safety rather than just concentrating on human anatomy and physiology. The last sentence also doesn't really add anything and if anything detracts from the final paragraph.

Conclusion: The essay is badly let down by a lack of a conclusion. This title required a critical analysis of the strengths of the two sides of the arguments in order to produce a well-constructed conclusion that answered the question, and this was not done effectively in this essay.

Language: There are some very elegant turns of phrase here however this sometimes results in a loss of focus. This doesn't affect the essay's readability (and therefore the language score) but indirectly affects the strength of the argument quite substantially. Nevertheless, as all parts of the question are answered to a sufficient level, it still scores a solid 3 for content/argument.

Score: A3

Example Essay 6

"Medicine is a science; not an art."

Explain what this statement means. Argue to the contrary that medicine is in fact an art using examples to illustrate your answer. To what extent, if any, is medicine a science?

This statement argues that medicine is more deeply rooted with the facts and set observations associated with scientific principles, and that no aspect of medicine is in itself open to interpretation; or art.
However many of the facets of medicine could very well be regarded as art. The manual dexterity and presision required by surgeons; particularly the visually aesthetic finish reconstructive or plastic surgery aims towards has a deep basis in artistic ability. Medicine as a career has a hugely important social aspect too; healthcare proffessionals are expected to be involved with communication and, when it comes to patients, even dealing with emotions has an interpretative aspect, and as such could be viewed as art. There is not one set approach to these situations; rather the outcome is reliant on a doctor's own personal judgement and choices as to the best course of action. Theoretical medicine, too, in terms of research may find success in new, recently discovered techniques; the development of which requires thinking 'outside of the box'. The clinical aspect of diagnostic medicine too is subject to unique approached, particularly when introducing extremely complex cases.

On the other hand, of course medicine involves most of the main scientific discplines. It is involved with biological structures, chemical processes and even principles of physics- generally all empirical; based on evidence and set in stone. In many cases there is a clear distinction between the right interpretation and thus course of action and the wrong one; for example when prescribing- for many patients (such as when allergies are involved) there are a whole host of drugs and courses of action that are unnacceptable. Therefore, the understanding of the human body and its inner workings that is so crucial for appropriate and successful medicine could well be argued to be science. Yet its application; its uses by doctors and other healthcare proffessionals is less impirical, more open to interpretation, and therefore more so an art. I feel that despite the fact medicine involves understanding and knowledge of the physiology and anatomy of the human body, it also involves the integral of caring for others, which is definitely more of an art form than a science, showing us that despite all the debates, medicine manages to combine art and science together resulting in the formation of a wondrous profession.

Examiner's Comments:

Introduction: An efficient introduction that explains the statement well. It could be improved by setting up the counter-argument.

Main Body: This is a good effort at a tricky essay – there is good breadth of argument with lots of examples like surgical precision and communication. There is also good consideration of the counter-arguments for why medicine is a science with good depth of argument e.g. drug prescriptions.

Conclusion: A strong conclusion that addresses the question well and summarises arguments from both sides concisely. It uses unnecessarily romantic language "wondrous profession" but nevertheless is an effective closing paragraph.

Language: Although there are a noticeable number of spelling mistakes (*"proffessionals"* and *"impirical"*), they do not detract from the flow of the essay, which is otherwise well written and fairly clear.

Score: B4.5

Example Essay 7

"The primary duty of a doctor is to prolong life as much as possible"

What does this statement mean? Argue to the contrary, that the primary duty of a doctor is not to prolong life. To what extent do you agree with this statement?

The most important responsibility of doctor is to cure diseases and extend a life of patients at his most ability acquired. The statement also states that doctors should try their best to prolong life of the sufferers as the first principle to consider. However, some might argue with this.

It is true to say that doctors are responsible for improving the conditions of diseases and alleviate the symptoms. This does not necessarily mean that is a major factor to tackle with each disease. Preventive care should be introduced at an early stage, and therefore the primary duty of medical professions, especially doctors should adopt this principle to be their main concern. For example, doctors should be aware of other health conditions of his or her patients that might be developed in the future regarding to patient's lifestyle or eating habits. To only address the present problems, relating to particular disease is not enough. Hence, prolonging life of patients is not the primary concern of doctors but to improve quality of life of the sufferers. By preventing the possible disease and acknowledging the patients are a proper most important role of a doctor.

Some people might argue with the statement as there have been a very controversial issue raised in recent years, euthanasia. Nowadays it is evidently seen that some patients in Switzerland and some other countries have their right to urge a doctor to help end their lives peacefully. Doctor may put an emphasis on methods or alternative ways to help prolong life of patient. Their prime concern is also finding the best beneficial treatment in order to fight with the disease unless there is a possible way. Therefore, putting patients at ease by ending their lives is also a primary concern to a doctor in some countries.

To some extent, I agree and support this statement as doctors have to delegate roles as healer to those who are in pain. Although it is illegal in some regions of the world to allow doctors taking life of a patient with their consent, it does mean that this method apart from prolonging life is one of the main duty for doctors to be well aware.

Examiner's Comments:

Introduction: This is a concise introduction that effectively just rephrases the statement rather than developing on it much – effectively not advancing an argument and thus wasting valuable space. A discussion of what a doctor's <u>primary</u> duty actually is would have been more appropriate here and would have set up the main body far better.

Main Body: This is a rather confusing and muddled answer The second paragraph is difficult to follow and strays from the real topic at hand. The student correctly identifies that a doctor's job is to improve quality of life and not duration. However, this isn't expressed with any degree of clarity or any examples. The third paragraph regarding euthanasia is better but again it is difficult to assess what point the student is trying to make.

Conclusion: Although the aim is to reach a clear conclusion supporting one side of the argument, you must consider and reflect on the counter arguments. Thus, it was necessary here to consider that both quality of life <u>and</u> duration are important. In addition, the euthanasia part is somewhat misguided – it is not necessarily true that doctors who do not prioritise duration of life are in favour of euthanasia.

Language: There are frequent grammatical errors e.g. "as there have been a very" which reduce the essay's fluency. The sentence phrasing also impacts the essay's overall readability, ultimately leading to a poor language score.

Score: C2

Example Essay 8

 "The primary duty of a doctor is to prolong life as much as possible"
What does this statement mean? Argue to the contrary, that the primary duty of a doctor is not to prolong life. To what extent do you agree with this statement?

A doctors' job is to cure disease through medical treatment to extend the life of the patient as much as possible. However, there is a certain limit to which doctors can go in order to prolong life as often quality of life is equally (if not more) important as quantity of life.

Doctors provide treatments to patients to help them overcome their disease so that they can live longer. This is also what the patients want. For example, for patients with kidney diseases, doctors will suggest them to have dialysis in order to remove the toxic substance in their body, which will kill them. Through dialysis, the patient's life will be extended and as this is the patient's will; a doctor's primary duty is to prolong life. People take preventative method, such as endoscopy of large intestine for symptoms of cancer, to avoid late discovery of disease, which will lead to a high chance of death. Therefore as a doctor has the skills to help people to extend their life, they should do as much as they can to fulfil the patient's wish, which is prolonging life.

However, doctors should also respect the patient's autonomy. If a patient doesn't want a treatment, even if the treatment is effective, doctors should not carry the treatment out, as everyone has the right to control their own life. Even when doctors want to help their patients, doctors should not over-ride the will of the patients.

Although one of a doctor's duties is to prolong life – this shouldn't be at the expense of quality of life. A doctor's primary duty is to offer the best possible medical advice and minimise suffering. Although the impact of this is usually to prolong life, in some cases, it may result in maximising quality of life.

Examiner's Comments:

Introduction: An excellent introduction that sets the scene very nicely for the main body and immediately conveys to the examiner that the student understands the essay of the essay.

Main Body: There are good points made throughout e.g. dialysis and the inclusion of preventative treatment shows a good insight into medicine. However, the student is not able to put together a substantial enough argument against prolonging life as they spend too much space discussing the reasons for prolonging life (which are less important). This is a perfect example of an unbalanced essay. In general, there is limited focus on the question and as a result, weak depth of argument.

Conclusion: A satisfactory conclusion that expresses the sentiments of the conclusion nicely and addresses the question well. It gives a clear position but also demonstrates insight into alternative points of view.

Language: The introduction and conclusion rescue this essay – as they convey a high degree of understanding in only a couple of sentences. The essay itself reads well and there are no obvious errors that reduce its fluency.

Score: A3.5

Example Essay 9

Animal euthanasia should be made illegal.

Explain what this statement means. Argue to the contrary that animal euthanasia should remain legal. To what extent do you agree with the statement?

The statement refers to the ethical dilemma of euthanasia. Euthanasia, synonymous of mercy killing, is the ending of someone's life because of a particular situation in which this living creature's future will be painful, sometimes short or it will ultimately be better to end it soon. The first argument for this statement is the fact that we as humans are generally kind hearted and benevolent: we see pain as a thing to be eradicated if not suppressed, and hence euthanasia could be seen as merciful, given this nature of ours. However, the natural argument against this is the undeniable fact that in the eyes of many, euthanasia is glorified murder, which goes against most people's morals since killing is seen as a negative thing in most societies nowadays. Another argument in favour of this process could be the fact that killing is not only condemned but also morally wrong since it involves us "removing" an otherwise healthy living being. The counter-argument illustrated by this statement is the fact that most people would look upon this as a grateful act of mercy, where although the consequences are taken into account, it is, morally, for some, the best thing to do, particularly if an animal, a source of fondness for some, is involved. The last argument to put forth is that animals are intelligent creatures capable of feeling pain like us and should hence receive the same mercy as is sometimes shown to humans. Naturally, people would say that in some cases this would be murder rather than mercy killing, eg the killing of old horses for glue rather than because of old age. I conclude this should remain legal since animals cannot voice their own opinion and hence give more weight to a decision.

Examiner's Comments:

Introduction: This is a good albeit somewhat long introduction. The student has defined euthanasia well and then established that there is an ethical dilemma surrounding it. However, the very long sentences make it needlessly difficult to follow.

Main Body: Whilst the writing style is excellent, there is a limited amount of content here. The student presents arguments for both sides simultaneously which sometimes makes it difficult to follow. This is made more confusing by the lack of paragraphs which means that the essay doesn't flow as well.

There are also some long and rambling sentences that detract from the clarity of the argument. The essay would benefit from a more focussed approach in which the student gets to the point. The point about killing horses for glue is also not as relevant to euthanasia outside a slippery slope argument (which isn't expanded upon).

Conclusion: The conclusion does not really build upon any of the arguments from the main body. This gives the impression that it was rushed with little planning.

Language: There are no obvious spelling errors but colloquialisms like 'nowadays' should be avoided. Overall, the student clearly understands the topic – but the essay is let down by a limited focus on the question and poor structure due to very long sentences and a lack of paragraphs.

Score: A3
Example Essay 10

Animal euthanasia should be made illegal.

Explain what this statement means. Argue to the contrary that animal euthanasia should remain legal. To what extent do you agree with the statement?

This statement means that the purposeful act of killing animals, carried out by veterinary practitioners, should be made against the law. Currently, human euthanasia is illegal and the introduction of legal euthanasia brings with it many potentially harmful implications such as the 'putting down' of healthy animals.

An ethical pillar of medicine is non-maleficence. By making animal euthanasia illegal we uphold this pillar and avoiding causing harm to potentially healthy animals. If animal euthanasia were to be legalised then some people may think it justified to slaughter animals for less than noble purposes.

However, there are many cases in which euthanasia may be the best way of progression in the medical treatment of animals. For example: if an injured or terminally ill animal has no chance of recovery and is suffering then euthanasia may be the most kind and compassionate thing to do e.g. if a horse has broken its leg and will never be able to walk again. In addition, if the quality of life of an animal is very low e.g. they have no home, are starving and there is nowhere for them to live, then euthanasia may also be the most compassionate course of action. This may be especially the case where an area is overpopulated with stray cats and dogs.

Another case where euthanasia seems the most beneficial course of action is if an animal has become infected with a disease that could spread to other animals and humans potentially causing widespread and significant harm e.g. if a dog becomes infected with rabies or a cow becomes infected with foot and mouth disease.

To conclude, I disagree with the statement as I think animal euthanasia should remain legal. It should however, only be carried out by a veterinary professional and only when the animal is undergoing significant suffering.

Examiner's Comments:

Introduction: A concise and focussed introduction that answers the first part of the question well and sets up the main body nicely.

Main Body: The student presents a sophisticated argument that addresses all the aspects of the question and uses good examples to back up their points. The arguments are well thought out and naturally follow on from each other. It would be better to argue why euthanasia should remain legal **before** giving reasons for it to be made illegal. This would help improve the flow.

Conclusion: A succinct and well-supported conclusion that ties together the major arguments in the main body. It also introduces a new idea —only vets should be allowed to perform euthanasia. This is a good point but should have been developed somewhat more.

Language: The student clearly understands the essay and puts together a strong essay. There are no glaring spelling or grammatical errors and it is easy to read and follow.

Score: A4.5

Summary

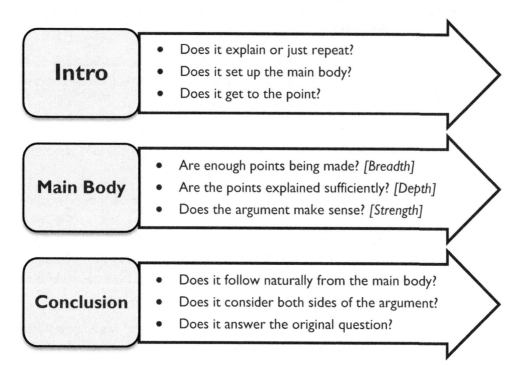

Intro
- Does it explain or just repeat?
- Does it set up the main body?
- Does it get to the point?

Main Body
- Are enough points being made? *[Breadth]*
- Are the points explained sufficiently? *[Depth]*
- Does the argument make sense? *[Strength]*

Conclusion
- Does it follow naturally from the main body?
- Does it consider both sides of the argument?
- Does it answer the original question?

General Advice

✓ Always answer the question clearly – this is the key thing examiners look for in an essay.

✓ Analyse each argument made, justifying or dismissing with logical reasoning.

✓ Keep an eye on the time/space available – an incomplete essay may be taken as a sign of a candidate with poor organisational or planning skills.

✓ Use pre-existing knowledge when possible – examples and real world data can be a great way to strengthen an argument- but don't make up statistics!

✓ Present ideas in a neat, logical fashion (easier for an examiner to absorb).

✓ Complete some practise papers in advance, in order to best establish your personal approach to the paper (particularly timings, how you plan etc.).

✗ Attempt to answer a question that you don't fully understand, or ignore part of a question.

✗ Rush or attempt to use too many arguments – it is much better to have fewer, more substantial points.

✗ Attempt to be too clever, or make up examples to support an argument – a tutor may call out incorrect facts etc.

✗ Panic if you don't know the answer the examiner wants – there is no right answer, the essay is not a test of knowledge but a chance to display reasoning and communication skills.

✗ Leave an essay unfinished – if time/space is short, wrap up the essay early in order to provide a conclusive response to the question.

ANSWERS

ANSWER KEY

Question	Answer	Question	Answer	Question	Answer	Question	Answer
1	A	39	D	77	B	115	C
2	C	40	A	78	D	116	B
3	A	41	B	79	A	117	C
4	A	42	B	80	B	118	D
5	C	43	E	81	E	119	A
6	D	44	B	82	B	120	B
7	D	45	D	83	C	121	D
8	A	46	E	84	C	122	C
9	A	47	B	85	D	123	D
10	B	48	D	86	C	124	B
11	D	49	B	87	C	125	A
12	C	50	D	88	A	126	C
13	D	51	A	89	C	127	E
14	A	52	B	90	C	128	C
15	D	53	D	91	A	129	E
16	A	54	A	92	A	130	C
17	B	55	C	93	D	131	C
18	B	56	D	94	D	132	A
19	A	57	C	95	C	133	B
20	B	58	A	96	B	134	B
21	A	59	D	97	D	135	C
22	C	60	D	98	E	136	D
23	C	61	D	99	D	137	D
24	A	62	B	100	B	138	C
25	B	63	C	101	D	139	C
26	A	64	B	102	B	140	B
27	D	65	B	103	E	141	C
28	A	66	D	104	E	142	E
29	A	67	E	105	B	143	B
30	B	68	C	106	C	144	D
31	A	69	E	107	E	145	B
32	C & E	70	D	108	A	146	D
33	B	71	F	109	A	147	E
34	B	72	B	110	D	148	C
35	D	73	A	111	A	149	E
36	A	74	C	112	C	150	C
37	A	75	D	113	D		
38	B	76	A	114	D		

Question	Answer	Question	Answer	Question	Answer	Question	Answer
151	D	189	D	227	C	265	C
152	B	190	C	228	E	266	C
153	E	191	C	229	D	267	A
154	E	192	B	230	C	268	C
155	D	193	C	231	B	269	B
156	E	194	C	232	E	270	E
157	E	195	C	233	D	271	D
158	A	196	D	234	A	272	C
159	C	197	C	235	C	273	B
160	D	198	E	236	A	274	E
161	C	199	D	237	B	275	C
162	C & E	200	C	238	C	276	C
163	D	201	A	239	A	277	E
164	C	202	A	240	B	278	B
165	B	203	C & E	241	E	279	D
166	A	204	D	242	C	280	C
167	C	205	D	243	E	281	D
168	B	206	B	244	B	282	E
169	C	207	D	245	B	283	B
170	D	208	A	246	C	284	C
171	B & C	209	B	247	C	285	C
172	B	210	C	248	C	286	D
173	B & D	211	B	249	E	287	B
174	D	212	C	250	C	288	E
175	E	213	D	251	A	289	C
176	D	214	D	252	C	290	A
177	D	215	C	253	A	291	C
178	B & D	216	C	254	B	292	C
179	B	217	C	255	B	293	D
180	C	218	A	256	E	294	A
181	C	219	C	257	E	295	D
182	D	220	A	258	C	296	D
183	B	221	E	259	B	297	C
184	C	222	C	260	C	298	C
185	B	223	B	261	D	299	C
186	E	224	B	262	C	300	E
187	B	225	C	263	B		
188	D	226	B	264	D		

Question	Answer	Question	Answer	Question	Answer	Question	Answer
301	A	339	E	377	E	415	C
302	D	340	E	378	A	416	B
303	B	341	C	379	A	417	A
304	A	342	E	380	D	418	E
305	C	343	B	381	C	419	A
306	C	344	E	382	B	420	E
307	D	345	E	383	E	421	D
308	B	346	B	384	A	422	E
309	A	347	A	385	C	423	B
310	D	348	C	386	C	424	E
311	D	349	C	387	E	425	D
312	A	350	B	388	E	426	A
313	C	351	D	389	E	427	D
314	A	352	A	390	E	428	D
315	E	353	B	391	A	429	C
316	D	354	C	392	C	430	E
317	A	355	C	393	B	431	E
318	D	356	A	394	D	432	D
319	D	357	B	395	B	433	D
320	B	358	C	396	E	434	E
321	A	359	B	397	B	435	C
322	E	360	C	398	D	436	A
323	D	361	E	399	D	437	B
324	E	362	F	400	D	438	E
325	E	363	A	401	D	439	C
326	E	364	A	402	E	440	D
327	D	365	A	403	E	441	C
328	C	366	C	404	B	442	C
329	E	367	A	405	B	443	E
330	E	368	A	406	D	444	C
331	D	369	E	407	A	445	B
332	C	370	E	408	B	446	B
333	A	371	D	409	E	447	B
334	E	372	B	410	B	448	E
335	A	373	D	411	D	449	D
336	A	374	A	412	A	450	C
337	C	375	E	413	A		
338	A	376	D	414	B		

Question	Answer	Question	Answer	Question	Answer	Question	Answer
451	E	489	D	527	C	565	C
452	A	490	E	528	C	566	B
453	D	491	E	529	C	567	D
454	E	492	A	530	E	568	C
455	E	493	B	531	A	569	A
456	C	494	C	532	C	570	C
457	D	495	D	533	E	571	C
458	E	496	E	534	E	572	C
459	D	497	E	535	C	573	B
460	D	498	C	536	D	574	D
461	E	499	C	537	E	575	E
462	B	500	B	538	E	576	D
463	C	501	D	539	B	577	D
464	E	502	E	540	C	578	B
465	D	503	E	541	B	579	E
466	E	504	D	542	B	580	E
467	C	505	D	543	C	581	B
468	A	506	A	544	D	582	C
469	E	507	D	545	C	583	A
470	E	508	C	546	B	584	C
471	E	509	D	547	A	585	B
472	E	510	C	548	A	586	B
473	E	511	D	549	D	587	B
474	D	512	E	550	A	588	C
475	E	513	B	551	B	589	C
476	D	514	A	552	A	590	A
477	E	515	C	553	E	591	C
478	E	516	C	554	D	592	D
479	B	517	E	555	A	593	C
480	A	518	C	556	D	594	D
481	C	519	B	557	D	595	A
482	C	520	B	558	E	596	C
483	D	521	E	559	B	597	B
484	D	522	C	560	C	598	B
485	E	523	E	561	B	599	A
486	B	524	E	562	C	600	C
487	A	525	D	563	A		
488	E	526	B	564	A		

PRACTICE QUESTION WORKED SOLUTIONS.

We didn't have room to include the worked solutions to all of our BMAT practice questions here (the book would be over 800 pages long if we did - but they are all available, online, for free. Just go to uniadmissions.co.uk/bmat-book to get detailed worked solutions to *every single question* absolutely free.

BMAT PAST PAPER WORKED SOLUTIONS

What are BMAT Past Papers?

Thousands of students take the BMAT exam each year. These exam papers are then released online to help future students prepare for the exam. Before 2013, these papers were not publically available meaning that students had to rely on the specimen papers and other resources for practice. However, since their release in 2013, BMAT past papers have become an invaluable resource in any student's preparation.

Where can I get BMAT Past Papers?

This book does not include BMAT past paper questions because it would be over 1,000 pages long if it did! However, all BMAT past papers since 2003 are available for free from the official BMAT website. To save you the hassle of downloading lots of files, we've put them all into one easy-to-access folder for you at **www.uniadmissions.co.uk/bmat-past-papers**.

At the time of publication, there are also too many past papers for us to fit answers to all of them in this book, so we've excluded the answers to the 2003 – 2012 papers. You can download these for free as a pdf from the address above, or they're available in print as part of *BMAT Past Paper Worked Solutions Volume I* from Amazon.

How should I use BMAT Past Papers?

BMAT Past papers are one the best ways to prepare for the BMAT. Careful use of them can dramatically boost your scores in a short period of time. The way you use them will depend on your learning style and how much time you have until the exam date but here are some general pointers:

- 4-6 weeks of preparation is usually sufficient for most students.
- Students generally improve their score for section 2 more quickly than section 1 so if you have limited time to prepare, focus on section 2.
- The BMAT syllabus changed in 2009 so if you find seemingly strange questions in the earlier papers, ensure you check to see if the topic is still on the specification. Often the questions no longer on the syllabus are marked with * but it is best to focus your preparation on BMAT papers from 2009 onwards.
- Similarly, there is little point doing essays before 2009 as they are significantly different in style. We've included plans for them in this book for completeness in any case.

How should I use this section?

This section is designed to accelerate your learning from BMAT past papers. Avoid the urge to have this book open alongside a past paper you're seeing for the first time. The BMAT is difficult because of the intense time pressure it puts you under – the best way of replicating this is by doing past papers under strict exam conditions (no half measures!). Don't start out by doing past papers (see previous page) as this 'wastes' papers.

Once you've finished, take a break and then mark your answers. Then, review the questions that you got wrong followed by ones which you found tough/spent too much time on. This is the best way to learn and with practice, you should find yourself steadily improving. You should keep a track of your scores on the next page so you can track your progress.

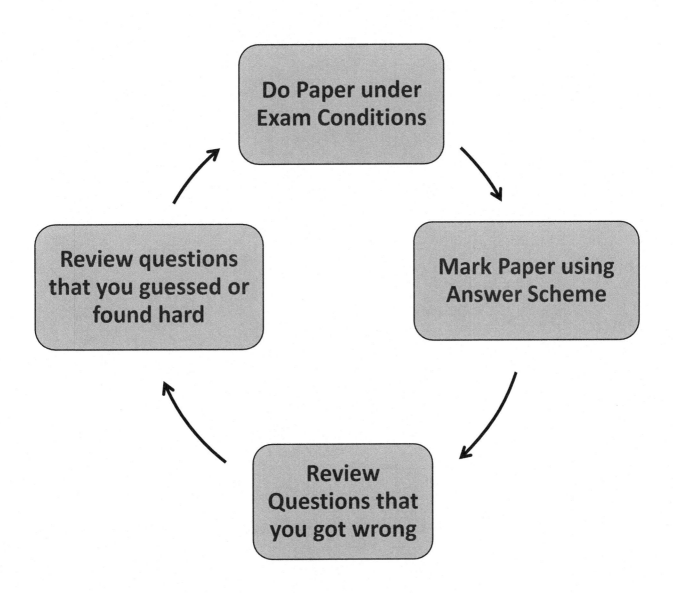

2003 - 2012

The solutions for these papers are all available for you, free, with the purchase of this book. You can access them all, online, at https://www.uniadmissions.co.uk/bmat-book

2013

SECTION 1

Question 1: A

This question can be worked out sequentially using the information given. Carla isn't working on Monday, which means Bob and Amy must be. That fills Bob's quota of 3 days in the week. This means Carla and Amy must be working on Wednesday and Thursday. Amy cannot also work on Tuesday, as that would make it 4 consecutive days, so Carla must work on Tuesday too.

Question 2: C

- **A** is a very bold statement and probably too strong to even be considered. The passage doesn't say that life can't exist on Kepler-22b and in fact even suggests that its new reclassification as uninhabitable may be inaccurate.
- **B** suggests the opposite of the background we are given. Cosmologists now suggest that fewer planets are habitable than previously thought.
- **C** summarises a main conclusion of the passage and is thus the best answer. The passage gives the example of Earth; that it is close to being outside the habitable zone but is robustly life-friendly, doubting the accuracy of the criteria.
- **D** cannot be inferred from this information alone. The passage doesn't describe a link between clouds and Kepler-22b.

Question 3: C

There are 6 combinations, and each one can be tested by determining the number of days between birthdays. If this equals a multiple of 7 then their birthdays are always on the same day of the week. 281 - 218 = 63, so Adam and Tara have their birthdays on the same day of the week every year.

Question 4: C

The main conclusion of the argument is easy to spot in this argument, as it is at the end of the passage and begins with the word 'therefore.'

- **A**, **B** and **D** are points which are mentioned in the text, but only as background information and they do not contribute to the conclusion of the passage.
- **C** paraphrases the conclusion that "the secret to losing weight is painfully simple - do more and/or eat less" and is thus the best answer.
- **E** is incorrect as the conclusion also describes that eating less can lead to calorie burn.

Question 5: D

Jason sold y Spruggles on day 1, and 2y on day 2. They cost £12 on day 1 and £9 on day 2, and he made £342 more on day 2. Thus, y x 12 + 342 = 2y x 9. We want 3y (how many were sold altogether).

18y = 12y + 342
6y = 342
3y = 171

Question 6: C

The main point of the Clovis-First theory is that the Clovis were *the first inhabitants* of the Americas.

C is the only answer that would seriously challenge this point, as it has a specific time linked to it. The Clovis first theory suggests that they arrived at -11500 BC, and if there was a human settlement present 500 years before this time then this disproves the theory. The other answers all link to the background information given, and could all be legitimate in a 'weaken' question, but none others seriously challenge that the Clovis were the first inhabitants.

Question 7: A

- Although elimination may seem quite a long process on the surface, it can be done rather quickly.
- Simon has 5 letters in his name so is limited to Hyde and Rush, and cannot be Rush because of the letter s. **Simon Hyde**
- Liam has 4 letters in his name so is limited to Doyle, Floyd and Shore, and must be Shore because of the letter l. **Liam Shore**
- Dylan has 5 letters in his name and must thus be Rush. **Dylan Rush**
- Eric must be Doyle or Floyd. It cannot be Doyle due to the letter e. **Eric Floyd.** Thus, Ian's surname must be Doyle. **Ian Doyle**

Question 8: D

D is clearly the correct answer, especially given that the question tells you that it is a sarcastic comment. If you find it hard to spot sarcasm, then we can rule out **B** and **C** as the comment links to the quote about children rather than the other 2, and the sarcasm of 'no' means we should agree with whatever the quote says, which is paraphrased in **D**.

Question 9: A

Answer **A** basically paraphrases this 'evidence' and is thus the correct answer. The statement does not relate to what wealth should bring, or anything about children. **D** assumes a causality which is not necessarily suggested by the statement.

Question 10: D

1. Kahneman suggests that the better you are at the job, the more time you must invest in it. However, this does not necessarily imply that people who work shorter hours will give more time to their children - they may use this 'extra' time in other ways.
2. The transcript talks about not getting happier as we get richer over a certain level. However, it does not suggest that wealth under this level will not cause stress.

Question 11: B

Anecdotal evidence is evidence based on personal accounts rather than facts or research, which this story clearly is. It is not necessarily conclusive without facts to back it up, there are no statistics, and it is relevant. We can argue that it is not hearsay as it is said that she is intimately involved with the family she describes.

Question 12: B

BEWARE that the symbol for Mercury and the symbol for Venus/Copper look very similar.

We can use the process of elimination here.

We can start with the first card, and at the top. There is only 1 equivalent to moon, which is silver on card 4, but the second item is different, so we can rule out cards 1 and 4 for having a pair.

The second card has 3 equivalents for the top item - cards 6, 7 and 8. The second item is only equivalent for card 7, and the last item for cards 2 and 7 are different, thus we can rule out all these cards as having a pair.

We are left with cards 3 and 5, and we can see that these are equivalent. Thus, we have 1 pair in total.

Question 13: D

The main conclusion of the argument is "*In the interests of providing the most desirable outcomes, it is clear that placebos should be used as a treatment offered by the NHS.*" Thus, if treatments (such as placebos) ensure better outcomes, they should be used, which is paraphrased in **D**.

A doesn't necessarily support the argument, as you do not know whether the placebo will work. **B, C** and **E** are unrelated to the fundamental point of the argument.

Question 14: B

We can start by ruling out 8 and 5, as this would break the alphabetical order rule. Let's call the missing digits x and y. From the information given in the text, we know that $4 + 0 + x + y = 8 +$ number of letters in x + number of letters in y. So, $x + y = 4 +$ number of letters in x and y.

Testing this rule out gives the exclusive answer of $x + y$ being 9 and 2. Here, it is important to re-read the question and make sure you give the number of letters that make up these digits, which is 7.

Question 15: B
A and **C** actually strengthen the argument as they back up some of the points made in the passage. **D** is a point against the argument, but doesn't weaken it, and is merely a statement saying the opposite of the passage. **B** is clearly the best answer as it directly contradicts the point that "*sport is what people do to counter the stress and pressures of work*" which is a key point in the author's argument that the growth of extreme sports is puzzling.

Question 16: C
- It is useful to quickly jot down the first letter of each month on a rough sheet of paper. J F M A M J J A S O N D. You can then determine which months the birthdays can occur on, along with the number of the month in the year.
- Jenny's and Alice's birthdays are 2 months apart - you can determine that this is only possible if Jenny's birthday is in June. It can't be in January or July because there would be no month 2 months away beginning with "A". Alice's birthday could be in August or April.
- Alice's and Michael's birthdays are 5 months apart - you can determine that this is only possible if Alice's birthday is in August and Michael's is in March, using the same logic as above. Thus, Jenny's birthday is in June, Michael's is in March, making them 3 months apart.

Question 17: D
This argument suggests that age makes us lack sleep and age makes us have impaired memory, so the lack of sleep must cause the impaired memory. This is an error of reasoning.
1 and 3 highlight different ways that these ideas may be connected, aside from lack of sleep causing impaired memory, highlighting weaknesses in the error of reasoning. 2 is unrelated to this error of reasoning and doesn't weaken the argument.

Question 18: B
Start by writing all the square numbers between 1 and 60. The month must be 09, the day can be between 1-30, the hour can be between 1-24 and the minute can be between 1-60.
There are 4 possible days: 1, 4, 16, 25. If we start with day 01, then there are 8 times: 4:16, 4:25, 4:36, 4:49, 16:04, 16:25, 16:36 and 16:49. There will be the same number of times for days 4 and 16, as you can have 2 different hours and 4 times per hour without including the same square number, giving 24 times from these particular days.
However, for day 25, there are more possible days, as you can have 3 different hours: 1, 4 and 16. These have 4 times per hour each, giving 12 times in total from this day.
24 + 12 = 36.

Question 19: B

There is no possible way that X can be green. The bottom left region must be green; it cannot be blue, yellow or red as there would be an edge with the same colour on both sides. The region to the top right of X can be red or yellow. When it is red, X can be yellow or blue. When it is yellow, X can be red or blue. Thus, the region can be yellow, blue or red.

Question 20: C

If this circle was smaller than a square, and contained within a square, it could be the opposite colour to the square and no rules would be broken. If it was not contained within a square, or larger than a square, then it would have to be a different colour to the black and white to prevent any edge having the same colour on both sides.

Question 21: A

This is probably best done by trial and error. Try using 3 lines in many different combinations, and you will eventually see that 2 colours will always be sufficient, as a segment will never share an edge with more than 2 other segments, both of which can be the other colour.

Question 22: B

We can think of the top and bottom as 2 separate circles (or any shape) with 5 separate segments. If we look at the circles individually, we can see that 2 colours cannot suffice, as there would have to be 2 adjacent. 3 colours suffice. When we superimpose this on another circle, 3 colours are sufficient to never have 2 adjacent faces the same colour.

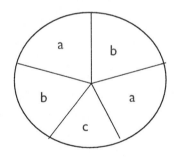 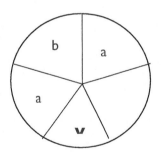

Question 23: D

Types of tiles:

- Fully black
- ½ black
- ¼ black

- ¾ black
- Black line outlines ¼
- Black line outlines ½

2 X black line outlines ½, ¼ black - this may be the one that people miss, as there are 2 forms of this shape, and one cannot be rotated to form the other. If in doubt, take a rough piece of paper, draw the shape, and try rotating it to see whether the tiles are equivalent.

Question 24: B

The main conclusion of this article is that "*to reduce the harm done by alcohol, it is vital to reduce consumption.*" The author argues that one of the best ways to do this is to make alcohol more expensive. Answer **B** suggests that cheaper prices have led to more consumption of alcohol, backing up the point that alcohol consumption is based on price.

Answer **A** is irrelevant to the point about reducing consumption and answers **C** and **D** do not strengthen the argument.

Question 25: B

- The clocks are 41 minutes apart. The hour change must between 19 and 20. Within the same hour there would be the same digit used, or with the same first hour digit. 23 to 00 cannot be used as the 0 is used twice and 09 to 10 also uses the 0 twice.

- **19:ab and 20:cd:** c cannot be 0, 1 or 2 as these numbers are already used. It cannot be 4 as cd could only be 40 or 41 (41 minutes apart), and the 0 and 1 digits have been used. Therefore, c must be 3, and a is 5.

- **19:5b and 20:3d:** b or d cannot be 1, 2, 3, 5 or 9 as these have been used. If b is 4, then d would have to be 5 which is not possible. b can be 6 or 7 and d can be 7 or 8.

- 19:56-19:57 and 20:37 and 20:38. There is no 4 used here.

Question 26: A

There is no easy way to do this- if you struggle with spatial awareness then this would be extremely difficult. Number 5 makes contact with P followed by 7 onto Q, 4 onto R and 1 onto S.

Question 27: C

We know Al is married and Charles is unmarried. Beth could be married or unmarried. **C** is the correct statement, as one way or the other, there was someone married looking at someone unmarried.

Question 28: D

The argument is *against* the idea that children should be exposed to harsh realities from a young age. Only **D** supports this, whereas the others suggest the opposite.

Question 29: E

Again, this was a difficult spatial awareness question and the only quick way to solve it would be to actually make the net in the exam. The rules don't explicitly state that you may not use scissors although you should certainly approve it with your exams officer before. Start by dividing the net into square shaped blocks. If spatial awareness isn't your forte – try ruling out the options that you can. Number 1 can be ruled out as, although there are 3 faces with one black block, the alignment is not correct in 1. Remember, there is no negative marking in the BMAT, so making educated guesses can be advisable.

Question 30: D

1 and 3 are fundamental principles of evolution, which can be assumed when the word evolution is used. The argument says, "*this characteristic must have evolved because it gave human beings a better grip underwater*" and this would not have 'evolved' if it were not advantageous. 2 is true from our previous knowledge, but is irrelevant to the argument and does not need to be assumed at any point.

Question 31: A

- First write down all the remaining numbers between 3 and 12, and cross each one out when used.

- **Bottom row:** 10+12=22, so the other 2 numbers must be added up to 7 to make 29. This means they must be 3 and 4, although we do not yet know in which order.
- **Top row:** 5+9=14, so the other 2 numbers must add up to 15. This can only be made from 7 and 8, although we do not yet know in which order.
- **Right column:** 11 must be above the 6, and we require 12 more. This means top right must be 8 and bottom right must be 4. This makes the person opposite 9 as 3.

Question 32: B

If 1% of non-cannabis users in the sample develop psychosis, and cannabis users were 41% more likely to have psychosis, 1.41% of cannabis users in the sample will have psychosis. 20% of young people report using cannabis, which is 2000 people.

2000 x 0.0141 = 2 x 14.1 = 28.2 users.

As we are referring to the **number of people**, the answer must be a whole number, so we round down to the nearest whole number, to give 28.

Question 33: A

Let's call the probability of psychosis for reasons other than cannabis use y. 80% of the population have probability y of getting psychosis, whereas 20% of the population have probability y plus the extra 41%.

We thus have 1.41 x 0.2y + 0.8y as the probability of getting psychosis and the percentage of getting it through cannabis use is:

$$\frac{0.41 \times 0.2y}{1.41 \times 0.2y + 0.8y} = \frac{0.082}{1.08} \approx \frac{0.08}{1.1}$$

This gives a percentage of slightly less than 8%, so **A** must be correct.

Question 34: B

The passage suggests that an increase in x (cannabis use) causes an increase in y (psychosis), whereas this answer provides an alternative link and suggests that y may also cause an increase in x.

A mentions age but this is not an alternative reason for the link. **C** suggests that the link may not be valid, which is irrelevant to what the question is asking. **D** is a point against an alternative link, suggesting more psychotic patients may have used cannabis than we think.

Question 35: C

Causal link is the key phrase here, and is a common theme in science and thus the BMAT. We need some evidence that A to B may be more than just a correlation.

- **A** may look tempting on the surface, but it doesn't prove causation. There may be a separate causal factor which affects you when young or old that links to cannabis and psychosis, rather than the cannabis itself.
- **B**, if anything, is against a direct causal link between cannabis and psychosis, highlighting that other factors may be involved.
- **C** suggests a causal link, because an increase in X (cannabis strength) **caused** an increase in Y (psychosis).
- **D** is irrelevant to showing a causal link.
- **E** is irrelevant to showing a causal link.

END OF SECTION

SECTION 2

Question 1: H

Both the nervous system and the endocrine system are involved in homeostasis. Some of the messaging takes place using chemicals and they can receive and send messages to and from the brain.

Question 2: D

This refers to the reactivity series. A displacement reaction can take place if the element in the salt is lower down in the reactivity series than the element it is being reacted with. This only applies to 1 and 4, where Al and Zn and higher in the reactivity series than Pb and Cu respectively.

Question 3: D

Both 1 and 2 are correct in their ability to damage. However, infrared does not cause damage when penetrating matter.

Question 4: A

$$\frac{4.6 \times 10^7 + 7 \times 2 \times 10^6}{4.6 \times 10^7 - 2 \times 2 \times 10^6}$$
$$= \frac{4.6 \times 10^7 + 14 \times 10^6}{4.6 \times 10^7 - 4 \times 10^6}$$
$$= \frac{4.6 \times 10^7 + 1.4 \times 10^7}{4.6 \times 10^7 - 0.4 \times 10^7}$$
$$= \frac{6.0 \times 10^7}{4.2 \times 10^7}$$
$$= \frac{6}{4.2}$$
$$= \frac{60}{42}$$
$$= \frac{10}{7}$$

Question 5: F

Protease would break down proteins into amino acids; lipase would break down fats into fatty acids and therefore lower the pH of the solution. However, carbohydrase would function to break up the carbohydrates and would produce the non-acidic sugar products, therefore not lowering the pH.

Question 6: B

This reaction is in equilibrium, with the greater number of moles on the left hand side of the equation than the right. This means that an increase in pressure would push the equilibrium to the right, therefore producing more T product. In addition, since the forward reaction is exothermic, a lower temperature shifts the equilibrium towards the products. Catalysts have no effect on the yield of product, just on reaction speed, and the addition of more reactants would obviously increase the product yield.

Question 7: H

The switch being closed has turned the circuit from a series to parallel which therefore has a lower overall resistance. Since total voltage is unchanged, current must increase in accordance with V=IR. Thus P increases. With the switch open, the voltage is shared across both resistors but with it closed, the second resistor can be bypassed (short-circuited) by the new branch. This means that only the full voltage is shared by the first resistor only. Thus, Q increases and R decreases.

Question 8: F

$$4 - \frac{x^2(1-16x^2)}{(4x-1)2x^3} = 4 - \frac{(1-16x^2)}{2x(4x-1)}$$

Thus: $\frac{8x(4x-1)}{2x(4x-1)} - \frac{(1-4x)(1+4x)}{2x(4x-1)}$

$= \frac{8x(4x-1)}{2x(4x-1)} + \frac{(4x-1)(4x+1)}{2x(4x-1)}$

Thus: $\frac{8x}{2x} + \frac{(4x+1)}{2x}$

$= \frac{8x+4x}{2x} + \frac{1}{2x}$

$= 6 + \frac{1}{2x}$

Question 9: F

Sensory neurons are the longest types of neurons as they must travel all the way to the spinal cord from receptors all over the body. The relay neurons are the shortest – they function to simply allow sensory and motor neurons to communicate with each other by connecting various neurons within the brain and spinal cord. An easy way to remember that motor neurons are medium length is **M**otor = **M**edium!

Question 10: B

This involves the equation $2Na + 2H_2O \rightarrow H_2 + 2NaOH$. You can therefore work out the moles of sodium by using $Mass = moles \times M_r$:

$Moles\ of\ Sodium = \frac{1.15}{23} = 0.5\ moles$

Since the molar ratio between sodium and hydrogen gas is 2:1, 0.25 moles of hydrogen are produced. Therefore, volume of hydrogen = $22.4 \times 0.25 = 5.6\ dm^3 = 560\ cm^3$.

Question 11: C

Remember that:

- Angle of incidence < Critical Angle: Light reflected back
- Angle of incidence = Critical Angle: Total internal reflection
- Angle of incidence > Critical Angle: Light leaves outside

Diagram 1 shows an angle below the critical angle. Therefore, total internal reflection does not occur and instead the light is reflected out. In diagram 2, the angle is greater than the critical angle. Therefore, total internal reflection does not occur and instead the light passes through.

Question 12: B

Label the corners of the square as A, B C and D and then see where they move in relation to the transformations performed. A reflection in the y axis therefore leads to the original orientation.

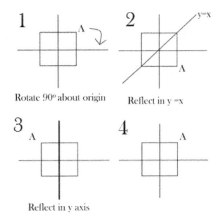

Question 13: C

The actual protein is not needed to produce it, as the intention is to allow the bacteria to produce the protein from the implanted DNA.

Question 14: A

In $MgCl_2$, the 2 valence electrons in Mg will each go to a chlorine atom, which will then mimic a filled orbital. The chloride atoms will both have the same electronic structure as argon, but when the Mg loses 2 electrons it will have the same electronic structure as neon, meaning the pair don't match.

Question 15: D

Don't get confused – this is actually easy, since background radiation has been corrected for!
Source X has a half-life of 4.8 hours and thus has 5 half-lives in 24 hours.

$Activity\ of\ X\ =\ 320\ x\ 0.5^5\ =\ 320\ x\frac{1}{32} = 10$

Source Y has a half-life of 8 hours and thus has 3 half-lives in 24 hours.

$Activity\ of\ Y\ =\ 480\ x\ 0.5^3 = 480\ x\frac{1}{8} = 60$

$Total\ Activity\ =\ 60\ +\ 10\ =\ 70$

Question 16: D

Start off by writing the relationships mathematically: $x\ \alpha\ z^2\ and\ y\ \alpha\ \frac{1}{z^3}$

Now make the powers equivalent so we can substitute: $x^3\ \alpha\ z^6\ and\ y^2\ \alpha\frac{1}{z^6}$

$So\ y^2\ \alpha\frac{1}{x^3}\ and\ x^3\ \alpha\frac{1}{y^2}$

Question 17: A

This question requires knowledge of somatic cell nuclear transfer. 2 is incorrect, because this procedure doesn't involve sperm cells. 4 is also incorrect, because the egg with the newly transferred nucleus must begin to divide, and not differentiate (point 1). The other 3 points are valid.

Question 18: E

$NaOH + HCl \rightarrow NaCl + H_2O$

The atomic mass of $NaOH = 23 + 16 + 1 = 40$

Theoretical maximum of NaOH in sample: $\frac{1.2}{40} = 0.03$

Moles of NaOH in sample that react is given by $n = cV$:

$\frac{50}{1000} \, x \, 0.5 = 0.025$

Purity: $\frac{0.025}{0.03} = \frac{5}{6} = 83.3\%$

Question 19: D

Recall that Power = IV = I²R. Since the resistors are in series, the overall current is given by: $I = \frac{V}{R1+R2}$

Thus, Power = $(\frac{V}{R1+R2})^2 R1 = \frac{V^2 R1}{(R1+R2)^2}$

Question 20: D
Smallest Cube:

We have 5 faces of the smaller shape, which is 5 x 1= **5 cm²**

Middle Cube:

Where the small cube joins the middle cube, we have a right-angled triangle with lengths x, x and 1. Using Pythagoras: $1^2 = x^2 + x^2 = 2x^2$

$x = \sqrt{\frac{1}{2}} = \frac{\sqrt{2}}{2}$.

Since the triangle makes up half the side, the total length of the side = $\sqrt{2}$

There are 4 faces fully uncovered and one face partially covered by the smaller shape. Thus, the surface area = $4 \, x \, \sqrt{2} \, x \, \sqrt{2} = 8$

Surface area of top face = $(\sqrt{2} \times \sqrt{2}) - (1 \times 1) = 1$

Surface Area of 2nd layer = 8 + 1= **9 cm²**

Largest Cube:

Using Pythagoras again: $\sqrt{2}^2 = x^2 + x^2 = 2x^2$

$2x^2 = 2$

Thus, $x = 1$. Since the triangle makes up half the side, the total length of the side = 2.

There are 5 faces fully uncovered and one face partially covered by the smaller shape. Thus, the surface area = $5 \times 2 \times 2 = 20$

Surface area of top face = $2 \times 2 - 2 = 2$

Surface Area of 3rd layer = 20 + 2 = **22 cm²**

Total Surface Area: 5 + 9 + 22 = 36 cm²

Question 21: E

The entire genome is found in every adult cell in the body (excluding red blood cells), hence 1 and 2 are correct. Starch is broken down before it reaches the liver so 3 is incorrect.

Question 22: C

There are 2 atoms of Cr in the equation, so **d** must be 2. The equation must balance for charge, so **b** must be 8. Comparing **a** and **c** shows that these coefficients change the number of oxygens only and do not affect the number of carbons or hydrogens. Thus, to have the correct number of hydrogens, **e** must be 4. **a** and **c** are 3, but this does not have to be deduced in this question.

Question 23: D

A. Rearrangement of $F = ma$

B. Rearrangement of $V = IR$

C. Rearrangement of $E_k = \frac{mv^2}{2}$

D. The relationship between wavelength and frequency is given by: $v = f\lambda$. If the y axis was wavelength, the axis should be $\frac{1}{f}$ i.e. the inverse of frequency (not a direct correlation).

E. Rearrangement of $Work\ Done = Force \times Distance$

Question 24: C

We need to calculate the probability of either:

- 2 blue balls and 1 red ball
- 2 red balls and 1 blue ball

Each combination has 3 permutations (i.e. the first combination can be BBR or BRB or RBB).

Probability of 2 Blue Balls = $\frac{8}{10} \times \frac{7}{9} \times \frac{2}{8} \times 3 = \frac{336}{720}$

Probability of 2 Red Balls = $\frac{2}{10} \times \frac{1}{9} \times \frac{8}{8} \times 3 = \frac{48}{720}$

Total Probability = $\frac{48}{720} + \frac{336}{720}$

$\frac{384}{720} = \frac{8}{15}$

Question 25: C

Let's call the dominant allele A and the recessive allele a. We're looking for the proportion of Aa cats. There are 50% in the first cross, and 67% in the second (dead organisms don't contribute to a population).

	A	a
A	Aa $(\frac{1}{4})$	aa $(\frac{1}{4})$
A	Aa $(\frac{1}{4})$	aa $(\frac{1}{4})$

	A	a
A	A (dead)	Aa $(\frac{1}{3})$
a	Aa $(\frac{1}{3})$	aa $(\frac{1}{3})$

Question 26: B

3, then 6.

As a catalyst is not used up in the reaction, we can see that using 3 then 6 uses NO to speed up the reaction but then replenishes it at the end. We can also see that a by-product which is not present in the net production, NO_2, is used up before the final products are formed. Finally, the reagents used up are SO_2 and $\frac{1}{2} O_2$ as seen in the reaction, and SO_3 is the only net product generated.

Question 27: E

The kinetic energy is given by $\frac{mv^2}{2}$ i.e. $\frac{4v^2}{2}$ = 1800.

Thus, v = 30 ms^{-1}

Using $F = ma$, the current acceleration is = $\frac{20}{4}$ = 5 ms^{-2}

Now calculate the velocity after 2 seconds of acceleration by the same force by using: $v = u + at$: $v = 30 + 2 \times 5 = 40$

Then calculate the final kinetic energy: $\frac{4 \times 40^2}{2}$ = 3200 J.

Finally, the extra kinetic energy is the difference between 3200 and 1800 = 1400 J

END OF SECTION

SECTION 3

"When you want to know how things really work, study them when they are coming apart."
(William Gibson)

- This statement suggests that the function of a system is not fully represented when it is working smoothly, rather it is only when the system is put under stress that the intricacies and factors in the system can be fully appreciated. This concept can be appreciated from a tangible example of a working system, such as the internal structures in a clock. When certain cogs in the clock fall out or stop working, it is easy to identify their place in the overall system. This follows the suggestion by the phrase that a functioning system can mask certain important properties that the system possesses. A system 'coming apart' allows identification of different parts, their purposes and their importance with regards to the function of the whole system. This can be represented by the mutation of certain genes leading to defects, such as in the case of cystic fibrosis where the mutation in the chloride channel shows its vital involvement in the secretion of mucus. In addition, 'coming apart' can help identify the original function of the system, as it will presumably be unable to perform it under those conditions.

- This method for studying systems does however have flaws. If the entire representation of the system is based on the lack of function of the system, this means that in some cases, the original function of cannot be understood. The use of such a principle therefore relies heavily on background knowledge, as without some understanding of the system, its lack of function could mean very little. For example, a certain part of a computer can malfunction, leading to a complete breakdown; however, without understanding the basics of what each part in the computer provides to the function, there will be too many unknowns to really build up an understanding. Additionally, if more than one factor leads to the coming apart of the system, a more subtle function of a part could be masked by a less subtle function of another.

- This method also underestimates the complexity of the system and the number of factors that may be interacting. It also does not allow you to appreciate the system as a whole.

- Overall, this method could be applicable in a simple system where there is already basic knowledge in place. However, for a more complex system that includes the involvement of a number of parts, or a system that has yet to be investigated, it may not accurately represent all the interacting features.

Good surgeons should be encouraged to take on tough cases, not just safe, routine ones. Publishing an individual surgeon's mortality rates may have the opposite effect.

- Tough cases present a challenge to surgeons, as increased difficulty of the procedure leads to an increased risk of mortality. It is therefore suggested that due to this increased likelihood of mortality, doctors may be reluctant to take tough cases on if the records were public as their reputation would be somewhat tarnished if the operations were unsuccessful. Public exposure of surgical league tables could also prevent consent from certain patients for certain surgeons to carry out procedures. In particular, certain cases of high mortality rates in surgeons could show that the surgeon is more experienced as they have performed in situations of increased risk. The surgeon could also feel increased pressure in certain procedures, leading to potential negative outcomes. This league table could also undermine the expertise of both the GMC and the hospital, which must make regular reviews of their doctors to make sure that they are working to the best of their abilities. In addition, mortality rates give very little information about the situation in which the death took place and in a number of cases, it could have no reflection on the surgeon at all.

- Publishing mortality rates could also have a positive impact on surgeons. I suggested above that this could put the surgeons under increased pressure, however all doctors should feel a certain pressure as they have a responsibility to the patient. In the case of the surgeon, they hold the patient's life in their hands and they should not forget that they have an obligation to perform to the best of their ability. Slightly contrastingly, a league table should have no impact on the performance of a doctor, as a doctor must always act with the patient's best interest at heart, rather than worrying about their reputation. Patients also have a right to know the performance record of a doctor, as they are putting their welfare into their hands, and therefore have a right to know all the factors and risks involved in making such a decision, including their surgeon.

- To use such a league table for the improvement of surgical performance would be extremely beneficial; however it is unlikely that this would be the case. It is more likely that there will be a negative impact on surgical performance, or even the decision to perform the surgery in the first place. In conclusion it is probably more beneficial for both the surgeon and the patient that the league table be kept private, as it enables the doctors to act as they see fit, potentially saving more lives due to the undertaking of riskier but more beneficial operations.

"Ignorance more frequently begets confidence than does knowledge: it is those who know little, and not those who know much, who so positively assert that this or that problem will never be solved by science." (Charles Darwin)

- Darwin is suggesting that people who know very little about science are very confident that they know the limits of the field and therefore do not think it is able to make progress in certain areas. This can be a valid idea in certain respects as little knowledge of a subject can cause a person to think they know the important aspects of a certain field. It might however actually be the case that the field has many more complexities that can offer very exciting potential developments, something which could be overlooked by an ignorant person.

- This concept can be represented by a self-diagnosis made by a patient online, they are only being exposed to very superficial (and potentially incorrect or over-simplified) areas of the field and this can lead them to incorrect conclusions.

- On the contrary this concept does not always apply. When looking at research environments, it is possible for a very knowledgeable researcher to have studied the subject for such an extended period of time that their sense of the bigger picture becomes clouded. This could mean that an ignorant person is then able to come and have a different take on the problem, as they have not developed the same mind-set of the knowledgeable researcher and they are less aware of possible limitations of the field, able to look 'outside the box'.

- This can be compared to a partnership between an engineer and a biologist in the development of medical equipment, the engineer might not know very much about biology, they are however able to give a different outlook on the problem. Similarly a scientist very knowledgeable in a field may not be able to accept that the field may be unable to progress, something that an ignorant person may be able to notice. There is also the possibility that an ignorant person may assume that science has no limits, which can be helpful and detrimental depending on the situation.

- Overall, this attitude does have some truth, I find however that it is cynical and a generalisation. It does not take into account that it is new and possibly uninformed minds that are pushing the boundaries of science, allowing it progress at a rapid pace.

In a world where we struggle to feed an ever-expanding human population, owning pets cannot be justified.

- This phrase suggests that the expense of owning a pet is unnecessary, and that instead the resources used to support a pet should be used to help support the over-expanding population. This argument is based on a number of assumptions, and I will attempt to sort through them here.

- On the one hand, it is true that the population is 'over-expanding' and we have a moral obligation to support those in need. It is also true that ownership of a pet requires a lot of resources. The pet must be fed, watered and housed appropriately, as well as veterinary care provided. This therefore suggests that the owner must have food and money available for this particular purpose. One could argue that the money spent on the pet, and the food that it consumes, could be better used elsewhere. If the resources used to support the pet were dedicated to supporting the needs of the expanding population, it is possible that a significant positive contribution could be made. In addition, one could argue that some expensive and exotic pets are an unnecessary luxury or too indulgent.

- On the other hand, the above statement assumes that the resources and food supplied to pets could be better used elsewhere. Owning a pet is an individual's decision, and people have free choice to spend their money how they please. There is also no guarantee that even if these owners did not own a pet, they would spend the equivalent amount supporting those in need, they might instead buy something equivalently recreational.

- An extrapolation of the above statement might lead one to say, "why don't we restrict television sales and other such luxuries as the funds could be better used elsewhere?" As further support against this statement, it is important to consider the importance of pet ownership, such as the life-saving duties of guide dogs, likely to be considered a worthwhile expense. In addition, pet ownership in some cases can reduce waste due to their consumption of leftover food, which would have been otherwise discarded.

- Additionally, the money spent on pets can be a form of support for the economy. A stronger economy allows the government to provide more assistance to those in need, and so encouraging pet ownership could in fact indirectly provide support.

- In conclusion, we have a moral obligation to help those in need, however, pet ownership and lack of resources for an expanding population are not necessarily related. Owning pets cannot account for the massive resource deficit. Making steps to prevent food waste or encouraging charitable donations may be a more worthwhile venture.

END OF PAPER

2014.

Fully worked solutions for the 2014 paper are released for every candidate to practice the BMAT. You can find them by going to this link: https://www.uniadmissions.co.uk/bmat-book

2015

SECTION 1

Question 1: D
Plotting the information:

Stuart > Ruth > Margaret

Tim > Adrian?

We don't know where Adrian sits in relation to Margaret, but we do know that Adrian is shorter than Tim, Ruth and Stuart. So, Adrian is shorter than Ruth and Stuart but not necessarily Margaret.

Question 2: E
E is the overall conclusion from the passage as it states, *"three quarters of all infections recorded last year were in people from deprived areas…and born outside"*, thus only a quarter were from more affluent populations and born in the UK.

A may be true but is posed as a potential reason to account for the weakened immune systems that may again be the cause of the increased incidence of TB in deprived areas, so it is not a conclusion. **B, C** and **D** may again be true but are not conclusions, they are possible reasons for the overall conclusion that TB incidence has increased but not in UK born affluent populations.

Question 3: A
The question shows the end of month readings so to find the greatest difference between September 1st and November 30th subtract the November readings from August (as August reading is end of the month, hence the value for September 1st).

The red van has the biggest difference, so the answer is **A**.

	August	November	Difference
Red	68 240	78 853	10 613
Orange	64 425	73 684	9 259
Yellow	71 302	81 163	9 861
Green	64 827	75 146	10 319
Blue	73 959	83 392	9 433
Indigo	68 623	78 229	9 606
Violet	63 088	72 826	9 738

Question 4: A
A (if true) would best express a flaw in the overall conclusion of the passage that physical attractiveness correlates with sporting performance, as it highlights the failure of the passage to account for other potential contributing factors to sporting success.

B is not true as an objective measurement has been taken using performance in a cycling endurance race. **C** may be true but the passage does not claim to extend the correlation to other sports, so it is not a flaw. **D** is never stated as no other sports are mentioned. Whilst if true, **E** would fail to support the findings of the research mentioned in the passage, it is not a flaw relating to the research that has been undertaken.

Question 5: D

The top 3 scores on either attempt were no. 4 1st attempt – 7.34m, no. 13 2nd attempt 7.29m, the 3rd would be no. 4's 2nd attempt (7.26m) but they have already qualified so then next best is no. 10 1st 7.17m. Anyone else within 50cm of third place 7.17m or 717cm qualifies, so anything above (717-50 = 667cm) 6.67m qualifies.

Competitor Number	1st	2nd
1		
2	✓	
3		✓
4	Already qualified	
5		✓
6		
7	✓	
8		
9		
10	Already qualified	
11		
12		
13	Already qualified	
14		✓
15		

Which totals 5 competitors in addition to the top 3, so 5 + 3 = 8, answer **D**.

Question 6: C

The passage is about the impending difficulties facing African and Asian agriculture due to future climate changes that will result in food shortages. It concludes that the development of seed banks with detailed catalogues about their trait so that farmers may be able to trial crops for the future, which is expressed in **C**.

A and **E** give the basis for the problem, rather than conclusions, and so are wrong. **B** is the hopeful outcome but not the conclusion. Also, the closing sentence states that seed banks are not the only answer, which **B** implies, so it is incorrect. The main conclusion is about creating seed banks, which would then allow for **D**, making it more of a secondary conclusion or suggestion rather than the main conclusion.

Question 7: C

Laying out share price for each day in a table:

Monday	Tuesday	Wednesday	Thursday	Friday
£1	£1.20	x	1.25x	£1

As Wednesday's price is unknown, use x. Thursday is 125% of x.

For Helen's shares:

Monday	Tuesday	Wednesday	Thursday	Friday
£1000	£1200	x	£1350	

This means on Tuesday the shares must have been worth £1200.

$1.25x = £1350$ so $x = £1080$ *(1350 ÷ 5 x 4 = 1080)*

Monday	Tuesday	Wednesday	Thursday	Friday
£1000	£1200	£1080	£1350	

Therefore, the price change between Tuesday and Wednesday is a decrease of 10% (1200 x 0.9 = 1080).

So, for Paul: £3600 x 0.9 = £2700

Monday	Tuesday	Wednesday	Thursday	Friday
	£3600	£2700		

He therefore made a loss of £300.

Question 8: B

In a direct comparison with non-custodial sentences the rate of reoffending was 22% whereas for custodial sentences it was 55%. This shows that the rate of reoffending was significantly lower for those with non-custodial sentences therefore which means that it would be a mistake, as stated in **B**, to give a custodial sentence when a non-custodial sentence is also appropriate as doing so would likely result in a higher chance or reoffending.

A cannot be reliably concluded as we have no specific data about serving half sentences.

That "70% of under 18s re-offend" would suggest that there are significant problems, but it does not fully support **C** that it is a mistake to send them to prison - we would need data comparing with those given a non-custodial sentence. **D** is incorrect as study 1 shows that only 72% reoffend, after 9 years. Whilst they all may do eventually, we have not been given the data to show this so it cannot be reliably concluded. For **E** we have no data on comparing rates of recidivism for short (less than 12 months) with over 12 months so again it cannot be reliably concluded.

Question 9: D

Of the 50 000 former prisoners:

- In year 1 44% reoffended = 22 000
- In year 5 66% had reoffended which is 33,000 (a further 11,000).

Question 10: A

The phrase "instead of" implies there is the choice of whether to give a community service rather than a prison sentence as they are eligible for both, which would account for the limited population where this was possible, where it reduces reoffending rates by 6%. The comparison of 55% versus 22% is simply a comparison of reoffending rates for all non-custodial versus custodial sentences but it may be that many of those given a custodial sentence were not eligible for a non-custodial so a direct comparison with the group where both were available cannot be made.

Question 11: E

E (if true) would strengthen the argument the most as it gives evidence compared with a control group that restorative justice is 20% more effective at reducing reoffending rates than just a community service order. This provides strong evidence that restorative evidence is effective and so would strengthen the argument that more offenders should be subjected to restorative justice to reduce reoffending rates.

A would not strengthen the argument as much as it does not provide evidence for the effects of restorative justice without a prison sentence, so there is no basis for comparison for the effects on reoffending rates.

B and **C** are moral and practical arguments that support the plan of sending individuals to **prison** rather than being subjected to restorative justice, thus they weaken the argument in question. **D** is a financial argument rather than being based on the evidence of the report so is not relevant here.

Question 12: D

Using the possible points of 9, 5, 3 and -2, it's possible to make:

- Crosswords: $9 + 5 + 5 + 3 = 22$
- Jigsaws: $9 + 9 + 5 - 2 = 21$
- Rubiks: $9 + 9 + 3 + 3 = 2$
- Tangrams: $9 + 9 + 9 - 2 = 25$

But it is not possible to make a score of 23 so the Solitaires' score must have been wrongly calculated.

Question 13: D

The argument is essentially a discussion comparing the benefits to an individual compared with the risks to society but eventually arguing that the risk of a negative message to society is not worth the risk of allowing a previously convicted criminal to return to a high-profile job. This makes the assumption that the rights of the individual are less important than the risks to society, which is stated in **D**.

Question 14: B

For the red paint, 20ml is left so 80 ml must be used. 20% is used as red, for 10% each purple and orange half of each mix is red, so a further 5% each, giving 10% in total. For brown, red is 1/3 of the 30%, so 10%. This means 20% + 10% + 10% = 40% of the mural is red from 80ml of paint, which means each 1% is 2ml.

For blue paint 10% is blue alone, for brown the blue makes up 1/3 of 30%, = 10%, and for 10% each green and purple half of each mix is blue, so a further 5% each, so 10% each. This totals 10% + 10% +10% = 30%, and each 1% is 2ml the blue paint used is (2ml x 30% =) 60ml. So, 40 ml of the 100ml of blue paint is not used.

Question 15: A

The passage concludes that human and primate brains are very similar such that we should consider research on primates to understand the effects of brain lesions, which in turn may help to develop treatments, which is most closely stated in **A**.

B, C and **D** are not mentioned in the passage. **E** would be too strong a conclusion based on the conditional tone of the passage that research 'may' help.

Question 16: D

Including Maisy there are 16 girls, and 10 boys, 26 children total.
There are no more than two children in each family.
So, the 3 older girls who have younger sisters = 6 girls.
The 2 girls with brothers (and vice versa) = 2 girls and 2 boys.
The two boys with brothers = 4 boys.
Which totals 8 girls and 6 boys with brothers and sisters, i.e., 14 children, so there are 12 of the 26 children without siblings.

Question 17: A

The argument states that the development of a new blood test will help reduce the rate of heart attacks recorded in women. It states that this will be because more sensitive detection will allow for earlier treatment. However, it fails to account for the fact that with more sensitive detection, the recorded rates of heart attacks in women are likely to increase, as those not previously picked up may now be, as stated in **A**.

Question 18: E

For the 720 points, the span goals make up 42 x 8 = 336, and beat 36 x 5 = 180, which together total 516. This leaves 720 – 516 points for tip goals, = 204 points total. This means that 102 tips were scored (as each tip scores 2)

This means that the points for span goals were roughly half, and beat and tips were roughly a quarter each, with beat being slightly smaller than tip, which is best represented in **E.**

Question 19: C

2 litres of sugary drink is 2000ml. 330ml contains 35g of sugar or 9 lumps. There is about 6 lots of 330ml in 2000ml. 6 x 9 lumps = 54 lumps

Question 20: C

Taking 100% of the daily-recommended sugar intake to be for example 100g. "Teenagers consume 50% more sugar on average, so 150% would be 150g. The information states that 30% is comprised of sugary drinks, which would be 30% of 150g = 45g. If this were to reduce to 1/3 as much, this would reduce by about 30g to 15g and so go from 150g to 120g, which would be the equivalent of 120% of the daily-recommended sugar, or 20% above the recommended level.

Question 21: C

If the tax resulted in a 10% reduction the 5 727 million litres of sugary drinks would reduce to about 5154 million (5727 x 0.9).

Question 22: A

If **A** were true, it would most weaken the argument for a sales tax rather than volume tax on sugary drinks. If retailers reduced prices to remain competitive this would negate the effects of a sales tax in the attempt to reduce consumption. The increase in price with the sales tax would be counteracted by the reduction in price by the retailers to remain competitive. This would not have the same effect with a volume tax as more expensive drinks would not be as heavily taxed so there would not be the same need to reduce price by the retailers. Statements **C, D** and **E** make no comparison of the advantage of the volume versus sales tax. **B** only states that most food is taxed by volume but does not give any arguments for or against this rather than taxing by price.

Question 23: E

As the tourist took the shortest route, it is logical that the attraction he did not visit would be either the Tower or the Palace as these are the furthest distance from all of the others. The shortest possible route is from the hotel to the courts (60m) which will definitely be taken so it makes sense to start with this. Logically the next step is to go towards the fountain as otherwise the tourist will get to the Palace and then need to either go to the Tower, which cannot be right as one of the Tower or Palace must be missed, or back to the hotel which would add unnecessary distance.

So, he goes to the fountain 80m, then the Arch 80m, the Castle 90m and then Tower 110m and finally back to the hotel 110m. This totals 530m, so the Palace is the attraction that is missed as it is not possible to do the other option of missing the Tower and doing the other 5 attractions in a 530m route.

Question 24: B

The argument states that banning caffeine drinks may backfire on schools' exam results due to the positive effects of caffeine on focus and short-term memory. However, if statements I and 3 were true it would weaken the argument as they suggest that caffeine that would have negative effects on sleep, which would result in a lack of focus and memory. Hence the negative effects of sleep deprivation would counteract the positive effects of caffeine itself on focus.

Question 25: F

Represent the six letters as: A-8-B-X-Y-Z.

We know that:

$$A8B \qquad \textbf{and} \qquad A8$$
$$+ \; XYZ \qquad\qquad\qquad + \; BX$$
$$\underline{\; 8\,0\,0\;} \qquad\qquad\qquad + \; \underline{YZ}$$
$$\qquad\qquad\qquad\qquad\qquad \underline{8\,0}$$

For the three-digit sums B + Z must = 10 due to the 0 digit below, the possibilities are 7 & 3 or 6 & 4 as 8 is already taken (and Y = I as the second column must also sum to 0 and it will have a I carried).

A and X must be 5 and 2 in either order as they must sum to 8 with the carried I. They cannot be 4 and 3 as B & Z must contain exactly one of the numbers 4 & 3 therefore making the pair incompatible. So, 8 + X + Z must sum to a number with 0 as the second digit (which cannot be 10 as the only way would be 8 + I + I which would use I twice and therefore not be allowed). Thus, it must be 20 which gives:

$$X + Z + 8 = 20 \quad \text{so} \quad X + Z = 12$$

From earlier we know that Z the possibilities were 7, 3, 6 or 4 and for X: 5 or 2. The only combination of these that sums to 12 is 7 and 5, so we now know that X = 5 and Z = 7. Which means that B = 3 and A =2. So, the full pass code is: 2-8-3-5-I-7

Question 26: C

The argument states that previous bad press surrounding saturated fats may have been based on misleading, now discredited data. The passage only discusses these particular studies but does not make any account for other studies that may have supported their results with more substantiated research, which is the assumption of the argument.

Question 27: C

As fewer than 5% owned neither device the total population that may overlap is about 95%. Plotting the potential overlap using the minimal overlap gives:

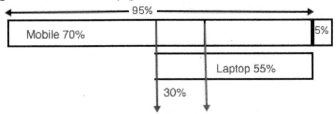

In which case the minimum overlap is 30% (70% + 50% -95%). It may be more than this if more of the mobile owners also own laptops, but this is the lower bound assuming the highest potential number of children who only own one device. However, because the percentage that owned neither is actually **fewer** than 5%, but by how much we do not know, the potential overlap is less than 30% as the laptop owners may move further into the 5% block and so the most correct answer from the options available is 25%.

Plotting the maximum overlap:

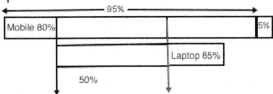

For which the overlap is 50% (80% + 65% - 95%). So, this must be between 25 - 50%.

Question 28: A

The argument states that people are different in whether they are more alert at night or in the morning according to levels of melatonin. Consequently, employers should adjust working hours to accommodate these sleeping patterns. If it were true, however, that ritualistic behaviour of for example staying awake at night increases melatonin levels, this would weaken the argument, as it would suggest that the behaviour controls the melatonin levels rather than vice versa.

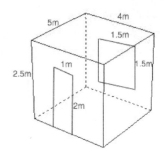

Question 29: A

Plotting the room:

The two larger walls are 5m x 2.5m = 12.5m², so for both walls 25m²

For these walls the high-quality paint would need to cover 50m² and the low quality 75m².

For high-quality this would need 4 tins as each will cover 15m², which would cost £60 (4 x £15). For low quality this would need 5 tins, which would cost £55 (5 x £11), so for the larger walls the low-quality paint is cheaper at £55.

The two smaller walls are 4m x 2.5m = 10m², so for both walls this gives 20m². The door is 2m² and the window 3m² so with subtracting these only 15m² needs to be covered. For these walls the high-quality paint would need to cover 30m² and the low quality 45m².

For high quality this would need 2 tins, which would cost £30 (2 x £15). For low-quality this would need 3 tins which would cost £33 (3 x £11) so for the larger walls the high-quality paint is cheaper at £ 30. Therefore, the total cost for the paint is £55 + £30 = £85.

Question 30: C

The argument states that, because working in A&E is not attractive to doctors, they will choose to work in other areas meaning that hospitals have to pay large sums of money for temporary staff. This leads to the conclusion that it is necessary to pay higher wages to A&E doctors as an incentive to work there. As this would cost the same as paying the extra temporary staff it would be of no more net cost to the health service. If it were true that many doctors work for agencies in A&E to supplement their salaries this would strengthen the argument, as it implies that doctors are incentivised by money to work in A&E. Thus, it would follow that if the wages were higher for working in A&E, more doctors would be happy to do so.

Question 31: B

From the information we know that 5 must be opposite the 6. This leaves either 2 or 1 opposite the 4 and 3.

This gives the potential combinations of the opposite sides and their totals:

- 4:2 = 6
- 4:1 = 5
- 3:2 = 5
- 3:1 = 4

Which shows that it must be 4:2 and 3:1 as the other options sum to the same totals.

Which give the pairs 5:6 = 11, 4:2 = 6, 3:1 = 4.

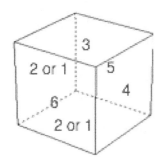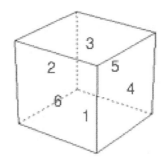

Looking at the potential answer combinations for the other die:

- A - 2:1 = 3, 3:6 = 9, so 5:4 = 9- not possible as two opposite sides would sum to 9.
- C – 4:1 = 5, 2:6 = 8, so 5:3- not possible as two opposite sides would sum to 8.
- D – 4:1 = 5, 3:6 = 9, so 2:5 = 7, not possible as 7 cannot be a total.
- E – 5:1 = 6, 2:6 = 8, so 3:4 = 7, not possible as 7 cannot be a total.
- F – 5:1 = 6, 3:6 = 9, so 2:4 = 6, not possible as two opposite sides would sum to 6.
- G – 5:1 = 6, 4:6 = 10, so 2:3 ≈ 5 not possible as the first die has 6 as one of its totals.
- B – 2:1 = 3, 4:6 = 10, so 3:5 = 8- this is the answer as it has all different totals to itself and to the opposite die.

Question 32: G

Statement 1: from 2011-2013, 4265 defendants were convicted. But each year had about a 98% success rate of prosecution so about another 2% a year were tried and subsequently cleared. The additional 6% of 4265 is about 255 which added together is more than 4300, so this is true.

Statement 2: in 2012 98% or 1552 were convicted, 1% = (1552 / 98= 15 remainder 82, so nearly 16) which means that the 2.1% acquitted was nearly 32, but definitely over 31, so this is true.

Statement 3: The total sentences handed out for includes all the suspended (220 + 178 + 140 = 538) plus the prison sentences (88 + 86 + 74 = 248). As 248 is less than half of 538, the prison sentences represent less than 1/3 of the total sentences, so the suspended sentences represent over 2/3 so statement 3 is also correct.

All three statements are correct.

Question 33: B

If the north of England had not risen by 6.6% to 566, the number of people convicted would have been (94% of 566 =) 532. If this had, along with the rest of the country, fallen by 11.7%, it would have decreased to (88% of 532 =) 469, which would have been 97 fewer convictions. 1371 − 97 = 1274.

Question 34: C

Statement 1: is not supported as the passage only mentions how many convictions there were in West Yorkshire for 2013, not for 2012. Even for 2013 it does not say whether this was the highest increase for any region for this year.

Statement 2: is not supported. Although it states the North of England rose by 6.6% between 2012 and 2013 this is for the North of England as a whole, not just for West Yorkshire, so it is not possible to confidently say that this was definitely the increase for West Yorkshire specifically, as other areas in the North may have changed by more or less and this is an average for the area.

Statement 3 is supported as in 2013 there were 1371 convictions, 566 of which were in the North which is over $\frac{2}{5}$.

So, only statement 3 is supported.

Question 35: B

The argument presented in the reader's comment is that rather than the north of England being particularly bad for cruelty they simply report it more as they are concerned about animal welfare. If it were true that a higher proportion of complaints resulted in conviction in the north than other regions, as in **B**, this would weaken the argument. This is because a higher conviction to complaint ratio would imply that there was not simply a higher proportion of complaints, as the reader would argue, but genuinely a higher rate of animal cruelty in the north. All of the other options fail to directly address the reader's argument that the rate of reporting is higher rather than the acts of cruelty.

END OF SECTION

SECTION 2

Question 1: E

The reflex arc after placing a hand on a hot object can be summarised as follows:

- Sensory neuron transmits impulse to the CNS – **A**
- Relay neurons pass electrical impulse from sensory to motor neurons – **D**
- Motor neuron transmits electrical impulse to muscle cells – **C**
- Muscle cell contracts – **B**
- So **E** is not part of the reflex.

Question 2: C

Alkenes will decolorise bromine water due to the presence of the C=C bond makes them unsaturated.

	Formula	Structure
1	C_2H_4	$H_2C = CH_2$
2	Polypropene is a polymer and has the repeating structure of $(C_3H_6)_n$	H H H H | | | | CH_3 CH_3 CH_3 CH_3
3	$CH_2C(CH_3)_2$	$H_2C = C$ with CH_3 and CH_3
4	CH_3CH_2I	$H_3C - CH_2 - I$

So only 3 and 1 have a C=C double bond, so only they will decolourise the bromine water. Hence, **C** is the answer

Question 3: B

Dark, matt surfaces are better at absorbing radiation than white surfaces. Black surfaces are also better at emitting radiation than white surfaces. White shiny surfaces are better reflectors of radiation so they will be better clothes in winter as they will reflect the radiation back into the person on the inside to keep them warm and radiate the heat away less on the outside. This is true because in winter the ambient temperature is likely to be lower than the body temperature, hence the answer is **B**.

Question 4: B

There are 3 black beads and the total number of beads in the bag is 8.

- The odds of picking a black bead the first time is $\frac{3}{8}$.

- Having picked this, the probability of picking another black bead is $\frac{2}{7}$ (the total being one less as the previous bead has been removed).

- To find the probability of picking two black beads multiply these odds: $\frac{3}{8} \times \frac{2}{7} = \frac{6}{56} = \frac{3}{28}$

Question 5: A

Anaerobic respiration is: Glucose $\Rightarrow$ lactic acid (+ little energy)

So there is no formation of carbon dioxide, use of oxygen or water formed, hence the answer is **A**.

Question 6: A

The energy change for the forward reaction is **a** as this is the overall energy change from reactants to products, hence the reverse reaction is **–a,** and so the answer is **A**.

Question 7: D

As the question says, a step-down transformer decreases the voltage of an alternating current (a.c.) electricity supply. Decreasing the voltage does not decrease the power, so it **stays the same**. But as the voltage goes **down**, the current goes **up**, hence the answer is **D**.

Question 8: E

Drawing the triangle PQR:

To work out the tangent of angle PQR we must first work out the length of the line joining P to the midpoint of QR (M).

Using Pythagoras:

- $a^2 + b^2 = C^2$
- $QM^2 + PM^2 = PQ^2$
- $4^2 + PM^2 = 6^2$
- $PM^2 = 6^2 - 4^2$
- $PM^2 = 36 - 16$
- $PM^2 = 20$
- $PM = \sqrt{20}$

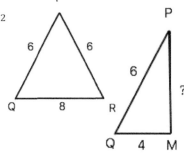
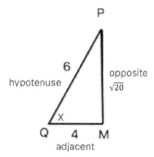

Using $Tangent = \frac{Opposite}{Adjacent}$

$Hence, \frac{\sqrt{20}}{4} = \frac{\sqrt{5 \times 4}}{4}$

$Hence, \frac{\sqrt{5}\sqrt{4}}{4} = \frac{2\sqrt{5}}{4}$

$= \frac{\sqrt{5}}{2}$

Question 9: D

The white mouse must be recessive, hence genotype cc.

The possibilities for the black mouse, 1 are CC or Cc

Punnet squares for both possibilities:

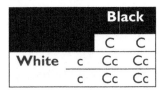

		Black	
		C	C
White	c	Cc	Cc
	c	Cc	Cc

Genotype 100% Cc + Phenotype 100% black

		Black	
		C	c
White	c	Cc	cc
	c	Cc	cc

Genotype 50% Cc, 50% cc + Phenotype 50% black, 50% white

As all of our first-generation offspring are black we know that the black mouse 1, is genotype CC, and mouse 2 is genotype Cc.

A cross of these mice:

		Mouse 1	
		C	C
Mouse	C	CC	CC
2	c	Cc	Cc

Genotype 50% CC, 50% Cc + Phenotype 100% black.

This gives 50% heterozygous (CC), black only offspring and heterozygous and homozygous genotype, hence the answer is **D**.

Question 10: D

Rubidium is an alkali metal and in Group 1. From the physical and chemical properties of group 1 we know that the Group 1 metals become more reactive as you move down the group, therefore is it logical to assume that being far down the group, rubidium is very reactive and would need to be stored under oil, **D**.

A. Is not correct as rubidium is more reactive than hydrogen and so electrolysis of rubidium chloride produces hydrogen not rubidium.

B. Is not correct as knowing the group 1 trends that melting and boiling points decrease moving down the group and so rubidium **does not** have a higher melting or boiling point than sodium.

C. Again, knowing the group 1 trends we know that rubidium reacts vigorously with water, increasing in reactivity down the group with rubidium being lower than sodium, so it is incorrect that rubidium reacts more slowly with water than sodium.

D. The chemical formula for rubidium sulphate is Rb_2SO_4, not $RbSO_4$ so this is incorrect.

Question 11: A

Statement 1 is correct – neutrons emitted in nuclear fission can cause further fission

Statement 2 states "*the half life of a radioactive substance is half the time taken for all its nuclei to decay*", which is incorrect. The half-life of a substance is **the time taken for the number of radioactive isotopes in a sample to halve**, which is different.

Question 12: B

As X is a whole number greater than zero, and most of the options are true for all values of X we can take X to be 1. Hence:

$$a = \frac{3}{5+1} = \frac{3}{6} = \frac{1}{2}$$
$$b = \frac{3+1}{5} = \frac{4}{5}$$
$$c = \frac{3+1}{5+1} = \frac{4}{6} = \frac{2}{3}$$
$$so \; a < c < b$$

To check that G is not correct try with another value, $X = 2$

$$a = \frac{3}{5+2} = \frac{3}{7}$$
$$b = \frac{3+2}{5} = \frac{5}{5} = 1$$
$$c = \frac{3+2}{5+2} = \frac{5}{7}$$

Where still $a < c < b$, hence **B** is correct

Question 13: B

On a hot day, a human would sweat more and lose more water, thereby making the blood plasma more concentrated. This would mean than more water is reabsorbed into the blood and the urine becomes more concentrated so there would be **less** water in the urine on a hot day compared with a cold day. Whilst the concentration of urea in urine would be affected by the volume of water, the mass of the urea will be unaffected by external temperature so the mass of urea in urine will be the **same** on a hot day compared with a cold day.

Question 14: C

Cycloalkanes are examples of alkanes, which only have single bonds, which means they are saturated, hence **C** is correct.

A- is incorrect as they have the general formula C_nH_{2n}.

B- they do not rapidly react with bromine water. Alkenes with double C=C bonds react to decolourise bromine water.

D- they burn in excess oxygen to produce carbon dioxide and water, not hydrogen, so this is incorrect.

E- is incorrect as they are part of a homologous series due to their similar properties and the same general formula.

F- cycloalkanes are not giant covalent structures.

Question 15: A

The vertical force upwards can be deduced by subtracting the downwards force 20N from the lift from the wings 25N = 5N.

Using the equation *force = mass x acceleration* and given mass = 2kg:

$5N = 2kg \times a$

Thus $a = 2.5\,ms^{-2}$ upwards

The horizontal acceleration can be deduced by subtracting the air resistance (drag) force 40N from the force from the engine 50N, (50N – 40N =) 10N.

$10N = 2kg \times a$

Thus $a = 5\,m/s^{-2}$ to the right

Question 16: E

- A : B : C
- 1: $\frac{2}{3} : \frac{4}{5}$
- C = £3000, which = $\frac{4}{5}$ so divide 3000 by 4 and multiply by 5 to find:
- 1, = £3750, which = A.
- B = $\frac{2}{3}$, so $\frac{2}{3}$ of 3750 = 2500,

Total amount collected by charity = $A + B + C$

$= £3750 + £3000 + £2500 = £9250$

Question 17: H

Process 1 is photosynthesis, which involves neither respiratory nor digestive enzymes. Process 2 is when plants are eaten by animals. Process 3 is when the carbon in animals is taken in by decomposers such as microorganisms by feeding, so both 2 and 3 involve digestive enzymes. Process 4 is when the decomposers respire, releasing the carbon as carbon dioxide into the atmosphere and so uses respiratory enzymes.

Question 18: C

The equation given is an exothermic reaction, as shown by the negative enthalpy change. This means there is a rise in temperature as the reactions transfer energy to the surroundings. Adding a catalyst would increase, not decrease the rate of reaction so **A** is incorrect. The state of chemical 'T' being a gas means the rate of reaction would be faster, as the surface area is increased, more particles are exposed to the other reactant, there is a greater chance of particle collision and there is a greater chance of the particles colliding. Hence **B** is incorrect. Increasing the temperature will increase the rate of reaction because at a higher temperature the reactant particles will have more energy and move more quickly and so collide more, and more collisions result in reaction, so **D** is incorrect. **E** is incorrect because the volume of gas will not change in this reaction – 1 mole of S and 2 moles of T = 3 moles total, which then forms 1 mole W and 2 moles of X = 3 moles total. S, T, W and Z are all gases, so we start with 3 moles of gas and end with 3 moles of gas. According to Avogadro's law, "equal volumes of all gases, at the same temperature and pressure, have the same number of molecules", which we can extrapolate to show that 3 moles of any gas will take up the same volume as 3 moles of any other gas at the same temperature and pressure. Thus, there will be no change in gas volume, and we cannot use changes in volume to monitor the rate of reaction. The activation energy is the energy required to start the reaction, the higher the activation energy the slower the rate of reaction. This is because a high activation energy will mean that a few particles will have enough energy to collide, slowing the reaction, hence **C** is correct.

Question 19: G

V and Y represent mass numbers, which is the number of protons plus neutrons. W and X are atomic numbers, which is the number of protons. Beta emission involves changes to the nucleus whereby a neutron is converted to a proton. This means from M to N the proton number will increase by 1, so W +1 = X. During alpha emission from N to Q the nucleus loses two protons and two neutrons, this means the mass number decreases by 4 and the atomic number decreases by 2. This means that Y will decrease by 4 when N decays to Q so Y = V-4.

Question 20: D

- m = mean, n = number of pupils, Therefore, the total score = mn

The expression for the number of pupils when another pupil is added is $n + 1$ and the mean is $m - 2$. The extra pupil scores n, so now the total score is $mn + n$.

Hence the expression for when the extra pupil takes the test and scores n:

$$\frac{mn+n}{n+1} = m - 2$$
$$mn + n = (m - 2)(n + 1)$$
$$mn + n = mn + m - 2n - 2$$
$$n = m - 2n - 2$$
$$3n = m - 2$$
$$n = \frac{m - 2}{3}$$

Question 21: A

The question concerns the nature of aerobic bacteria and white blood cells.

1- the structure of their DNA is a double helix, so this is correct
2- they both do not possess a cell wall, so this is incorrect
3- they both do not possess a nucleus, so this is incorrect
4- they both do possess a cell membrane, so this is correct

Hence, 1 and 3 are correct.

Question 22: C

The lower number is the mass number, which is the number of protons and electrons in an uncharged particle.

Thus, $^{35}_{17}Cl^-$, $^{40}_{18}Ar$ and $^{39}_{19}K^+$ all have 18 electrons in the arrangement 2, 8, 8 so they are the same.

	Number of Electrons
$^{35}_{17}Cl^-$	18
$^{35}_{17}Cl^+$	16
$^{40}_{18}Ar$	18
$^{39}_{19}K^+$	18
$^{40}_{20}Ca^+$	19
$^{41}_{19}K^-$	20

Question 23: D

The reaction time is 1.4s (0.7s x 2).
The speed of the car is 20m/s
- Distance = speed x time
- Distance = 20 m/s x 1.4s, = 28m travelled in the reaction time.

For the braking distance the car is decelerating at a constant speed from 20 to 0 in 3.3s so can take 10 m/s for this time as this will be the average speed.

$Distance = 10 \, x \, 3.3 = 33m$

Adding the braking distance and reaction time distance: 28m + 33m = 61m

Question 24: D

$$= \frac{2x+3}{2x-3} + \frac{2x-3}{2x+3} - 2$$

$$= \frac{(2x+3)(2x+3)}{(2x-3)(2x+3)} + \frac{(2x-3)(2x-3)}{(2x-3)(2x+3)} - \frac{2(2x-3)(2x+3)}{(2x-3)(2x+3)}$$

$$= \frac{4x^2 + 6x+6x+9+4x^2-6x-6x+9-2(4x^2 + 6x-6x-9)}{(2x-3)(2x+3)}$$

$$= \frac{4x^2 + 6x+6x+9+4x^2-6x-6x+9-8x^2 - 12x+12x+18}{(2x-3)(2x+3)}$$

$$= \frac{9+9+18}{(2x-3)(2x+3)}$$

$$= \frac{36}{(2x-3)(2x+3)}$$

Question 25: G

Looking at the columns of chromosomes:

XAA = X:A 1:2 (or 0.5:1) so XAA is male, which makes answers **A, B, C** and **D** incorrect.

XYAA , the Y is irrelevant to the sex of the fruit fly so the ratio of X:A is again 1:2, or 0.5:1, so XYAA is also male, which makes answers, **E** and **F** incorrect.

At this point we are left with answers **G** and **H**, for which notably XXAA and XYAA are given as the same sex so to save time it would be wise to skip to the final column, XXYYAA.

XXAA has ratio of X: A, 1:1, so it is female

XXYAA has ratio of X:A , 1:1 so it is female.

XXYYAA has ratio of X:A, 2:2, 1:1 so it is also female, so **H** is incorrect, **G is the correct row**.

Question 26: B

To work out the excess of oxygen first work out the moles of CH_4 and CO_2.

$$Moles = \frac{Mass}{Molecular\ Mass}$$

Molecular Mass of:

- $CH_4 = 12 + (4 \times 1) = 16$
- $CO_2 = 12 + (16 \times 2) = 44$
- $O_2 = 32$

Moles of $CH_4 = \frac{1.6}{16} = 0.1$

Moles of $CO_2 = \frac{4.4}{44} = 0.1$

So, we know that the number of moles of CH_4 to make CO_2 is 0.1.

The ratio of O_2 to CH_4/CO_2 is 2:1, so therefore we must have twice as much oxygen as CH_4/CO_2.

0.1 x 2 = 0.2 so there are 0.2 moles of O_2.

$$Mass = Molecular\ Mass \times Moles$$

$$0.2 \times 32 = 6.4g$$

So, the mass of O_2 used is 6.4g which means 1.6g of the 8g is left unreacted.

Question 27: D

Considering each statement in turn:

1. force = mass x acceleration

   $5.0N = 4.0 \text{ kg} \times 1.25ms^{-2}$

2. speed = wavelength x frequency

   $5.0 \text{ ms}^{-1} = 1.25m \times 4.0 \text{ Hz}$

3. voltage = current x resistance

   $4.0 \text{ V} = 1.25A \times 5.0 \text{ } \Omega$

So, 1 and 2 are true.

END OF SECTION

SECTION 3

"Computers are useless. They can only give you answers." (Pablo Picasso)

- In the statement, Picasso argues that computers are useless because they are only able to supply answers for the questions posed to them. A calculator, as an example of a computer, is entirely useless until one enters a sum for it to solve, but it requires the human input to perform any sort of function.

- As an artist Picasso must have thought computers to be against his industry of creativity and indeed at the time they would have been able to contribute very little to the arts. He may have been making a wider statement about the evolution of creativity, perhaps that it will never be possible for mechanical devices to be truly creative- a thought still held by many.

- This argument, however, while debatable in its truth at the time when Picasso made it, is almost impossible to defend today. Computers in their many forms are almost essential to modern day life in the western world. Most people in developed countries own mobile phones, along with a plethora of computer devices; laptops, tablets, even some cars have computers.

- We rely on computers for a huge number of daily functions that those who lived before them could not have anticipated. From maintaining contact with people across the world to simply arranging meetings, from organising our online banking to ordering clothes and food on websites we use computers in almost every aspect of modern life.

- Even in medicine, computers are often used to store patient notes and be used to keep track of lab results and as a viewing platform for medical scans. Even stripped down to the bare essentials of search engines providing answers to questions, as Picasso would argue, they are still useful to us in many areas. From searching locations, to medicine where the new recommendations for treating conditions can be found via computers.

- The real limits of technology are changing all the time. The evolution of technology has developed quickly, from simple calculator arithmetic functions, to now virtual realities and artificial intelligence that rival our human abilities. Computers are on the brink of being used to drive cars more safely than humans, and being able to design and create in the way that Picasso would have never thought possible. It is possible now to interact with robots in a way that could change the way we function in the future, with computers even becoming more efficient than humans in many jobs, industrial labour for example.

- There are still limitations of technology – for instance in the medical world, computers cannot replace doctors as of yet. The human ability to be able to look at another human and assess how unwell they may be is a long way off being mechanised. If one was to search flu symptoms into an online search engine, a brain tumour may appear as one of the possible diagnoses. It takes a person to ask the right questions, even if computers may be able to help with the answers.

- In conclusion, Picasso's argument may have been relevant to him and his contemporaries in the context of the era and abilities of computers at the time. However, it would have been difficult for him to anticipate the exponential increase in computer use and elevation of their status to now almost essential devices in everyday life.

"That which can be asserted without evidence, can be dismissed without evidence."
(Christopher Hitchens)

- Hitchens means that things that are said to be true but have no physical evidence, can be dismissed as they cannot be proven. He probably is referring to religion, the proof of the existence of which has no material evidence that can be replicated and documented in a scientific fashion. In the world of science, for example, something wouldn't be simply accepted because of folklore and people believing it to be true. The very heart of the Christian religion, the existence of God, for example, has not been positively proven.

- Arguing against this, however, there are many domains where beliefs are held without material evidence. In a court of law for example a person's testimony is often held without argument. It is not often the quality of the evidence but the person who gives it that is taken into account. Why then, with religion, do we not believe the thousands of people, both today and historically, who we would judge to be honest caring people who deeply believe that their god exists? Especially when we will sometimes take one person at their word for deciding whether someone should be convicted or exonerated for a crime.

- It also brings into question what we consider to be 'evidence'. Does something have to be proven by multiple people, reportable and repeatable as with scientific research? There are many scientific principles held to be 'proven with evidence' that only have one experiment to support them. The MMR vaccine being linked to an autism scandal, for example, was previously supported by evidence until it was later discredited. Bloodletting was the commonplace of medicine centuries ago and was believed to have evidence supporting its efficacy, whereas now it is unthinkable. How many things then do we hold to be true today that will later be disproven as scientific techniques evolve?

- Equally there are probably many things that we do not as yet have evidence for that may emerge over time. We do not yet know the full mechanism of some drugs such as paracetamol for example. Do we dismiss its worth because we cannot prove it? Hitchens' statement would argue that we should, when clearly this would be a mistake.

- I somewhat agree with this statement. It depends on the context to which it refers. It is potentially valid for example with new medicines. We would be wrong to accept their safety and efficacy without solid scientific evidence.

- However, there are also many areas of medicine that cannot be dismissed just because they don't have solid material evidence. Many personal accounts of out-of-body and near-death experiences where they have the option to 'move towards the light' cannot be simply dismissed because they have no physical proof. It may be that such things will never be proven, or indeed disproven in the same way that we take paracetamol to be effective simply on anecdotal evidence.

- Ultimately it depends on where the burden of proof lies. Is it with those who propose the hypothesis that something may be true, to then confirm their theory with evidence? Or is it for those who doubt their statement to disprove it? Hitchens would argue the former, but many issues such as widely held concepts of faith and religion are arguably the latter.

When treating an individual patient, a physician must also think of the wider society.
- The statement means that doctors should not only consider their patient's needs but also that of the other people within society when considering treatment. It would be ridiculous, for example, for a doctor to spend all day by a patient's bedside if that may be in their specific best interests, when they have their other patients to consider as well. This also applies in the wider context of finite resources within the NHS such as medicines, hospital resources such as radiological scans and even human tissue such as blood and organs. It is not feasible in a resource-limited system such as the NHS to prioritise any individual patient with no regard to the expense of other patients and wider society.

- It is, however, a doctor's role to place the needs of their patients above all else. Doctors have historically prioritised patients even over the law. Helping patients addicted to drugs or alcohol for example, whilst reporting them and forcing them to change their lifestyle might be what is the best for society, but doctors will often ignore these societal priorities and treat the patients regardless.

- Equally, in a society where we have a growing population and many children in foster care or waiting for adoption, it is not in society's interests to treat an infertile couple with IVF, where otherwise they may adopt these children and accept their infertility. But it would be against the foundations of medicine to begin to deny patients the treatment that was available and best for them to force them to accept conditions such as infertility, merely to satisfy the needs of society.

- There are many times where the patient's interests can conflict with those of the population. For example, antibiotics will become increasingly precious as resistance to them is increasing. It would be in society's interest to ration these for only the very unwell that won't survive without them. However, this would mean that many patients suffer for longer with illnesses that could easily be treated with antibiotics that are being saved for the future society's benefit.

- Equally, with treating individuals who are dangerous, such as serial killers, it could be argued that treating these patients does not benefit society. However, this would be completely against the ethical principle of non-maleficence and against doctors' nature and medical training.

- Vaccination is another key example. Often, it is not in the individual patient's interest to receive a vaccination - they may be more likely to experience side effects from the vaccine than to acquire the infection it prevents. Despite this, it is necessary to vaccinate everyone to protect the most vulnerable members of society, such as the young, elderly and immunosuppressed via herd immunity.

- In conclusion, there are times when the patient's best interests contradict that of society, and doctors must make a jugement call regarding the best way to spend their limited time and resources. It would for many go against the foundations of medicine to deny some patients the treatment they need, in order to satisfy the needs of society. It may, however, end up as a necessity, to ration some resources, such as antibiotics, for those most at need in order to prevent the rise of resistance, which would make them useless to everyone else in society.

Just because behaviour occurs amongst animals in the wild does not mean it should be allowed within domesticated populations of the same species.

- This statement suggests that there are differences in the behaviours of domesticated animals to those living in the wild. It consequently argues that just because a certain type of behaviour exists in wild animals, that does not mean it should be automatically allowed in domesticated animals as well.

- It could refer to the hierarchical nature of wild animals that operate in a 'survival of the fittest' modality. It is common for animals to fight each other in the wild, for example, for simple status such as with wolves to be head of the wolf pack. This behaviour would not be acceptable in domesticated animals. It would be very difficult, for example, if every time someone attempted to walk their dog in the park it attacked other dogs to assert status, and this was allowed or even encouraged as the behaviour was fitting with its analogous species in the wild. Quite apart from the danger this might pose to the other dogs if this behaviour was encouraged, it could even lead to a change in behaviour of the fighting dog such as aggression towards humans that would make it not only impractical but dangerous to keep it as a pet.

- Equally, with hunting where animals such as wolves and big cats would hunt their food in the wild, it would be obscene to allow the domesticated versions to keep attacking other people's house bunnies or guinea pigs to satisfy their carnal nature.

- Arguing against the statement however, one could claim that if we are having to continually disrupt our domesticated animal's behaviour to fit with what works practically within our society, then it may be best to not keep these animals as pets. Changing the nature of these animals through behavioural modification, known as taming, could be considered immoral and unfair to these animals. Training dogs to be more docile and to repress their animalistic instincts on the face of it seems unethical. One could argue that just because we can train these animals to change their behaviours does not mean that we should in order to keep them as pets.

- With breeding programmes, for example, our pets have been slowly bred to be less like their historic ancestors, and to be more aesthetically pleasing. Whilst the taming of certain behaviours to prevent harm to other animals and even humans can be justified, it is hard to accept the alteration of animals' natural breeding behaviour for aesthetics. This has even gone to extremes where selective breeding of pugs and other dogs have been bred to their detriment, as they may experience breathing and other health problems.

- However, it is also impossible to avoid the fact that domesticated animals would ultimately not be kept as pets had their behaviour not been altered through time. So, it could be argued that these changes are necessary for their existence, and so merely a form of survival adaptation. Indeed, they have been so successful that they don't rival the top of the food chain humans and are even protected by them.

- In conclusion, while modifying the behaviours of domesticated animals is ideologically difficult, it is vital to continue to allow their existence as a valued and cared for part of our society.

END OF PAPER

2016

SECTION 1

Question 1: D

In order to solve this question, we have to fill in the blanks in the table. Years 1 – 3 are easy as there is only one gap to fill, but years 4 and 5 are a little more difficult. From the totals we can calculate the total number of boys and girls in all 5 years.

We can calculate the total number of students in year 4: 120-24-16-40-24=16

We also know that the probability of a boy being in year 4 is 1/12, applying this we can calculate the number of boys in year 4: 72/12=6.

Since we know that there are 16 students in year 4 and 6 of them are boys, the probability of a student being a boy is 6/16 or 3/8.

Year	Boys	Girls	Total
1	18	6	24
2	16	24	40
3	8	8	16
4	6	10	16
5	24	0	24
Total	72	48	120

Question 2: B

Answer **A** is incorrect as the forest fires in Indonesia represent only a fraction of the net allowance of CO_2 emission and there is no information on further culprits, therefore failing as a valid conclusion.

Answer **B** is correct as the text specifically mentions that only CO_2 from burnt plant material is taken up by new vegetation. Since peat is being burnt in the forest fires, this leads to the conclusion that some of the CO_2 will not be taken up by regrowth.

Answer **C** is incorrect as it has no backing from the text presented in the question.

Answer **D** is incorrect, as it basically ties in with Answer A; they both ignore other contributing factors to CO_2 emission.

Question 3: B

This question is pretty straightforward. First, calculate the price per stay and then compare.

Stay 1:

Hotel: $50 + $50 + $40 + $40 + $40 = $220

Car: 5 x $5 = $25

Total: $25 + $220 = $245

Stay 2:

Hotel: 8 x $40 = $320

Car: $25 + $5 = $30

Total: $350

Difference: $350 - $245 = $105

Question 4: B

Answer **A** is incorrect as this is a direct rephrasing of the second sentence, and is therefore not a flaw. Answer **B** is correct, as if the ability to synchronize to particularly slow music was an innate ability of musicians that occurs naturally, there's a chance that non-musicians would have the same trait, despite not having taken up music. Answer **C** is incorrect as it is not supported by the text and is also irrelevant. Answer **D** is incorrect as whilst it might generally speaking be true, the text passage here deals with synchronization to slow music, not with any other abilities.

Question 5: A

The first challenge in this question is to correctly identify the axes of the diagram.

Looking at the distribution of results, the Y axis must represent the written paper in steps of 10 and the X axis must represent the score of the practical paper in steps of 5. Both start at 0.

Moving from left to right, the dots therefore represent the following students:

Ina – Liz – Els – Joe – Fio – Gho – Amy – Kai – Ben – Den – Haz.

Con would fall between Fio and Gho.

Question 6: D

Answer **A** is incorrect as the new changes do not make it more difficult to seek justice per se, but only for claims of £200 000 or more.

Answer **B** is incorrect since fees associated with the court cases have been in place for a long time and the text only takes issue with the recent rise, not fees as a whole.

Answer **C** is incorrect for similar reasons as answer A. The statement is too general for a rather specific answer.

Answer **D** is correct since it is in keeping with the text as a whole.

Answer **E** is incorrect as they address an issue beyond the scope of the text.

Question 7: D

To answer this, we need to use a two step approach. First, it is easiest to work backwards from the answers provided to determine the amount of taxes Paul would be paying for the respective incomes. We know that Paul must be paying $4800 ($5600 - $800). In addition to that, we also know that Paul makes more money this year than he did the year before, but pays less tax. This is only possible by crossing the 50 years of age boundary. For this reason, the correct answer is **D** rather than C, despite the net tax payment being the right amount in both cases.

Income	Tax if under 50	Tax if over 50
$24000	$3000	$2000
$29000	$4000	$3000
$33000	$4800	$3800
$38000	$6500	$4800
$43000	$7700	$5800

Question 8: C

For this question we have to look at figure 1 as well as figure 3. From Fig 1 we know that there were 800 units of products 1, 2 and 3 sold in April to June and none of products 4, 5 and 6 since they have not been released yet.

Looking at Fig 3 we can determine the cost for each set of **100 units**:
Income from product 1 was 8x£1500=£12000
Income from product 2 was 8x£2000=£16000
Income from product 3 was 8x£1500=£12000
Total Income: £40000

Question 9: D

From the diagram in Figure 2 we know that in the 4 months that product 2 has been on sale, by June, a total of 1700 units had been sold. We also know from the question that the 900 units sold in March represent 2/3 of the sales for the months of March to May.
If we define X as the number of units sold from March to May, we can express this as
900 = X x 2/3, solving this for X gives X = 1350.

Therefore, the number of units of product 2 sold in June must be 1700 – 1350 = 350.
We know from Figure 3 that 100 units of product 2 are worth £2000, which means that in the month of June the sale of product 2 generated £7000 income (£2000 x 3.5 = £7000).

Question 10: D

From figure 1 we know that there were 600 units of product 6 sold in November and December. Using the data from figure 3 this equates to an income of 6x£4000=£24000. From Figure 2 we can see that the income produced from product 6 in December was £6000 and therefore the income from November must be £24000-£6000=£18000. This value is equivalent to 450 units of product 6, since £18000/£6000=4.5 and since prices are given for batches of 100 units, the result is 450 units sold in November.

Question 11: E

To solve this, you will have to calculate the total number of units sold in a year and divide this by the months since release. For product 1 this is 12, product 2 it is 10, product 3 it is 8, for product 4 it is 6, for product 5 this is 4 and for product 6 this is 2.

Product	Total sale since release	Average monthly sale since release
Product 1	3000	250
Product 2	3200	320
Product 3	2600	325
Product 4	1200	200
Product 5	1400	350
Product 6	600	300

Question 12: D

When Helen goes to bed it is 21mins to 2300, meaning it is 2239. When Helen wakes up it is 23mins to 0400, meaning 0337. Therefore Helen has slept from 2239 to 0337 which means she has slept 4hrs and 48 minutes.

Question 13: A

Answer **A** is correct since sports and entertainment are provided as specific exceptions to high IQ professions that are well paid by the text.

Answer **B** is incorrect since IQ is independent of education.

Answer **C** is incorrect as the statement misinterprets the argument of the text that specifically states that the research did not consider parental profession when analysing IQs.

Answer **D** is incorrect as it argues a point beyond the scope of the text.

Question 14: D

The easiest way to approach this question is to calculate the balance per month.

Month	Balance	Month	Balance
January	$1300	July	$1100
February	$1100	August	$1300
March	$1300	September	$1200
April	$1300	October	$1500
May	$1700	November	$1400
June	$1500	December	$1400

Sam has **over $1300** in her account in May, June, October, November and December.

Question 15: B

Answer **A** is incorrect since the text makes no claim about unhealthy lifestyle accelerating synapse loss in old age.

Answer **B** is correct since the text claims that healthy lifestyles can produce additional synapses meaning that after age related synapse loss there are a higher number of synapses still available.

Answer **C** is incorrect as it has nothing to do with the text.

Answer **D** is incorrect since the link between quality of life and existing synapses is not mentioned in the text.

Answer **E** is incorrect since it does not provide a relevant conclusion to the topic addressed in the text.

Question 16: A

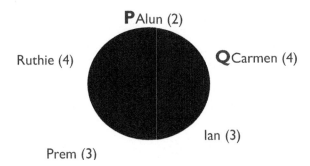

After Alun's throw, he has 0 coins, Ruthie has 7 coins and Carmen has 7 coins.

After Carmen's throw, she has 2 coins, Alun has 1 coin and Ian has 7 coins.

After Ian's throw he has 2 coins, Carmen has 4 coins and Prem has 6 coins.

After Prem's throw she has 1 coin, Ian has 3 coins and Ruthie has 9.

After Ruthie's throw he has 4 coins, Alun has 2 coins and Prem has 3 coins.

Question 17: A

Answer **A** is correct since only the regulation of use of these new antibiotics will maintain their effect on bacteria.

Answer **B** is incorrect since it neither strengthens or weakens the argument but rather provides more background information.

Answer **C** is incorrect since it has nothing to do with the topic discussed in the text.

Answer **D** is incorrect since it has little relevance to the incentive provided to pharmaceutical companies to research new antibiotics.

Question 18: D

To solve this, it is easiest to calculate the interest for the 1st year for the individual mortgages this person qualifies for.

Since they already have £25000 and want a house worth £150000, they need to cover approximately 83% of the house through a mortgage. This means they do not qualify for Mortgages 1 and 2.

To calculate the interest for year 1, we need to apply the respective percentages to the mortgage amount of £125000.

Mortgage No	Year 1 Interest	Arrangement Fee	Year 1 Total Cost
3	£6250	£500	£6750
4	£3750	£2000	£5750
5	£5000	£1000	£6000

Mortgage 4 is the cheapest in year 1, resulting in answer **D**.

Question 19: C

To calculate this, we need to calculate the growth for Texas, where the number of new wells in 2012 represents 40.11% of the wells drilled from 2005 – 2011.

40.11% of 2694 = 1081 = answer **C**.

If you struggled with this multiplication, or are short of time, finding 40% of 2694 will give 1077.6, which is roughly equal to answer **C.**

Question 20: C

To answer this, we need to calculate the net water use per well. This is easiest using approximate numbers:

For Louisiana: 12000/2500= 4.8 million gallons per well.

For Utah: 600/1200=0.5 million gallons per well.

Difference between Louisiana and Utah: 4.8 – 0.5 = 4.3 million gallons per well. Since we used approximation we will use the next closest answer, which is **C**.

Question 21: E

Statement 1 is correct, since the risk is specifically mentioned in the text. According to the information in the table, Texas used 110000 million gallons of water, equating to 99.2% of the fracking fluid. The penultimate sentence in the paragraph titled 'Chemicals used' tells us that the remaining 0.8% is made up of a mix of chemicals. Therefore, they must have used 880 million gallons of chemicals.

Statement 2 is incorrect since the total amount of pollution produced equates to 28 coal power plants per the text which means that one power plant equates to 100,551,000/28 = 3,591,107 tonnes of carbon dioxide, not 36,000,000 tonnes carbon dioxide.

Statement 3 is correct since the text states that 26 000 million gallons are enough to supply 200,000 Colorado households for a year meaning that each household uses
26,000 million/200,000 = 130 000 gallons of water.

Question 22: H

Statement 1 does not provide an explanation for the difference in water consumption per well since the amount of water required for the process is independent from the amount of water available as mentioned by the text quoting the drought in Texas.

Statement 2 does not provide an explanation as the development in technology as it does not explain the degree of variability in water consumption, even if we assume that the fundamental understanding of technological progress is to equate in a reduction of water consumption.

Statement 3 does not provide an explanation since the per well water consumption for states varies, as demonstrated by question 20.

Question 23: D

Answer **A** is incorrect: it ignores the square, demonstrated by the proximity of the right and down pointing diagonal to the circle.
Answer **B** is incorrect since the angle of what is supposed to be the larger triangle is incorrect.
Answer **C** is incorrect since it misrepresents the position of the circle on the different cards.
Answer **D** is correct.
Answer **E** is incorrect since it ignores the square card.

Question 24: C

Answer **A** is incorrect since it is irrelevant as the text has already demonstrated the effectiveness of high impact walking irrespective of age and gender.

Answer **B** is incorrect since it is too specific and does not account for a general weakening of the argument.

Answer **C** is correct since it provides a general explanation for why the weight loss effect of playing sports is less marked.

Answer **D** is incorrect since it is irrelevant for the argument in the text that addresses the effect on BMI rather the motivation for performing a certain type of exercise.

Question 25: D

To answer this question, we have to calculate the overall time that elapses between the start of the showings at 1015 and the end of the showings at 2245 which is 12.5hrs or 750mins.

Since the time between films is the same throughout the day, we can simply add up the duration of all the showings: $2 \times (117 + 109 + 119) = 690$mins

The total amount of time spent on breaks therefore is $750 - 690 = 60$mins.

Since 2245 marks the end of the 6th showing 60mins has to be distributed over 5 equal rest intervals: $60/5 = 12$mins.

Question 26: B

Answer **A** is incorrect since it has no relevance to the text that addresses the effectiveness of the mentioned self-help books not the attainability of a happy life.

Answer **B** is correct since this is the population that would most benefit from the purchase of a self–help book about happiness.

Answer **C** is incorrect since this possibility is already accounted for by the text through the use of the phrasing 'are more **likely** to be anxious…'.

Answer **D** is incorrect since it has no relevance to the text that specifically addresses the connection between self-help books and anxiety and not the lack of a connection.

Question 27: F

There are two ways to answer this question.

Firstly, you can approach this from a theoretical perspective. In order to maximize the difference between systole and diastole, we have to have a very high systolic reading or a very low diastolic reading. Therefore, you can scan the diastolic column for the lowest readings (78 and 81) and find the combination with the highest associated systolic reading. To then ensure that you have not missed anything, check your result against the highest systolic reading.

The second option is to work backwards from the solutions looking at the differences for all the days of the respective pulse number presented in the answer.

Question 28: D

Answer **A** is incorrect since the text specifically mentions a degree of uncertainty in the last sentence.

Answer **B** is incorrect since it oversimplifies the issue with regards to the information presented in the text.

Answer **C** is incorrect since the text is not about cancer prevention but about what brings cancer about.

Answer **D** is correct since only with some influence on pollution and stress can a patient take control over them developing cancer as claimed in the last sentence.

Question 29: B

In order to find the solution for this, it is easiest to follow the instructions step by step drawing them out:

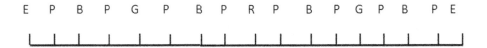

The second approach is to use the folding pattern for help. We know that the number of strands doubles on each fold, moving from 1 to 2 to 4 to 8. Since the green mark is added at the 2-strand-stage, half of the marks on the respective strands will lie between the two green marks. This means, that we will add 2 blue marks as there are 4 strings and 4 purple marks as there are 8 strands. Then, we need to add the single mark of red since it will provide the mid-point of the overall string that must necessarily lie between the two green marks.

Question 30: D

Answer **A** is incorrect since it is an absolute value, ignoring the concept of the text itself.

Answer **B** is incorrect since the text does not aim to provide a time frame but merely highlight the possibility of consciousness developing in machines.

Answer **C** is incorrect for similar reasons as Answer A.

Answer **D** is correct since the connection between consciousness and brain function is unclear even when dealing with living creatures such as animals.

Question 31: D

The best way to approach this is to set up an equation.

X = points achieved by the Argents.

Z = number of fesses achieved by the Sables.

Y = number of fesses achieved by the Argents.

X = 11Y since the Argents achieves twice as many pales as fesses and X is total number of points.

X − 1 = 9Z since the Sables achieved the same number of fesses as pales.

Z − 1 = Y since the Argents achieved one less fess than the Sables.

Z − 1 = Y since the Argents achieve one less fess than the Sables.

Therefore Z = Y + 1.

Combining the points achieve we get 9Z + 1 = 11Y.

Insert the value of Z we have calculated above:

9Y + 10 = 11Y and solve for Y to get 5 = Y.

Since Y is the number of fesses scored by the Argents, and they achieved twice as many pales as fesses, they must have scored 10 pales which is answer **D**.

Question 32: A

The answer from this can be found in the Table in the column giving total amount of polyunsaturated fat. For Olive Oil this is 10% and for Sunflower oil this is 65.7%. Since they both occur in a 50:50 mix the easiest way to find the solution is to add half of the content of each meaning 5% from the Olive Oil and 32.85% from the Sunflower Oil This leads to 37.85% which can be rounded up to 37.9%.

Question 33: A

Answer **A** is correct since it specifically says so in the text (in the second paragraph in the last two lines).

Answer **B** is incorrect since the constituents of diets in hunter-gatherer societies around the world vary according to climate.

Answer **C** is incorrect since the third paragraph specifically mentions that the fall in heart disease is associated with the consumption of polyunsaturated fats.

Answer **D** is incorrect since the text specifically contradicts this by pointing out the cardioprotective nature of high polyunsaturated fat oils.

Answer **E** is incorrect since the heart protective effect has improved with a higher ratio.

Question 34: B
According to the text the average adult ingests 143 grams of fat of which 30% are vegetable oil which equates to 43g.

Since 100g of Canola oil contain 0.6g of Erucic acid, this equates to roughly 0.26g (260mg) of Erucic acid which equates to just over 50% of the daily suggested intake.

Question 35: D
We know from the table that Sunflower oil does not contain any omega 3 so can only contribute Omega-6.

Similarly, we know that the ratio of Omega-6 to Omega-3 in flaxseed oil is 0.2:1. For this is easiest to work with 100g of flaxseed oil and work out how much sunflower oil has to be added to achieve the 2:1 ratio.

Since Flaxseed oil contains 53.3g of Omega-3 we need to achieve 106.6g of Omega-6 in the mix.

12.7g of that are already contained in the flaxseed bolus, meaning we need to add the equivalent of roughly 94g of Omega-6 from sunflower oil. The amount of sunflower oil is (94/65.7) × 100g = 143g.

This means the mixture of oils will weigh 243g. Looking at this we can approximate that the flaxseed oil representing slightly less than half of the mass. The closest percentage to this is 41% from answer **D.**

Calculation: 100/243 = 0.411 = 41%.

END OF SECTION

SECTION 2

Question 1: D

Answer **A** is incorrect since the vessels are named incorrectly.

Answer **B** is incorrect since the vessels are named correctly, but the content of urea is not allocated correctly.

Answer **C** is incorrect since the vessels are named incorrectly.

Answer **D** is correct since the structures are named correctly and so is the urea concentration.

Answer **E** is incorrect since the structures are named incorrectly.

Answer **F** is incorrect since the structures are named incorrectly.

Question 2: F

Statement 1 is incorrect - no element is in both period 3 and group 12.

Statement 2 is correct since the element has up to 3 electrons to donate and oxygen can take up 2 with the smallest common denominator being 6.

Statement 3 is correct since Br can take up one electron and the compound has 3 to donate.

Statement 4 is correct since elements will contain the same number of protons as they contain electrons meaning that the atomic number must be 13.

Statement 5 is incorrect since alkali metals are only in group 1.

Question 3: D

From the diagram, we know that two pieces of the material displace 100 cm³, therefore 1 piece must displace 50 cm³.

The weight difference from no pieces of material to one piece of material is 300g. Therefore, the correct formula is 300g/50cm³.

Question 4: A

First, we need to establish the slope of the line that passes through the two points given in the question using the $y = mx + b$ equation:

$m = \frac{(y_2 - y_1)}{(x_2 - x_1)}$ Meaning m $= \frac{2}{3}$

To determine b, we need to insert one of the two points into the equation: $y = \frac{2}{3}X + b$:

Using $\left(\frac{6}{9}\right)$ this gives $9 = \frac{2}{3} x 6 + b$; solve for b to find $b = 5$.

Therefore, the equation is $y = \frac{2}{3}X + 5$

Since two lines are parallel if they have the same slope, the only possible answer is **A** where the slope (m) is 2/3.

Question 5: B

W must be a chromosome since it is the origin of the removed DNA.

X must be a restriction enzyme since it cuts the DNA strand to remove the DNA fragment.

Y must be a restriction enzyme since it cuts a gap into the DNA of the other organism.

Z must be a ligase since only ligase enzymes can fuse DNA strands which will result in insertion of chromosome DNA into the other organism.

Question 6: E

Answer **A** is correct since removing water from a solution of calcium carbonate and water will leave solid calcium carbonate.

Answer **B** is correct since the two substances will have different vaporisation points due to different molecular size.

Answer **C** is correct since silicon dioxide is a solid that dissolves poorly in water allowing filtration as an effective means of separation.

Answer **D** is correct since the vaporisation point of sodium chloride and water is different.

Answer **E** is incorrect since ethanol and water form a solution that cannot be separated by mechanical means but only by thermic separation techniques.

Question 7: D

Statement 1 is correct since the difference in weight is due to different numbers of neutrons in the core that will not change the chemical properties of the element.

Statement 2 is correct since the atomic number is equal to the number of protons.

Statement 3 is incorrect since all the isotope nuclei contain 28 protons but a varying number of neutrons.

Question 8: B

The easiest way to solve this is via an equation where we assume that N consists of a number of people all weighing exactly 75kg and that Jim, Karen and Leroy each weigh 90kg.

$$78 = \frac{[75N + (3 \times 90)]}{N + 3}$$

Solving this equation for N, leads to N = 12, which is solution **B**.

Question 9: F

From statement 1 we know that the enzyme must be in the stomach as the stomach is the only place in the human body with a pH of below 4. From statement 2 we know that the enzyme must be a protease as it digests proteins into amino acids. From statement 3 we know that the enzyme must be inside the body, as the body's core temperature is ~37 degrees.

Answer **A** is incorrect as amylase digests sugars (carbohydrates).

Answer **B** is incorrect as amylase digests sugars (carbohydrates)

Answer **C** is incorrect as lipase digests fat.

Answer **D** is incorrect as lipase digests fat.

Answer **E** is incorrect since the pH in the small intestine is above 4.

Question 10: F

To solve this, we need to count elements and add the respective molecular weight.

N = 14, there are 2 in the equation leading to 28.

H = 1, there are 20 in the equation leading to 20.

Fe = 56, there is 1 in the equation leading to 56.

S = 32, there are 2 in the equation leading to 64.

O = 16, there are 14 in the equation leading to 224.

In total this adds up to 392, which is answer **F.**

Question 11: B

Rotating the coil at a faster constant speed will accelerate the voltage change thereby increasing the amplitude and since the coil will also complete its turns faster, the frequency of the waves will increase as well. Only answer **B** ticks these boxes.

Question 12: D

This is a two-part question. First, we need to find the radius of the arterial lumen and then apply the equation defining the area of a circle to this.

Since the overall diameter of the artery is 1.6cm and the walls are 1mm thick, the inside diameter is 1.4cm, leaving the radius to be 7mm.

Since $A = \pi r^2$, $A = \pi \times 7^2 = 49\pi$ mm^2

Question 13: B

Carbon dioxide is only produced in aerobic respiration since it requires oxygen which per definition is absent in anaerobic respiration. Glucose represents the energy substrate in either type of respiration. Lactate is only produced in anaerobic respiration.

Question 14: C

Answer **A** is incorrect since the cathode would form hydrogen as the calcium is more reactive.

Answer **B** is incorrect as the anode would give oxygen since the nitrate is a complex ion.

Answer **C** is correct.

Answer **D** is incorrect since products are allocated to the wrong electrodes.

Answer **E** is incorrect since molten NaCl does not contain any hydrogen.

Question 15: D

Firstly, all waves named in the question travel at the speed of light which is defined as 3×10^8 m/s or 300 000 km/s.

Since the distance from satellite to transmitter and receiver is 45,000 km each, the waves must travel 90 000 km, which will take $= \frac{90,000}{300,000} = 0.3s$

Secondly, we know from the question that the waves have a frequency of 1.5×10^{10}Hz, which defines them as microwaves, which leads to answer **D.**

Question 16: C

To solve this question, we have to work in several steps as we need to calculate both the length of RS and the length of PQ.

To calculate RS we need to know that tan is defined as opposite/adjacent, in this case RS/6. Since the question gives us the tan as 4/3, we know that RS must be 8.

RS/6 = 4/3 = 8/6.

Applying this we can calculate the area to be $A = (5x8) + \left(\frac{6x8}{2}\right)$

$= 40 + 24 = 64cm^2$

Question 17: G

Statement 1 is correct, this is known as a gain of function mutation.
Statement 2 is correct, this is known as a loss of function mutation.
Statement 3 is correct, this is known as a loss of function mutation.
Statement 4 is correct, this is known as a gain of function mutation.

Question 18: E

We know from the question that the acid is able to donate two protons therefore we know that we must use twice the volume of NaOH, which is able accept 1 proton.

Since the concentration of NaOH is half of that of the diprotic acid, we need 4 times the amount of the solution provided in the question to neutralize the acid.

Since the volume of the acid is given as 30cm³, we need 120cm³ (4x30cm³) of NaOH.

Question 19: D

Since we don't know the resistance of the body, the best formula to use here to find the current is I = Coulombs/time in seconds.

First, we need to calculate the charge in Coulombs (1 Coulomb = 1 joule per volt):

$\frac{125J}{500V} = 0.25$ C

Therefore $I = \frac{0.25C}{0.01s} = 25A$

Question 20: C

The main point here is not to lose the overview of the different components of the equation. Then there is no real trick to this, other than going through it step by step:

$$\frac{a}{b} = \frac{c}{d} + \frac{e}{f}$$
$$\frac{a}{b} = \frac{fc+de}{df}$$
$$dfa = bfc + bde$$
$$dfa - bfc = bde$$
$$\frac{f(da-bc)}{da-bc} = \frac{bde}{da-bc}$$
$$f = \frac{bde}{da-bc}$$

Question 21: A

Statement 1 is correct as it basically aims at the doubling of the genetic material which happens through the process of mitosis.

Statement 2 is incorrect since growth happens through the production of proteins which does not occur during mitosis.

Statement 3 is incorrect since again this happens through protein production, which is halted during mitosis.

Statement 4 is correct since stem cell division follows the same principle as asexual reproduction.

Question 22: B

Due to the increased concentration of the acid (twice that of the original sample), the reaction speed will be higher. However, as the net amount of the acid in the sample is the same as in the original experiment (half the mass at twice the concentration), the equilibrium will be the same. Only line B fits this bill.

Line **A** equates to higher concentration and higher net amount.
Line **C** equates to lower concentration of acid and the same net amount.
Line **D** equates to a lower net amount of acid.
Line **E** equates to a lower concentration as well as a lower net amount of acid.

Question 23: E

From the information given in the question we know that the object must weigh 1.5kg on earth (15N at g=10N/kg).

The mass of the object will remain unchanged as it is transported to the planet. The difference will be due to the different gravitational fields.

Since we know from the text that the object has a weight of 3N on the planet, it will have a kinetic energy of 30J after a fall of 10m.

Question 24: D

This question deliberately aims to confuse you with additional information. The population distribution of the different blood groups is not needed to answer this question. There are 4 different blood groups A, B, AB and 0. One individual can have one of those 4 blood types. Since the question asks about the probability of the one criminal having both A and B antigens, he/she must have blood group AB, the likelihood of which is 25% or ¼.

Question 25: H

From the diagram, we can deduce that a grey coat colour must be dominant, and a white coat colour must be recessive. Therefore, any grey mouse can potentially be heterozygous. There are 12 grey mice in the diagram, leaving answer **H**.

Question 26: D

From the equation, we know that the volume of the final product will be equivalent to the volume contribution of X. Since we use 100cm³ of reagent X and the reaction occurs under exclusion of air and to completion, the final volume in the syringe must be 100cm³.

Question 27: C

The easiest way to answer this question is to calculate backwards to determine the wavelength of the light through air.

We know that 360nm equates to ¾ of the wavelength through air; if X = wavelength through air: 360nm = X × 0.75

X = 480nm. If Y is defined as the wavelength of light through glass, we can express this as

Y = 480nm x 2/3 which gives Y = 321.6nm.

Rounding down, **C** is the closest answer to the nearest 10 from our calculation.

END OF SECTION

SECTION 3

'You can resist an invading army; you cannot resist an idea whose time has come.' (Victor Hugo)

- In this question, Hugo basically evaluates the strength of physical confrontation and violence versus intellect and words when it comes to changing the status quo. This position has to be connected to be taken in connection with his time. It has to be taken into account that he lived in France just after the Revolution and during the reinstitution of the monarchy in France. You can also remove the quote completely from the historical background and look more at the conflict between forced change and passive change with the invading army representing forced change and the idea a more passive and natural form of change. The forced change does not necessarily have to come in the form of an invading army consisting of battalions of soldiers but can also represent something as simple as a prescribed mind-set or political conviction that is propagated by a political regime.

- Arguing against this statement, it is obvious that sufficient force seems to be able to suppress most ideas, especially if execution of opposing violence is well-publicised and very large scale. One example of this would be the wide-ranging suppression of individuals in dictatorships such as North Korea where dissidence is not only punished physically but also enforced by a high degree of isolation. In the end, if it is possible to enforce a perception that any form of dissent will result in personal injury and in the injury of loved ones, violence can suppress ideology, especially if punishment is severe.

- A further point to consider when arguing against the statement in the question is the definition of an "idea whose time has come". Considering the vagueness of this point, it provides a good target for arguing against the statement as it renders the statement as a whole rather moot.

- The idea of physical and psychological violence versus ideas is also discussed in other works of literature such as Orwell's 1984; some of the means by which thought control can be achieved in the book provide good examples to add to an essay.

- Addressing the point of the power of ideas, there are many examples in history than can be used for this. The American Revolution presents one of them, as does the resistance against Nazi occupation in France or other countries during the Second World War. In all cases ideology refused to be intimidated by violence. In this context however it is essential to highlight the fact that ideas usually achieve power by encouraging to violence thereby almost resulting in a circular argument. An obvious exception to this rule is Ghandi's peaceful disobedience promoting Indian independence without use of violence.

- In this question you could take multiple routes, either staying rather close to historical examples or moving in a more philosophical direction, looking at the issue with a broader perspective in mind and questioning the statement of 'an idea whose time has come'. The challenge of the latter option is that it will be easier to lose track of your arguments and more difficult to maintain relevance of the points you make with regards to the original question.

- This is a very interesting question as it challenges the idea of progress in general to a certain extent as it attempts to quantify the influence of eternal stressors on the intellectual development of a society.

Science is not a follower of fashion nor of other social or cultural trends.
- The statement attempts to explain and unpack the reasoning behind scientific progress and the interaction between science and culture and other influences on social progress. The basic assumption is that science develops independently of other social driving forces that may or may not be subject to trends and temporary interests.

- In order to write a good essay, you have to be clear about what you understand by science, fashion and social/cultural trend.
  - Science could be defined as the endeavour of searching for new information and truths in order to widen the horizons of our knowledge and out understanding of the world around us. In addition to that, science also has a clear set of rules that define the value and the truthfulness of the information that is being found.
  - Fashion as well as cultural trends both essentially go in the same direction. They both describe temporary and variable perceptions and interests within society. What is important in this is that trends as well as fashions tend to be variable from individual to individual or from group to group. This makes it different from science which claims to hold an overarching always valid truth, external to social norms or trends.

- The basic idea when arguing to the contrary is to provide arguments of why science is indeed influenced by fashions and trends. There are several ways this can be done. On one hand, you can argue from a social perspective putting fashion and trends into context with social interests. This is a good starting point as it makes it clear how science and the areas of research that individuals are interested in are influenced by what fascinates the masses at any given point in time. One example for this is for example the great degree of progress in military technology in the early 20th century when Europe was ravaged by war and there was a great degree of militarism throughout many levels of society.

- On the other hand, you can argue in a different direction illustrating the relationship between science and trends by starting at scientific discovery and arguing its influence on trends and fashion. One example for this would be the progress in nuclear technology in the 1950s when society began dreaming about nuclear powered cars etc and striving for new technology and the idea of progress through technology was very widespread, in particular in the US.

- Other examples include the great social interest in geography and the natural sciences in general during the age of discovery of the Americas. The idea of widening frontiers and pushing civilization onwards became a great social driving force contributing to large amounts of funding as well as large movements of populations to these newly discovered lands. This also directly reflects social issues in the European heartland that made a new life abroad more favourable and attractive.

- When phrasing your agreement or disagreement, it pays to be direct with your opinion since this will make it easier to present your point to the examiner in an efficient fashion. However, be sure to have argued your points appropriately before voicing your opinion to avoid jumping to a conclusion too hastily.

The option of taking strike action should not be available to doctors as they have a special duty of care to their patients.

- In general, when arguing this point you have to be aware of the implications of this question. You have to keep in mind that the essay you write will be shown to any admitting university that may or may not use your essay during the interview. For this reason, it may be wise to be a little careful with what you write in this type of essay.

- Now, when it comes to arguing this particular essay, there are several bases you have to appreciate. Firstly, there is the nature of the relationship between doctors and their patients. As doctors we have a complicated relationship with our patients in the sense that we know many intimate details about our patients but at the same time we need to keep a professional distance to make the decisions that we deem to be in the patient's best interest. This tight bond between patients and doctors makes for a unique relationship between the healthcare provider and those receiving a priceless service that is not found outside the health care setting. Secondly you have to consider the purpose of strike action.

- Doctors going on strike is never an easy decision, particularly due to the sometimes life and death nature of health care. With doctors not working, the care that vulnerable patients require may not be deliverable. This can be argued to be a direct violation of the basic ethical pillar of 'do no harm', since the doctor knowingly and without direct external pressure decides to withhold treatment that may alleviate the patient's suffering. The other ethical pillar that can be argued to be violated by doctors going on strike is the idea of beneficence, since it is the doctors duty to always act in the patient's best interest, which withholding of treatment most definitely is not.

- The most significant justifying reason for supporting doctors' strikes is the fact that poor working conditions will necessarily have a direct influence on the doctors' performance which in turn will have negative and detrimental effects on patient care. Good examples of this are fatigue due to long working hours or unsafe levels of staff due to general unattractiveness of the profession or widespread symptoms of exhaustion in the work force.

- Another point to consider is the reason why doctor strikes are an issue in the NHS and not as much in other countries and health care systems. This will give you a good chance to demonstrate that you have an understanding of how the NHS works and how health care is provided in different countries.

- You can also consider addressing the reasoning for strikes in general and why every work force should have the right to protest unjust working conditions by going on strike. This will necessarily take you away from the question somewhat which can be a challenge but may lead to an interesting essay. In this context it is absolutely paramount to not lose track of your argumentative chain and to relate your main body back to the question in your conclusion.

- You should definitely consider using the 2016 junior doctor strikes as the pivot point of the essay. It provides a good real-world example to organize your essay around and it will also give you scope to write about how doctor strikes can be made as safe as possible and how potential harm to patients can be reduced as much as possible.

If we truly care about the welfare of animals, we must recognise them as fellow members of our communities with their own political rights and status.

- This can be a tricky topic to argue since the question covers a wider range of different points that need to be addressed or at least be clear in your head in order to write a good essay.

- The first point to define is the general idea of animal welfare. What do we understand by animal welfare and what implications does this have for the interactions between humans and animals? At this point it would be helpful to be aware of what rules exist already and what defines our current considerations of animal welfare. Having a basic knowledge of this will strengthen your essay. However, if you do not have any idea of animal welfare, you can always stick to making points of a more general and philosophical variety.

- The second point to be clear about is the meaning of political rights and status. Make sure you have a vague idea of what this would mean and also how this could be put into action, in particular since you are required to argue that we do not need to allocate animals political right and status. Points to consider in this context are that with rights always come duties and think through how these can be applied to animals.

- In the context of this question it can also be helpful to consider the idea of conscience and intellect in animals. Try to connect this to the idea of political rights and status and the attached duties and possibilities. Taking the right to vote for example, how are cats and dogs supposed to understand what politicians say and even if they were to understand what they say, how do we know if they have the intellectual faculties to make a decision about such matters.

- Another avenue you might want to explore is the matter of ownership. When it comes to animals, whether they are pets or animals that themselves are a resource such as in the milk and meat industry, what role does the owner of the animal play? If we make them members of our community with political rights and status, does that make them ownerless and what implications does this have? Or do they retain their ownership and if so, how does that then impact the de facto role they take in society? Since animals will always require some degree of human support either through for example access to medical care or in order to gain access to food, how does this arrangement work form a legal perspective?

- In general it could be helpful for this essay title to imagine animals as humans that are somewhat limited in their ability to communicate with the rest of society and how this would shift the role they play as members of society.

- When it comes to political institutions, again it would be useful to have a general understanding of what institutions already exist that deal with animal welfare and animal rights and the protection and care of animals. Examples could be NGOs such as Greenpeace or organisations such as the RSPCA.

- Finally, much like the previous topic, keep in mind that this is a difficult and somewhat charged subject and that your admitting university will receive a copy of your essay that may well be discussed, at least on a theoretical level, in your interview. Things you should avoid are extreme claims such as animals not needing special rights since they have no independent mind and are just things or property etc.

END OF PAPER

2017

SECTION 1

Question 1: C

Start by converting everything to ml so that you don't get confused. Hugh had 900ml of yellow paint left, meaning he used (1500 – 900 =) 600ml of it. This means that he also used 600ml of red paint to create the orange paint, as they were mixed in a 1:1 ratio. If the room was 40% orange and represents (600 + 600) = 1200ml of paint, this means that the 60% pink represents 1800ml of paint. A quarter of the pink paint was red, meaning 450ml of red paint was used to make the pink.

600ml (in the orange) + 450ml (in the pink) = 1050ml of red pain used overall

1500ml – 1050ml = 450ml of red paint left

Question 2: D

A is definitely not true; there is definitely no causal evidence from this statement alone that use of antidepressants leads to *worse* outcomes for patients (i.e. we cannot know if the disability claimants are taking antidepressants or whether they represent a different population). There is also no evidence from the information that there is short term efficacy of this medication.

B is not necessarily correct; we don't know if doctors are 'increasingly encouraged' to prescribe these drugs. There may be other reasons for the increase in numbers seen over time; there may be more people suffering with their mental health in the present, there may be more people seeking help for it than previously, or more people aware of the claims there are entitled to make.

C is also not necessarily correct. There may be other reasons for the mental health figures getting worse over time, including an increasing incidence of mental health issues in society. There is a causal link between the prescriptions and claimants in this answer, which cannot be drawn from the figures alone.

D can be drawn as a conclusion from this stem alone. The figures do indeed show that on a population level, the long-term mental health in the UK is declining, and that this is not improving despite increasing use of antidepressant drugs. This answer doesn't assume any causal link between the drug prescriptions and disability claimants, instead just focuses on the numbers.

Question 3: D

A satisfies the bedrooms, garage, garden, distance to grocery store, but not distance to sports facilities. 4/5

B satisfies the garden and distance to sports facilities, but not bedrooms, garage, or distance to grocery store. 2/5

C satisfies the bedrooms, garage, garden, and distance to sports facilities, but not distance to grocery store. 4/5. However, this is not within their budget of $900,000, so we can strike it out as an option.

D satisfies the bedrooms, garage, garden, distance to grocery store, but not distance to sport facilities. 4/5

E satisfies the bedrooms, distance to grocery store and sports facilities; but not garage or garden. 3/5

Only options A and D satisfy 4/5 options and are within budget. To find the 'lowest price' in terms of dividing the cost of the house by the number of bedrooms, we do 2 simple equations:

A = 825,000 / 5 = 165,000
D = 640,000 / 4 = 160,000

So, **D** is the house they should choose,

Question 4: A

The key argument against nuclear power in this passage is that it 'poses an unacceptable risk to the environment and to humanity' and that stations 'create tens of thousands of tons of lethal, high-level radioactive waste.' Thus, the phrase that would weaken this argument the best would minimise these concerns.

A would weaken the argument well; it highlights that there is a way to counteract the unacceptable risk to the environment/humanity that the author talks of.

B is incorrect; the passage suggests that we should be using less nuclear power so the author would be satisfied with this statement. It doesn't tackle the issues that are raised against nuclear power.

C gives an argument against wind power, but doesn't counter the key arguments raised against nuclear power highlighted above.

D is not correct; the argument actually acknowledges that nuclear power is 'less air-polluting that fossil fuels' and this is not a key reason for concern for the author.

Question 5: D

This question is obviously quite hard to explain with text alone. These questions often require trial-and-error and visualisation in your mind of how each shape would appear in different orientations. Check if you are allowed to bring scissors to the exam and cut shapes out, although that may

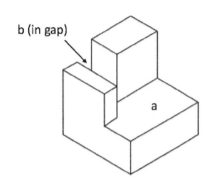
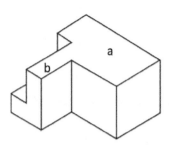

not help so much with this question compared to previous ones regarding cube folding, etc.

As it is a trial-and-error type of question, it is a good idea to find the options that you can instantly rule out. For example, we can quickly discard A or C, as they contain no thin 'b' protrusion which could slot into the 'b' gap of the original structure.

Look at the shape of 'a' on the diagram. It is nearly a rectangle with one width slightly smaller than the other. If both 'a's were abutted directly there would be a perfect fit, whereas none of the other shapes given would be able to do this.
Also, when looking at the gap made by 'b' in the diagram, you can see it would need to be filled by a thin part of the block, which **D** is able to do.

Question 6: C

For any disease, there are 'risk factors' that increase your likelihood of developing it. For example, other risk factors for prostate cancer include increasing age, having a family member with the disease, or being black. However, even with no risk factors, there is still a chance (albeit smaller) that one may develop the disease. Conversely, even with every risk factor for a disease, there is no guarantee that one will definitely develop the disease.

The key aspect of this argument, and simultaneously the flaw, is the suggestion that men can prevent the development of prostate cancer by minimising weight gain, as highlighted by **C.**

A and **D** address other cancers which is not an issue tackled by this argument.
B doesn't address the causal link between changes to diet and exercise and prostate cancer itself.

Question 7: B

Adding up all the students at the school gets to 144, therefore there are 72 girls.
None of the girls swam and 40 girls played rounders.
28 pupils ran, of which 14 were girls.
Therefore 72 girls in total – 40 (rounders) – 14 (running) = 18 girls who played football

Question 8: C

The reconviction frequency rate is the number of offences per 100 offenders in the cohort. Thus, over a 9 year follow-up period, there were 1057.5/100 = ~10.6 reconvictions per person in the cohort. However, the question asks specifically for the frequency of those who reoffended in the first place (74%), so we must do 10.6/0.74 = 14.3.

(Alternative method if it helps you understand it better: 42, 721 offenders in total, of which 74% were reconvicted, which is 31613 reconvicted. There were 10.6 reconvictions per offender in the total cohort, so 10.6 x 42721 = 452842 for total number of reconvinctions. If we want a figure for the number of reconvinctions per re-offender, 31613 reoffended, so we need to do 452842/31613 = 14.3)

Question 9: E

1 is true. The information explains that the figures are cumulative. 43% were reconvicted in the first year and cumulatively 55.2% by the second year, so the proportion convincted in the first year is 43/55.2 = ~0.78 = ~78%. (If you'd rather work with numbers, 43% is 18370 and 55.2% is 23582. 18370/23582 = ~78%)

2 is not true. As the figures are cumulative, 43% were reconvicted in the first year, and 61.9% by the third year, meaning only 61.9% - 43% = 18.9% were reconvicted in years 2 and 3. This is less than a third of the 74% that were reconvicted over 9 years. (Again, working with numbers, 31613 were reconvincted overall, 18370 over the first year, 26444 over three years; therefore 8074 over years 2 and 3. This is less than a third of 31613.)

3 is true. The reconviction frequency rate was 185.1 per 100 offenders at this time, meaning 1.851 per offender. There were therefore 42721 x 1.851 = 79077 reconvictions.

Question 10: C

The key information for this question is: 'offenders who received sentences of less than 12 months... committed 39% of all offences that led to a conviction in the first year of the follow-up.'

As shown above there were ~79077 reconvictions in the first year. 39% of these were by those who received sentences of less than 12 months.
79077 x 0.39 = 30840 = ~31000

Question 11: G

The wording of the question, 'might feasibly,' means that any reasonable suggestion that could account for a decline is acceptable; it doesn't need specific evidence or to be supported by the data.

1 makes sense because out of those who are going to eventually be reconvincted each year, a proportion had already done so and were in prison, so there is a smaller cohort each year.

2 is correct because we are looking at a specific cohort only (i.e. nobody can be added to the population size) and **reconvictions** rather than any new convictions. Naturally, some of the cohort will pass away.

3 could be true; if there was indeed harsher sentencing, this could feasibly deter those who would have likely reoffended with lighter sentencing.

Question 12: D

Unfortunately, there is not always a neat mathematical way to solve this type of question; trial and error is probably best.

A cannot be true because the passage states that 'the performer who plays the role of Gracie cannot play any other characters.'

B cannot be true because both Teddy and Guard 1 are in scene 1.

C cannot be true because of the statement 'the performer who plays Rose also plays Guard 1' which would mean one performer would have to play Rose, Guard 1 and Sarah. Rose and Sarah are both in scene 10.

E cannot be true because Sarah needs to be played by a female, and Graham needs to be played by a male.

F cannot be true since the performer would have to play Rose, Guard 1 and Guard 2. Both guards are in scene 1.

Question 13: B

This is a classic BMAT correlation does not necessarily equal causation question. The passage even states that the drop in selenium levels *correlates* with the three extinction events.

A is nothing to do with extinction, which is the main topic of the passage.

B is correct because it acknowledges that we do not know the exact causation but that the factors could be linked.

C is incorrect because there is no mention of evidence regarding other trace elements, so we do not know how they compare in importance.

D is incorrect because it implies definite causation.

Question 14: B

The Citrons gained 57 seats and lost 29 hence now have $80 + 57 - 29 = 108$. $108/240 = {\sim}0.45$

The Jonquils gained 26 seats and lost 80 hence now have $126 + 26 - 80 = 72$. $72/240 = {\sim}0.3$

The Saffrons gained 51 seats and lost 25 hence now have $34 + 51 - 25 = 60$. $60/240 = {\sim}0.25$

The pie chart that shows this best is B (the part which is exactly a quarter is probably most helpful, as we know this corresponds to the Saffrons' seats).

Question 15: A

The argument presented is in the following format:

- There are two potential hypotheses for an observation, one of which is correct.
- One reason can be ruled out based on objective evidence.
- Therefore, the other reason must be correct.

A is presented in this way.

B doesn't give evidence to rule out the first potential hypothesis, only evidence to support the other reason.

C doesn't give 2 potential hypotheses for an observation.

D doesn't rule out one of the hypotheses objectively, only subjectively.

Question 16: D

All the digits add up to 38, and each digit is different. There are 8 digits in total, so each number except one from 1-9 are used. Now use trial and error to see which digit can be excluded to get to a total of 38.

A $2 + 3 + 4 + 5 + 6 + 7 + 8 + 9 = 44$

B $1 + 2 + 4 + 5 + 6 + 7 + 8 + 9 = 42$

C $1 + 2 + 3 + 4 + 6 + 7 + 8 + 9 = 40$

D $1 + 2 + 3 + 4 + 5 + 6 + 8 + 9 = 38$

E $1 + 2 + 3 + 4 + 5 + 6 + 7 + 8 = 36$

Question 17: D

The argument is very specifically suggesting that being a childhood star can cause problems in later life, so do not extrapolate this argument to include any other groups of people.

1 is not correct because the author doesn't ever argue that break-ups and mental breakdowns are exclusive and unique to childhood stars; instead, they are merely suggesting that the *risk* would increase as a childhood star.

2 is incorrect because the author does not suggest any link between being adored as a child and addictions/broken relationships. The author uses the word 'adored' to emotionally charge the argument rather than as a logical explanation.

Question 18: B

Cocoa powder used = 100 + 85 = 185g, 315g remaining.
Eggs used = 2, 10 remaining.
Sugar used = 330 + 200 = 530g, 70g remaining.
No lemons used, 5 remaining.
Milk used = 250ml, 2250ml remaining.
Butter used = 400 + 150 = 550g, 50g remaining.
Flour used = 400 + 225 = 625g, 375g remaining.
There is only 50g of butter left, and 8 pancakes requires 50g of butter, so this is the maximum that can be made (there are no other limiting factors at this point).

Question 19: B

The claim in the headline is that a fifth of the papers contained errors in the spreadsheets. Therefore, if they are not actually errors, the support that the data gives to the claim would be weakened - this is why **B** is correct.

A is not correct because the headline isn't referring to any other scientific fields.
C is not correct because even if the results were successfully replicated, that wouldn't discount the fact that there were errors.
D is not correct because there is no way of knowing whether this would actually reduce the number of papers containing errors.

Question 20: B

1 is incorrect, we don't have data for total number of papers published for an individual year so we cannot make this conclusion.
2 is correct; looking at the graph the figure was 50 for 2009 and just above 100 for 201.
3 is incorrect; Nature published 23 with errors and BMC Bioinformatics with 21.

Question 21: D

The papers with a higher % than the average (1st graph) are: Nature (23 papers), Genes Dev (55), Genome Res (68), Genome Biol (63) Nature Genet (9), Nucleic Acid Res (67) and BMC Genomics (158). Adding these together gets to 443 papers, and out of 704 this represents 63%.

Question 22: C

'The authors found that the number of genomics papers packaged with error-ridden spreadsheets increased by 20% a year over the period, far above the 10% annual growth rate in the number of genomics papers published'

Often with these proportion questions, it is easier for the human brain to convert into figures first, which doesn't actually change the calculation. So, imagine that in 2015, there were 100 papers published and 80 were affected.

2016: $80 \times 1.2 = 96$ papers affected. $100 \times 1.1 = 110$ papers published
2017: $96 \times 1.2 = 115$ papers affected. $110 \times 1.1 = 121$ papers published
2018: $115 \times 1.2 = 138$ papers affected. $121 \times 1.1 = 133$ papers published
Therefore, every paper would be affected by 2018.

Question 23: E

Looking at the shaded area, we can work out that the puzzle pieces that would fit would require 4 bits that stick out to fill the gaps and 2 that go inwards. Additionally, in the middle to join the 2 pieces, there would need to be a bit that stuck out and a part that stuck in. Therefore, in total we should expect 5 parts that stick out and 3 parts that go inwards. This is only fulfilled by **E**.

Additionally it may be useful to instantly discard option **A** from the start, as we can see that one of the pieces has no parts that go inwards.

The diagram above, using x and y, shows one way that the pair of pieces could fit into the jigsaw puzzle.

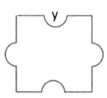 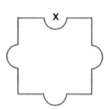 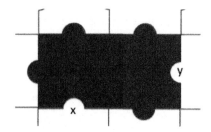

Question 24: E

The argument suggests an exclusive, causal inverse link between the number of children a woman has and their life expectancy based on a correlation. Only **E** acknowledges that there could be other factors that influence this correlation.

A is not correct because the argument doesn't assume or mention women's recognition of the toll of childbirth.

B is incorrect because the argument actually acknowledges that it is referring to women in rich nations.

C is incorrect because the argument is not about infant mortality.

D is incorrect as the argument doesn't suggest this, only that the *risk* is increased with multiple pregnancies.

Question 25: E

The distance for the original journey between the two towns is not known, so let's call this x. We also don't know the time taken for the original journey, so let's call this y. Obviously, distance = speed x time.

Original journey: x = 30y

The distance is increased by 4km in the new journey, the speed is reduced by 3km/h and the journey time increases by 25%

New journey: x + 4 = 27 x 1.25y

Solve the simultaneous equations. x = 30y, and x + 4 = 33.75y so 4 = 3.75y and y = 1.067. x = 32

New journey is x + 4 = 36km

Question 26: C

The passage says that the government suggests the health checks **could** equate to prevention of over 2,000 heart attacks and strokes. However, even though the researchers identified many patients with risk factors, they need to follow up long-term to see whether there is actually prevention. No firm conclusions can be drawn from the health checks yet, which is why **C** is correct.

A is not correct because we don't know if this number of heart attacks and strokes have actually been prevented.

B cannot be concluded; we don't know whether patients' health-related behaviours changed or not.

D may be subjectively true but is not specifically implied by the argument here, without the context of 'health checks'.

E is incorrect, there is no suggestion to think this from the information given.

Question 27: B

The lowest common multiple of 400g, 300g and 200g is 1200g therefore Charlie would have ordered 1200g of each fish. This would make 24 plates. 3 packs of prawns are needed, and the third is free. Therefore $4.08 x 2 = $8.16. 4 packs of squid are needed, and one is free. Therefore $4.08 x 3 = $12.24. 6 packs of cockles, whelks and smoked salmon are needed and 2 are free for each. Therefore $4.08 x 3 x 4 = $48.96 → 8.16 + 12.24 + 48.96 = $69.36 in total. This is making 24 dishes, therefore 69.36/24 = $2.89 each.

Question 28: E

1 is true. The second line, especially the phrase 'encourages the police officers to better regulate their own behaviour', suggests that the author thinks the 'violence' and thus level of force used can often be reduced.

2 is incorrect. The first line states that police departments across the world should use body-worn cameras, but it doesn't assume that there is any interaction between police departments in different countries.

3 is correct. The last two lines highlight this. The argument is suggesting that police-worn cameras are better because there is always a clear warning about filming from the start, which implies the assumption that those filmed by bystanders are not always being consensual.

Question 29: E

The die, with the orientation on the left, is missing a dot from its 4, and a dot from its 6. The die, with the orientation on the right, is missing the same dot on the 6 but we do not know which of the faces with 2 is missing a dot to make it a 3.

If we look at the orientation of both dice, the face to the left of the three dots on the 6 face must be the 5. This means the face to the left of the two dots on the 6 face must be 2 to make opposite faces add up to 7. Therefore, the die without missing dots is as in the diagram above. Therefore, on the right die, the bottom face is a 1, and the face that cannot be seen on the right is the 4 (with a missing dot). If you rotate this, you can get **E** but none of the others.

Question 30: C

This is yet another correlation does not equal causation question. Stronger brain connectivity is linked to 'positive' lifestyle traits; however, the causation may be in the other direction, or there may be confounding variables which influence both factors in the same way. Only **C** addresses this flaw.

Question 31: D

You can build a circle based on working out the preference of whoever is adjacent to someone. For example, on the left of Jess is someone who likes cricket and rugby, who is the same person who is on the right of David. You can go around the circle like this. Also, it helps to know that Jess is next to 2 people that share a common interest; so, she must like 1 of cricket and rugby, and 1 of football and golf.

```
                    George
                    (F&S)
        David                    Eli
        (C&H)                   (H&S)
        Amir                    Peter
        (C&R)          Jess     (F&G)
                      (G&R)
```

Question 32: C

This question should hopefully be straightforward, with the use of the word 'could'. There is no evidence required but clearly both of the answers have the potential to increase the problem of London's road congestion.

Question 33: E

6 underground strikes at £10 million each = £60 million. GVA due to congestion = £5.5 billion = £5500 million.

5500/60 = 92x greater cost

NB a conservative estimate is one in which you are cautious and estimate a low amount, which is more likely to be lower than the real amount rather than higher. So, in this instance, you need to use the lowest estimate

Question 34: C

1 is not true. The decline of car ownership in London has occured at the same time as a reduction in London's road congestion problem, but you cannot infer a causal link. 2 is not true, there is no evidence or suggestion provided this could be the case.

3 is true. This is heavily suggested by the third paragraph and confirmed by phrases such as 'demand for the bus service has started to decline.'

Question 35: B

To increase the GVA, there must be an increase in the production of goods and services (as per the definition).

If the widening of the congestion charge zone somehow did not increase the production of goods and services, then the proposal would not have the desired effect. With **B**, there may be less of the public visiting profitable retail areas (deterred by the congestion charge), meaning that the GVA may not increase – hence this is correct.

END OF SECTION

SECTION 2

Question 1: A
P is the gallbladder, which releases bile.

Q is the stomach, which secretes acidic HCl (thus hydrogen ions) and proteases.

R is the pancreas, which secretes insulin and enzymes to digest proteins, fats and carbohydrates (protease, lipase and amylase respectively).

Question 2: E
The anode and cathode are made of copper therefore this will be the reacting species. Remember that oxidation occurs at the anode and reduction occurs at the cathode. Therefore, there is the loss of electrons at the anode to generate the copper ion, and gain of electrons at the anode to convert the copper ion into the element.

(OIL RIG = Oxidation Is Loss, Reduction is Gain)

Question 3: A
1 is incorrect; microwaves have a larger wavelength (the value is **LESS** negative) than visible light. They do have a smaller **frequency**, but this is not what is stated.

2 is incorrect; all electromagnetic waves travel at the same speed through a vacuum. (This speed value is the 'c' seen in the equation $E = mc^2$).

3 is incorrect; Gamma rays have the smallest wavelength of any electromagnetic wave (it is the **MOST** negative).

4 is incorrect; X-rays are used in hospital radiography to look for broken bones. In contrast, radiowaves are more commonly used in communication.

Question 4: F
$$(\sqrt{5} - 2)(\sqrt{5} - 2) = 5 - 2\sqrt{5} - 2\sqrt{5} + 4 = 9 - 4\sqrt{5}$$

Question 5: G
This question is about the conversion of DNA into proteins, which requires both transcription and translation.

Humans have 2 alleles of every gene, one from the mother and one from the father. These can be the same, or different, and can be either recessive or dominant. The particular allele represents the exact DNA code; that is the order of each nucleic acid base (A, C, G, T) in the DNA. This is converted to mRNA by transcription, and then each triplet of bases in the mRNA code is read and converted into a specific chain of amino acids. The exact amino acid encoded, as influenced by the order of bases, can determine things like the tertiary structure of the final protein and thus the shape of the active site of the enzyme. In this disease, any of these steps could go wrong (and thus be different to normal) to give a faulty enzyme which is unable to produce healthy white blood cells, and as such all the answers are correct.

Question 6: D

1 contains 16 protons (atomic number), therefore 18 neutrons (to give a mass of 34) and 18 electrons (to give a 2- charge).

2 contains 17 protons, therefore 20 neutrons and 18 electrons.

3 contains 18 protons, therefore 22 neutrons and 18 electrons.

4 contains 19 electrons, therefore 20 neutrons and 18 electrons.

5 contains 20 electrons, therefore 20 neutrons and 20 electrons.

Question 7: F

1 is correct. The rate of evaporation increases with the temperature of the liquid because molecules have more kinetic energy and move faster on average, thus more can escape.

2 is incorrect. Still air would cause the air above the puddle to become saturated. Windier conditions allow more evaporation because a gradient is maintained between high water concentration and low water concentration above the puddle.

3 is correct. The increased surface area means that more molecules can escape at the same time.

Question 8: A

We want to pick a patient suffering from a migraine twice, so we use AND and multiply the 2 probabilities.

The probability of picking the first patient is 5/20; if this occurred the probability of picking the second would be 4/19.

Therefore $5/20 \times 4/19 = 20/180 = 1/9$

Question 9: E

There is water in all the experiments and glucose in 2, 3 and 4. Even with no/minimal gradient, random entropy means that both water and glucose will move through the partially permeable membrane, and there will be some (~equal and opposite) movement in the other direction to equilibrate.

Question 10: C

This is a classic metal + acid → salt + hydrogen reaction

$Mg + 2HCl \rightarrow MgCl_2 + H_2$

As the reaction occurs, the magnesium chloride and hydrogen are formed (the latter causing the bubbling) which reduces the concentration of the reactants. Therefore, only 2 is correct. The activation energy is irrelevant (this regards the start of the reaction) and the particles do not have less energy, it is just a concentration issue.

Question 11: C

Remember, in series, the current is always the same at each resistor (hence 4 is true), whereas the total voltage and the total resistance is additive. 2 is correct because using $V = IR$, doubling the resistance at R_1 with an equal current means that the voltage must be doubled here. (In parallel circuits, the voltage is constant, and the current is additive.)

Question 12: F

Firstly, you can rule out A/B/C as PS is clearly over 1.8cm!

Thanks to QT and RS being parallel, the triangles PQT and PRS are proportional to one another. That means the ratio of PT:QT is the same as PS:RS. Let's call PT x and the proportion between the sides y. $x = 0.3y$ and $1.8 + x = 1.5y$ therefore $1.8 + 0.3y = 1.5y$ and $1.8 = 1.2y$. Thus, y is 1.5 and x is 0.45. PT is 0.45cm therefore PS is 2.25cm.

Question 13: E

1 is true. The diploid cell had 54 chromosomes therefore the haploid gamete cell must have contained half as many, which is 27.

2 is true; early embryonic cells need to be able to produce every different cell type in the body from an un-differentiated state, and they are therefore regarded as stem cells.

3 is not true. The gametes that are used at the start must have been produced by meosis to become haploid (and then become diploid again upon fusion).

Question 14: E

1 $Fe\ (0) + CuCl_2\ (+2; -1\ -1) \rightarrow FeCl_2\ (+2; -1, -1) + Cu\ (0)$
2 $Cu_2O\ (+1\ +1; -2) \rightarrow Cu\ (0) + CuO\ (+2; -2)$
3 $Cl_2\ (0) + H_2O\ (+1, +1; -2) \rightarrow HCl\ (+1; -1) + HClO\ (+1; +1; -2)$
4 $BaCl_2\ (+2; -1, -1) + Na_2SO_4\ (+1, +1; +6; -2, -2, -2, -2) \rightarrow BaSO_4\ (+2; +6; -2, -2, -2, -2) + 2NaCl\ (+1; -1)$
5 $Hg_2Cl_2\ (+1, +1; -1, -1) \rightarrow Hg\ (0) + HgCl_2\ (+2; -1, -1)$

As you can see, there is reduction and oxidation of Cu in 2, Cl in 3 and Hg in 5

Question 15: F

Nuclear fission comes up quite regularly in BMAT questions and you just need to memorise how it works.

When an atom of U-238 is exposed to neutron radiation, its nucleus occasionally captures a neutron, making it U-239. It then needs to undergo two rounds of β decay which converts 2 neutrons into 2 protons, now giving the atomic number for plutonium.

Question 16: A

Just be careful with using standard form, and simplify your calculations as much as possible (e.g. 3.6/7 → 3.5/7 to make 0.5) and you should be fine with this question. Firstly, you can rearrange to get $M = gR^2/G$.
$R^2 = 6 \times 10^6 = 36 \times 10^{12} = 3.6 \times 10^{13}$
$gR^2 = 10 \times 3.6 \times 10^{13} = 3.6 \times 10^{14}$
$gR^2/G = 3.6 \times 10^{14} / 7 \times 10^{11} = \sim 0.5 \times 10^{25} \sim 5 \times 10^{24}$

Question 17: G

This question refers to coronary arteries on the surface of the heart.

1 is not true; glucose cannot freely diffuse from the blood stream; it requires *facilitated* diffusion with specific transporters.

2 is true, arteries carry blood away from the heart at high pressure.

3 is true, there are smooth muscle cells in the middle section of the arterial vessels (also known as the tunica media, although this is not in the specification so don't worry about learning it!) to allow vasodilation and vasoconstriction to occur.

Question 18: C

You can count each element on both sides of the equation to see if it is balanced, which works for **A-C**, whereas **D** and **E** have other problems.

A: 7C 12H 5O 1Mg → 6C 12H 5O 1Mg

B: 4C 6H 5O 1Mg → 4C 7H 5O 1Mg

C: 7C 12H 7O 1Mg → 7C 12H 7O 1Mg

D: The formula for propanoic acid is not correct – remember, propanoic acid always contains 3 carbon atoms only! In this example, 4 C atoms are present, so it is incorrect.

E: Hopefully intuitively you can see that $Mg_3C_3O_2$ is not a compound likely to form; it would require a C_3O_2 ion to be 6-...

Question 19: E

It takes 0.01s longer for the sound wave to travel to the far wall than the near wall. The microphone is placed next to the sound source, so the sound must travel to the wall and back before it is detected.

Distance to be reflected from near wall: 2 x 2 = 4m

Distance to be reflected from far wall: 8 x 8 = 16m

Therefore, it takes 0.01s to travel an extra 12m. 12/0.01 = 1200m/s

Question 20: E

Make the denominators the same so that you can add and subtract. Times the top and bottom of the first by x-1, the second by 2x, and the third by 2(x-1)

$$\frac{1}{2x} + \frac{1}{x-1} - \frac{1}{x} = \frac{x-1}{2x(x-1)} + \frac{2x}{2x(x-1)} - \frac{2(x-1)}{2x(x-1)} = \frac{x-1+2x-2x+2}{2x(x-1)} = \frac{x+1}{2x(x-1)}$$

Question 21: D

Males and females are affected similarly, ruling out this being an X-linked inheritance pattern.

A female and male with freckles are able to produce offspring without freckles, showing that it cannot be inherited recessively. Thus it is inherited in an autosomal dominant fashion.

Take F to be the allele for freckles and f the allele for no freckles.

Parent 1 is Ff and 2 is ff, so the offspring have a 50% chance of being either Ff (freckles) or ff (no freckles).

Parent 5 is Ff and 6 is Ff, so the offspring have a 25% chance of being FF (freckles), 50% chance of Ff (freckles) or 25% chance of ff (no freckles).

Question 22: B

Mr of hydrated copper (II) sulphate is $64 + 32 + (16 \times 4) + 10 + (16 \times 5) = 250$

Mass/Mr = $10/250 = 0.04$ moles of hydrated copper sulphate in $100 cm^3$

Therefore, in $1 dm^3$ (which is $1000 cm^3$), there would be 0.4 moles.

Concentration = moles/volume = 0.400/1, thus the concentration is 0.400 mol/dm^3.

Question 23: F

Newton's third law states that forces come in pairs, if object 1 exerts a force on object 2, then object 2 exerts an equal and opposite force on object 1. So, if the floor is exerting force P on the table, there must be an equal and opposite force which would be the force that the table exerts on the floor.

Question 24: C

Length of the third side of the triangle using Pythagoras's theorem is $\sqrt{(9^2 - 6^2)} = \sqrt{45} = 3\sqrt{5}$

Area of the triangle is $\frac{1}{2} \times 6 \times 3\sqrt{5} = 9\sqrt{5}$

Area of the circle is $\frac{1}{4}\pi 6^2 = 9\pi$

Therefore, together $9\pi + 9\sqrt{5}$

Question 25: G

1 is not correct. The gene is inactive at warmer temperatures, but this is not the same as it denaturing. (Remember denaturing is generally irreversible so if this gene denatured, it would never be active again and the cat could never get darker).

2 is correct. In the colder environment, the enzyme is active which causes the coat colour to darken, which is why the extremities of the body are darker.

3 is correct. The gene confers the potential to have a darker coat, and whether the gene is active or not depends on the environment.

Question 26: B

$2Na + H_2O \rightarrow Na_2O + H_2$

Mass/Mr = $0.23/23 = 0.1$ moles of sodium, hence 0.05 mol H_2

Volume = $n \times 24 = 0.05 \times 24 = 0.12 dm^3$

Question 27: C

For a straight-line graph, $y = mx$ where m is the gradient. y is the kinetic energy and x is the square of the speed which we can sub in below.

KE = $0.5mv^2$. We have the mass to sub in as 2.5kg, so KE = $1.25v^2$ and y=1.25x, so the gradient is 1.25.

END OF SECTION

SECTION 3

'He who has never learned to obey cannot be a good commander'. (Aristotle)

- This quote refers to the quality of leadership. Some may believe that good leaders were 'born to lead' and have always had the relevant characteristics and strengths to become an effective, charismatic leader. However, Aristotle argues that the best of leaders are those who have learnt to lead by developing their attributes and assets based around other successful leaders.

- As with any profession or role in life, natural talent is not enough for success. A talented sports player will need to follow the advice and potentially strict training and rules given by their coach if they are to thrive and potentially become a professional. A 'gifted' musician or writer will still need to work hard and learn from other successful pioneers in their profession in order to prosper. Leadership is an art form like any other and without learning from others talented in this art it would be difficult to succeed.

- Moreover, if one has not experienced something for themselves, they are less likely to be able to be good at it. For example, it is hard for someone who has never learned how to ride a bike to teach someone else, or someone who has not had children to give advice about parenting. In a similar vein, until one has experienced good leadership and has understood what it feels like to be led well, it would be difficult to truly know whether one's own leadership is good or not. Also, it is important to understand what it feels like from the 'other side'; that is, the insight gained from previously being commanded should help foster empathetic leadership.

- One could argue that there are other factors that are more important in a leader than learning how to follow, and that these can be developed independently. For example, you could elaborate on some of these points:
  - Ability to inspire confidence and provide direction
  - Planning, organising and setting targets
  - Listening, supporting and giving constructive criticism if relevant
  - Accepting responsibility for mistakes
  - Being assertive and always looking forward
  - Managing time, risk and people well

- You could link this quote back to medicine in the conclusion; and although they can't mark you down for an opinion, possibly err on the side of arguing that learning to follow is key to be a good leader. As a doctor, one starts with foundation training where every decision is reviewed and checked by a senior doctor, and where it is vital to follow the experience of someone who as worked for many years more. This 'leader' should hopefully also teach many attributes of the job that cannot be taught from a book at medical school. In order to thrive as a doctor, and then become a good leader in the future as a consultant, one must take on board all the advice that is given to them early on in the job.

The only moral obligation a scientist has is to reveal the truth.

- At its most basic level, science can be considered as the pursuit of truth. The scientific method involves systematic observation, experimentation and the formulation, testing and modification of hypotheses; all scientists should use this empirical technique if they are to be successful at their role. This statement suggests that using scientific research in order to advocate for a particular way of thinking, or other roles such as educating the public, are not the moral obligations of a scientist.

- It is clear cut that the role of a doctor is to look after their patients, the role of a lawyer is to represent their client in legal matters; similarly, the role of a scientist is considered to be using experimentation to acquire knowledge. One could argue that the role of a scientist is solely to reveal the truth, as this is what their expertise is, and then the role of others with varying professions, such as policy makers and politicians, to decide how that truth is applied and used. Moreover, some may believe that advocacy may hurt the credibility of scientists; by having a preconception of how things should be, they may consciously or unconsciously bias their experimentation in a particular way (although strictly speaking, using strong scientific methods should minimise this, such as the use of double blinding and randomisation in a clinical trial).

- There are events in history whereby successful scientists have used their power and privileged position to go beyond their pursuit of truth and harm others. A few examples:
  - Dr Reiter and Dr Wegener (both who have diseases named after them) were German physicians who committed war crimes under the Nazi regime by authorising medical experimentation on concentration camp prisoners

- One could argue that a scientist has the moral obligation to advocate for a particular scientific truth, especially when there could be a major danger to society. An example could be the case of global warming. Despite much rigorous scientific data showing that increasing carbon dioxide levels are causing atmospheric temperature increases, many politicians and global influencers deny the changes. As such, many countries and governments are not taking steps to reduce their emissions, which will make things worse and may eventually cause the destruction of the planet as we know it. If scientists publically came out to promote and support changing our lifestyles to save the Earth it may significantly improve things for generations to follow; that is, the potential for making the world a better place means acting on an ethical obligation.

- Also, one could argue that a scientist has a moral obligation to educate others and/or lead. They have likely developed a thorough, deep understanding of the science in their area of expertise and it would benefit society for them to spread this knowledge. Alternatively, one could argue that if their research is publically funded, i.e., paid for by via taxation, they have an obligation to give back to the public via education.

- You could conclude by arguing that the baseline moral obligation of a scientist is to reveal the truth and to be a good scientist (especially if this is their sole job description), but that the best and most influential people are those who are able to successfully apply (or at least think of how to apply) their knowledge, and those who are willing to educate the public.

The health care profession is wrong to treat ageing as if it were a disease.

- The statement is arguing that because ageing is a natural, normal and inevitable process which everyone must go through, whereas disease is an abnormal deviation from the norm which only affects a proportion of the population, we should not equate ageing and disease. That is to say, something that is universal cannot be abnormal.

- Instead of considering ageing as a disease, it may be more useful for the health care profession to consider ageing as a risk factor for chronic diseases, and then target the specific pathology caused by that disease (which is associated with but not solely caused by ageing).

- Another reason to not consider ageing as a disease may be due to its certainty. Some may believe that treating ageing as a disease would lead to the misallocation of limited resources to a futile cause, which may just prolong periods of pain and/or illness towards the natural end of life.

- In contrast: disease is defined as a disorder in structure or function of a bodily system causing harm to the affected individual, and with this definition, ageing does seem to fit the bill (if a healthy adult human is considered 'normal'). For example, ageing is associated with:
    - Bones become more brittle, muscle mass decreases, joints degenerate and become less flexible
    - Wrinkling of the skin, greying of hair, hearing loss, loss of near vision
    - Cognitive decline, potentially leading to dementia, cardiovascular changes, increase risk of cancer
    - Inability to control bowel and bladder movements

- Moreover, if we treat ageing as a disease that can be treated and prevented, we can use scientific research to target fundamental pathways in this process, and potentially find methods to halt or slow its development. That is, treating it as a disease legitimises medical efforts to try to cure it or eliminate other conditions associated with it. Before the advent of vaccinations and antibiotics, it was rare to live over forty years old; by targeting ageing itself there is the potential to make us even healthier.

- Targeting fundamental ageing mechanisms rather than 'age-related disease' may be beneficial to medicine; it may be futile to fight chronic diseases without striving to first understand their ultimate cause. Like with other diseases, modern experimental techniques have found molecular targets and deleterious changes related to ageing: particular genes involved in the process, and structures at the ends of chromosomes called telomeres which act as biological clocks for human cellular ageing. People's bodies age at different rates according to an interaction between genes and the environment.

- The goal of biomedical research is to enable people to be as healthy as possible, for as long as possible. With modern technology and methodology, it should be possible to better delineate the exact pathways involved in ageing. If we found more substantial evidence that ageing is a preventable or curable process, then it would make a lot of sense to treat it as such and fund and develop procedures to slow the process. However, in contrast, if the evidence suggested that resources were better spent directly targeting specific pathologies in age-related conditions, then it would not be worthwhile to treat ageing as a disease.

END OF PAPER

2018

SECTION 1

Question 1: B
As there is a seat every 400m for 3.2 km and a seat at the end of the path, there are 9 seats along riverside walk. Therefore, there are 18 bins situated around seats. There are 8 gaps of 400 m between seats, each with 3 bins (one every 100m), so there are 24 bins between the seats. In total, there are 42 bins along Riverside Walk.

Question 2: B
As someone with the MV combination has died from the disease, having one V variant of the gene cannot guarantee resistance to vCJD.

Question 3: D
It would take too long to calculate exactly how much time each employee spent in the office between 10:00 and 12:00 but this question can be answered using estimation and elimination.
Phil left the office for 39 minutes between 11:03 and 11:42 but was in the office for the remainder of the time.
Quentin was only in the office for about 23 minutes between 11:23 and 11:46 plus 5 minutes from 11:55 to 12:00. Eliminate Quentin.
Rob left the office for 9 minutes between 10:17 and 10:26 and 38 minutes from 11:00 to 11:38. He spent more time out of the office than Phil and is therefore eliminated.
Sanna left the office for 15 minutes, 6 minutes then left again at 11:50. She was out of the office for 31 minutes between 10:00 and 12:00. Eliminate Phil.
Theresa was out of the office between 10:02 and 10:42, much longer than Sanna and can also be quickly eliminated.

Question 4: C
The reasons stated by the author of this passage for being against cutting funding to state nurseries are all based upon their experience of sending their child to a state nursery. This is a flawed reasoning strategy because it is based on a generalisation; we have no way of knowing if this is a typical state nursery experience and if others consider state nurseries to be of similar importance.

Question 5: D
This pie chart consists of 2 small equally sized segments, 1 medium sized segment and 2 large equally sized segments. Only Dolly's average usage in minutes per day fits this pattern – 20, 20, 40, 80, 80.

Question 6: C

The conclusion to this passage is stated in the final line, 'Social medial has become a vehicle for spreading untruths, and has thereby undermined democracy'. **C** is essentially a rewording of this sentiment.

Question 7: D

The doors have just closed at floor 11 and the lift must go to floors 1, 4, 6, 15 and 24. It will go in the order 15, 24, 6, 4, 1. When it is at floor 15, it goes to floor 24 next rather than floor 6 even though they are equidistant because it will continue in the direction which it was previously travelling.

It takes 4x3 = 12s to get to floor 15, where it waits for a minimum of 9s. It then takes 27s to get to floor 24, where it also waits for a minimum of 9s. It then takes 54s to get to floor 6, where it will wait for at least 9s. Finally, it will take 6s to get to floor 4.

Total = 12 + 9 + 27 + 9 + 54 + 9 + 6 = 126

Note: there is a quicker albeit less intuitive way of answering this question.

You know that the lift will travel from floor 11 to floor 24 then back down to floor 4, i.e. it will traverse 33 floors, giving 99s travel time. It will also stop at 3 floors for at least 9s each, i.e. at least 27 stationary time. 99 + 27 = 126s.

Question 8: B

This question can be answered quickly using estimations. In the '25 and under', '36 to 45', '46 to 55' and '56 and over' categories, women did about 50% more hours of unpaid work than men. In the '26 to 35' category they did almost double.

Question 9: B

1) Incorrect. According the first table, women spend almost (but not quite) twice as much time doing housework compared to men.
2) Correct. Women spend on average about 6 times longer per week doing laundry than men, which is much higher than any of the other ratios.

Question 10: C

C is not a plausible explanation. Using public transport to commute to work would fall under the category of non-leisure travel. If lower income people are more likely to use public transport to commute to work without the option of driving, they are actually likely to spend more time on non-leisure travel. The other 3 options are all plausible.

Question 11: C

We are told that total unpaid work in the UK in 2015 had a value of £1000 billion and that women carried out an overall average of 60% more unpaid work than men.

Using ratios is the best way to get to this answer. If women are carrying out 60% **more** work unpaid than men, we can say that the ratio for women:men is 160:100 (as 160 is 60% more than 100).
Adding 160 and 100 = 260.
Divide 1000 billion by 260 → this will be roughly equal to 1000bn/250 = 4bn (the actual value will be slightly below this). This number represents the value of the unpaid work done per unit time.
Now multiply up by 160 to find the total value of unpaid work done by women.
4bn × 160 = 640bn.
As we know that 4bn was a slight overestimate, we can conclude that 640bn will also be a slight overestimate. Thus, **C** is the answer, as it is the closest answer to our estimate and is on the smaller side, which balances the overestimation that affected our previous answer.

Question 12: C

Customers 3, 9 and 10 were delayed by > 20 minutes
Customer 4 was delayed by 10-20 minutes
(3x6) + 4 = $22

Question 13: D

1) If this were true, it would weaken the counterargument that young people could channel aggression in a controlled manner via a conduit less dangerous than rugby, therefore strengthening the above argument.
2) If this were true, it emphasises the need for young people to channel aggression in some way.
3) Irrelevant – we are only concerned with the benefits and risks of rugby here.

Question 14: F

The easiest way to do this is by process of elimination. Answer **A** cannot be correct because both Roger and Sam would both have lied about both but the question states that only one person lied about both. Continue through the answers until you find one that fits with all the conditions. Unfortunately, in this case it happens to be the final one you get to.

Question 15: A

1) This passage states that 'Newspaper ownership in Britain is concentrated in the hands of a few businessmen' and, in the second sentence, refers to the 'political Right's promotion of the freedom and prosperity of business'. It follows that the Right promotes the interests of the newspaper owners in Britain.
2) While this passage suggests that political opinion presented in newspapers is biased in favour of the Right, it does not give information about the Press as a whole.

Question 16: A

Half of the small container is filled at a rate of 6 l/min and half of the container is filled at a rate of 12 l/min. Therefore, it must spend 2y time being filled at 6 l/min and y time being filled at 12 l/min.

As the containers swap pumps when the small container is half full, the large container is filled at 6 l/minute for y time and 12 l/minute for 2y time.

Forming expressions for container volume…

Small: $(6 \times 2y) + (12 \times y) = 24y$

Large: $(6 \times y) + (12 \times 2y) = 30y$

30 is 25% greater than 24, so A is the correct answer.

Question 17: A

This passage states that if the prince's whereabouts had remained secret, the risk could be managed, but as the whereabouts were not secret, the risk could not be managed. This is a logical flaw because it may still be possible to manage risk even if the whereabouts are not kept secret. Similarly, in option **A**, it may still be possible to reach the island if the tide isn't out.

Question 18: E

Number of combinations with 2 yellow and 4 red:

$C(6,2)$ = The number of combinations with 4 yellow and 2 red is also 15.

Number of combinations with 3 yellow and 3 red:

$C(6,3) = 15 + 15 + 20 = 50$

There are 50 possible combinations.

Question 19: A

Look for the phase of flight with the smallest ratio of fatal crashes to onboard fatalities. 8% of crashes occur during the cruise phase of flight, but these crashes account for 16% of onboard fatalities (double the mean average per crash). There is no phase of flight where the percentage of onboard fatalities is double the percentage of fatal crashes.

Question 20: C

As per the estimations in table 2, 27% of a 1.5-hour flight is spent in descent, initial approach, final approach and landing. 27% of 1.5 hrs is 24.3 minutes.

Question 21: D

The answer to this question can be found by elimination. **A** is incorrect; it is easy to tell that the proportion due to pilot error in the 1990s is similar to in the 1960s. **B** is incorrect – the proportion of crashes due to mechanical failure in 2010s is similar to that in the 2000s. Continue like this until all the options but **D** are eliminated.

Question 22: C

The final paragraph says that although deaths from air travel were quite high in 2015 at 560 fatalities, air travel has never been safer because a majority of these fatalities were due to 2 crashes that were deliberately caused by sabotage. However, this is not a fair assumption because sabotage may be becoming more common and the incidents are resulting in more fatalities, contributing to a lack of safety associated with air travel.

Question 23: B

1st	2nd	3rd	4th
n	n+1	x	x+1

The quickest way to do this is intuitively, by trial and error. If the digits in the PIN were large, there's no way that multiplying two of the digits would get the same value as adding two.

If we start off by trying 1234, we get 1+ 3 = 4, 2 x 3 = 6. The multiplied digits are still too great, so we have to go with smaller numbers. The first 2 numbers cannot be made smaller, so we make the final two digits one less.

1223

1 + 3 = 4

2 x 2 = 4

Question 24: C

This passage concludes, with the Canberra study as its only evidence, that domestic cats should be encouraged because they have a positive effect on native wildlife. This is flawed because the conclusion is not valid if the results of the Canberra study do not hold true for the general environmental effects of keeping domestic cats.

Question 25: B

In order to split the constituent charities into the smallest possible amounts, we must first give the maximum of $100 to Charity 1. If the remaining $100 is shared between the other 4 charities relatively equally but with all charities still receiving different amounts, STARS will receive $27 and charities 3, 4 and 5 $26, $24 and $23 respectively. The same answer can be gained by giving $98 to charity 1, $27 to STARS and $26, $25 and $24 to charities 3, 4 and 5 respectively.

Question 26: D

This passage ends by saying that more action is needed in the developing world to promote breast feeding 'if we are to help women and avoid infant ill health and mortality'. There is evidence in the passage for the benefits of breastfeeding to infants, but this argument could be strengthened with evidence that breast feeding would also 'help women'. This ensures that both criteria (i.e. mother's health as well as infant's health) are satisfied in the concluding statement at the end of the passage.

Question 27: A

	Type	Number of extra toppings	Cost of order ($)
India	vegetarian	none	5
James	ham	one	8
Keira	ham	none	6
Lance	ham	one	8
Maddie	Paul's special	two	12
Nellie	vegetarian	two	9
Ollie	Paul's special	one	10

At this point, we do not know for sure which of these 2 options would be cheaper:

Option 1: using the 'buy any four pizzas, get another one for free' promotion once with one 'buy one pizza, get any other one for half price' promotion.
Option 2: using the 'buy one pizza, get any other one for half price' promotion 3 times and paying for one pizza without a promotion.

Option 1:
Buy the cheapest 4 pizzas for $27 (5 + 6 + 8 + 8) and get the $12 pizza for free.
Pay $9 for Nellie's vegetarian special with two toppings and pay half price for Ollie's $10 pizza. Total: $41.00

As the total for option 1 is the lowest of the possible answers available, we can be confident that **A** is correct without calculating what the minimum cost would be using option 2. However, if you were to calculate it, you would find that it is $42.00, confirming that our answer is **A**.

Question 28: D
This passage argues against the effectiveness of universal screening programmes saying that 'A study of routine health screening programmes found no consistent evidence that they improved health or reduced death rates' but also refers to the 'value of targeted screening programmes in geographical areas where levels of disease are more prevalent'. From this passage, it is therefore safe to conclude that targeted screening programmes would be more effective than universal programmes.

Question 29: C

The die on the LHS must be the '2 to 7' die because if it were the '0 to 5' die the sides with '3' and '2' would be facing each other, which they are not.

Equally, the die on the RHS must be the '0 to 5' die because if it were the '2 to 7' die the sides with '4' and '5' would be facing each other, which they are not.

Option	Eliminated/answer	Reason
A	Eliminated	Not the LHS dice because if the LHS dice were flipped such that the '3' side were on top, the side revealed would have a '4' rather than a '6' because it is opposite a '5'. Not the RHS dice because one of the sides has a 6.
B	Eliminated	Not the LHS dice because the '4' and the '3' would be opposite each other. Not the RHS because '3' square would be a '0' because they both share corners with '4' and '2' and are therefore opposite.
C	Answer	Not the LHS dice because the side with a '1' would be opposite the '2'. Could be the RHS dice – '1' opposite '4', '3' opposite '2'.
D	Eliminated	Not the LHS dice because although the other side sharing a corner with the '2' and the '5' should be a 6 (as it's opposite the '3'), the dots on the '2' side are in the wrong place. Not the RHS dice because one of the sides has a 6.
E	Eliminated	Not the LHS dice because the '4' side should be a '7' as it is opposite the '2'. Not the RHS dice because the side with the '3' should be a '2', the same as the example view of the dice.

Question 30: D

1) If this were true, there would be increased food availability. This would strengthen the argument that we should transition to growing biofuels.

2) If this were true, it would make it easier to transition from fossil fuels to biofuels, strengthening the argument that this avenue should be pursued.

Question 31: A

If the largest difference between 2 scores was between design and evaluation, these must be the highest and lowest scores. We already know that design and construction were the 2 best sections, so design must be the best section, followed by construction, research and finally design.

Section	Score
Design	x+11
Construction	x+9
Research	y
Evaluation	x

x + 11 + x + 9 + y + x =48

3x + y + 20 = 48

$x \leq 9$ or else the score for design would exceed 20.

Trial and error:

If x=4, y=16. This cannot be the case because a score of 16 in research would be greater than the score in construction (13).

Therefore, x must be greater.

If x=7, y=7. The score for research cannot equal the score for design so this cannot be correct.

If x=6, y= 10. Design: 17, Construction: 15, Research: 10, Evaluation: 6.

This meets all the conditions of the question.

Question 32: D

The passage says that phenylalanine can be converted to tyrosine and that methionine can be converted to cysteine. However, these conversions rely on the presence of other molecules. Therefore, tyrosine and cysteine can be made sometimes but not others – they are conditionally indispensable.

Question 33: B

Table 1: 0.8 g/kg recommended daily intake of protein, 56 g for a 70 kg man.

Therefore, 4x56 = 224 mg of Try recommended.

Table 2: 5mg/kg body weight Try recommended. For a 70 kg man, this is 350 mg of Try.

350-224 = 126 mg

Question 34: A

Newborn: 560g breast milk at 2.5% protein – 14 g protein per day
55mg/g protein Iso, 14 x 50 = 700 mg per day

8-week-old: 700g breast milk at 1% protein – 7g protein per day
55mg/g protein Iso, 7 x 50 = 350 mg per day

Question 35: B

To work this out, first find the ratio between Thr (human milk) and Thr (hen's egg).

This ratio is 44:19, which we can roughly simplify to 2.25:1

Now, multiply divide 21 (His (human milk)) by 2.25 – which is equal to multiplying by $\frac{4}{9}$, giving us $\frac{84}{9}$.

We can see that $\frac{84}{9}$ is less than 10 ($\frac{90}{9}$) and more than 9 ($\frac{81}{9}$).

As we are asked to estimate the answer, we just need to decide if $\frac{84}{9}$ is closer to 10 or 9. As 84 is closer to 81 than to 90, it is easy to see that $\frac{84}{90}$ will be closer to 9 than to 10, thus the answer is 9mg.

END OF SECTION

SECTION 2

Question 1: H

Although these enzymes can be produced at multiple sites, all three of them are produced by the pancreas and should therefore be included in this medication.

Question 2: C

Q is magnesium, a metal which forms an ion with a 2+ charge.
Z is fluorine, a halogen which forms an ion with a 1- charge.

To cancel out the charges, the compound must contain a ratio of 1Q to 2 Z. The bonding between metals and non-metals is ionic.

Question 3: E

1: incorrect – all light travels at the same speed in a vacuum (the speed of light)
2: correct – see diagram
3: incorrect – light travels fastest in a vacuum
4: correct – see diagram

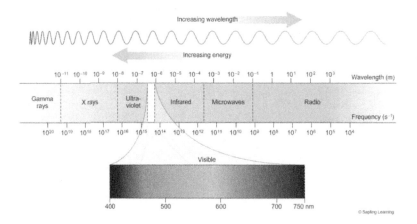

Question 4: B

1) Divide the numerator and denominator by $8m^3p$
2) When a factor is moved from the bottom to the top of a fraction its power is multiplied by -1.

Question 5: A

Denitrification, which is generally carried out by bacteria in the soil, is the process by which nitrogen is released from nitrates and returned to the atmosphere in the form of nitrogen gas.

Question 6: C
1: Incorrect - fractional distillation is used to separate particles with different boiling points. No bonds should be broken or formed. Fractional distillation breaks the weak intermolecular forces which keep the substances in a liquid state, not the covalent bonds.
2: Correct- the particles cool down between positions 2 and 3 and therefore become closer together.
3: Incorrect – it takes energy to turn a liquid into a gas so the process at position 1 is endothermic. Because it absorbs energy from the outside environment, it is endothermic.

Question 7: E
Wave speed = frequency x wavelength = 50 Hz x 0.4 m = 20 m s^{-1}
Use the equation speed = distance / time → 20 = 100 / time
Therefore, the wave takes 5s to travel 100m through this material.

Question 8: C
Probability of picking two 50p coins in a row: 2/7 x 1/6 = 1/21
Probability of picking two 20p coins in a row: 5/7 x 4/6 = 10/21

The probability of not picking the exact money to buy the item is 1/21 + 10/21 = 11/21
The probability of picking one 50p and one 20p is 1 - 11/21 = 10/21

Question 9: B
Cells can divide by mitosis over and over again whereas meiosis produces gametes, which are incapable of dividing further.
$4 \times 2^5 = 128$, the cell number is doubling each time, which is consistent with mitosis.

1: correct – mitosis produces identical daughter cells.
2: incorrect – meiosis, not mitosis, produces gametes. Gametes are haploid male or female germ cells.
3: correct - mitosis produces identical daughter cells.
4: incorrect – this is mitosis rather than meiosis for the reasons above.

Question 10: C
In this reaction, Fe^{3+} is reduced to Fe and the CO is oxidised to CO_2. As Fe_2O_3 is donating oxygen to the CO, it is acting as an oxidising agent in this reaction.

A: incorrect, Fe^{3+} gains negatively charged electrons to become Fe.
B: incorrect, each Fe has a charge of 3+ and each oxygen has a charge of 2-.
D: incorrect, ΔH is negative which means energy is released and this is an exothermic process.
E: incorrect, electrolysis is a type of reaction in which an electric current is passed through a solution containing ions to cause chemical decomposition.

Question 11: E

Velocity and momentum are vector quantities, the value of which changes with direction. Conversely, kinetic energy and speed are scalar quantities – they are only associated with magnitude, not direction.

Question 12: F

Length of rectangle: 9cm
Width of rectangle: 3cm
Area of rectangle = 27 cm²

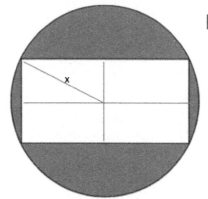

[diagram not to scale]

X= radius
$X^2 = 1.5^2 + 4.5^2$
$X=\sqrt{45}/2$

Area of circle = 45π/2
Shaded area = area of circle – area of rectangle = 45π/2 - 27

Question 13: H

All specialised cells, such as epithelial cells, mature red blood cells and muscle cells, come together to form tissues.

Question 14: B

Isotopes have the same number of protons but different numbers of neutrons. As they have different masses, they will show up as different peaks on mass spectrometry.

The peaks at 35 and 37 represent individual atoms of X. The peaks at 70, 72 and 74 represent diatomic molecules.
70: 35X + 35X
72: 35X + 37X
74: 37X + 37X

Question 15: E

$v^2 = u^2 + 2as$

v: final velocity

u: starting velocity

a: acceleration (acceleration due to gravity is about 10 m s^{-2})

s: displacement

$v^2 = 0^2 \times 2 \times 10 \times 1.8 = 36$

Velocity $= \sqrt{36} = 6\ \text{ms}^{-2}$

Note – the mass of the bar (200kg) is an irrelevant value in this calculation and does not need to be used!

Question 16: B

The difference between the rth triangular number and the triangular number 3 along in the sequence is the sum of the difference between r and r+1, r+1 and r+2 and r+2 and r+3. Since the difference increases by a regular increment (=1), the difference between r+1 and r+2 = 126/3 =42. Therefore, the difference between r and r+1 =41. Therefore, r is the 40th triangular number.

Question 17: F

During inspiration, the diaphragm contracts to flatten (rubber diaphragm pulled down), which increases the volume of the thorax (the volume of the bell jar). If the volume of the thorax increases, there are fewer gas particles per square volume so the pressure in the thorax decreases.

Question 18: E

The double bond breaks and each of the carbons involved form single bonds with other monomers. Every carbon atom in the chain will be bound to a -CH$_3$ group and a H.

Thus, every C atom on the polymer chain should be bonded to one methyl group (CH_3), as is seen in option **E.**

Question 19: B

According to Ohm's law, V=IR.

If the variable resistor has a resistance of 2.0 ohms, the total resistance of the circuit is 12 ohms. I=6/12= 0.5 A

If the variable resistor has a resistance of 20 ohms, the total resistance of the circuit is 30 ohms. I=6/30 = 0.2 A

0.5-0.2 = 0.30 A

Question 20: A

The total mass of sweets in the bag of 20 must be between 200 and 210 grams.

The 16 sweets already in the bag have a total mass of 152g.

The remaining 4 sweets must have a total mass between 48g and 58g.

48/4 = 12

58/4 = 14.5

Each sweet must be >12g because 'the mean mass of a sweet in a bag of 20 must be greater than 10 grams' i.e. not *at least* 10 g.

However, they must be 'not greater' than 10.5 grams, so the upper limit is included.

Therefore, x, the mass of each of the remaining sweets, must be greater than 12 and less than or equal to 14.5g.

Question 21: A

Glucose uptake is by active transport, which is an active process, thus requires energy to facilitate. Thus, there is no net movement of glucose molecules into the cell when respiration is inhibited.

Water molecules move into the cell by osmosis, a passive process, which does not require energy. Thus, osmosis continues to occur when respiration is inhibited.

Question 22: C

105g of the compound contains 57g of fluorine and 48g of oxygen.

Element	Fluorine	Oxygen
Mass (g)	57	48
Ar	19	16
n	57/19 =3	48/16 = 3

Therefore, there is a 1:1 ratio of fluorine to oxygen in the compound.

Empirical formula: OF

As the relative molecular mass is twice that of its empirical formula mass...

Molecular formula = O_2F_2

Question 23: B

When one alpha particle is emitted, the molecular mass decreases by 4 and the atomic number decreases by 2.

When one beta particle is emitted, the atomic number increases by 1 and the molecular mass is unchanged.

As the molecular mass decreases by 24, 6 alpha particles must have been emitted.

Emission of 6 alpha particles would result in a reduction in atomic number of 12. However, atomic number has only decreased by 8. Therefore, 4 beta particles must have been emitted.

Question 24: B

Factorise the denominator of the second fraction $\frac{x}{x-1} - \frac{x^2+3}{(x+3)(x-1)}$

Put both fractions over a common denominator $\frac{x(x+3)}{(x-1)(x+3)} - \frac{x^2+3}{(x+3)(x-1)}$

Combine $\frac{x(x+3)-x^2-3}{(x+3)(x-1)}$, expand the brackets $\frac{x^2+3x-x^2-3}{(x+3)(x-1)}$, and combine like terms $\frac{3x-3}{(x+3)(x-1)}$.

The numerator factorises to $3(x-1)$ and the '$(x-1)$' brackets cancel.

Question 25: C

B = brown body colour allele
b = black body colour allele

We can make a punt square to determine the possible genotypes, and then use this to determine the ratios of phenotypes.

	B	b
B	BB	Bb
b	Bb	bb

Possible genotypes: BB, Bb, bb (3)
Possible phenotypes: brown, black (2)

Question 26: C

3 mole of F-F bonds break, requiring 474 kJ
The overall energy change is -330 kJ/mol
804 kJ of energy is released when one mole of xenon hexafluoride is produced (with 6 Xe-F bonds per molecule).

The Xe-F bond energy is 804/6 = 134 kJ/mol

Question 27: E

The ratio of carbon-14 to carbon-12 has decreased from $1000:10^{15}$ to $100:10^{15}$.

Number of half-lives	Amount of carbon-14 to 10^{15} carbon-12
0	1000
1	500
2	250
3	125
4	62.5

The material has gone through between 3 and 4 half-lives and is therefore between 18,000 and 24,000 years old.

END OF SECTION

SECTION 3

'Liberty consists in doing what one desires.' (John Stuart Mill)

Explain the reasoning behind the statement. Present a counter-argument. To what extent do you agree that freedom is doing what you want?

Structure your answer such that you have one paragraph relating to each point in the question.

- Liberty is a state in which a person is free to live their life without oppressive restrictions. This statement argues that in a state of liberty, a person will do what they desire because they are free to do as they like.

- It can be argued that freedom of speech is one of the most important examples of liberty – it allows us to say what we desire, and therefore we have to liberty to express our views without fear of sanction. Without freedom of speech, societies would not be able to express opinions on, for example, government policies. Without this, we do not have the liberty to change our own society for the better.

- This statement assumes that the main driving factor behind human behaviour is desire. However, in a state of liberty, a person may be driven to do things other than what they desire, driven by other factors, such as their conscience. Even with the metaphorical chains removed, rather than what they desire, an individual may do what is morally right or what others want. Arguably, in some cases, this may be due to deprivation of a different type of liberty e.g. a feeling of constraint to social norms.

- A specific example may involve the avoidance of fast fashion, which is becoming popular across all generations. The desire of an individual may be to acquire clothes at a cheaper cost, as this is in their best interests. However, an awareness of the background of fast fashion, and the conditions workers face, often changes the behaviour of the individual. Many choose to buy more expensive clothing which is more ethically produced, despite this not really being in the individual's best interest – they do not have a specific desire to spend more. This is an example of a person having the liberty to balance between what they desire to do for themselves, and what they morally want to do.

- An alternative example can be seen in social media platforms – where certain posts are restricted or deleted, due to containing worrisome messages or hate speech. If hate speech was not controlled, i.e. we had full freedom to say and do what we desire on the internet, other aspects of liberty could be compromised, such as the liberty for individuals who are the targets of these hate crimes to live a safe life. Humans need to have a cut-off to their own liberty in order to create liberty for different groups of people.

In the final paragraph, introduce more personal opinion.

- On one level, liberty is about having autonomy over our actions – the ability to do what we want, but not necessarily what we desire. However, it could be argued that true liberty is a theoretical state in which we are free from negative consequences and therefore able to act in a way that fulfils our every desire.

  At the end of this essay, it is important to come to some sort of conclusion as to what extent you agree with the statement.

Rosalind Franklin said that science gives only a partial explanation of life.

Explain what you understand is meant by her statement. Argue to the contrary that science can give a complete explanation of life. To what extent do you agree with Franklin's statement?

- There are several ways to interpret this statement:
1) Science *currently* gives only a partial explanation of life, there is so much more to be known.
2) Science *is only capable* of giving a partial explanation of life and some of life can only be explained elsewhere (e.g. by religion/spirituality) or could never be explained.

- Science is an all-encompassing term that has its roots in the Latin word 'scire' meaning 'to know'. It includes ways of increasing our understanding of any aspect of life at both a subatomic level and the level of our planet and beyond. The ultimate aim of science is to explain every aspect of our surroundings, including life. Therefore, if given the opportunity to progress indefinitely and get infinitely closer to the truth about life, it may someday be able to provide an explanation.

- I agree with Franklin's statement to some extent. There is so much to life that we cannot currently explain with science and see no hope of explaining in the near future. Progress within science is restricted by elements such as the capacity of humans to interpret and understand information presented to them. Perhaps science has already given us an explanation to life but we haven't been able to understand it? On the other hand, if we think of science as a concept, rather than focus on the practicalities it is confined by in its current reality, I believe that (at some point) it could fully explain life.

In the age of modern healthcare, every time a patient dies after a routine operation or procedure, it's a case of medical error.

Explain the reasoning behind this statement. Argue that there can be reasons other than medical error behind such deaths. To what extent do you agree with the statement?

- This statement suggests that in the age of modern healthcare, the only reason for routine operations or procedures to result in fatalities is medical error. It suggests that, due to advances in fields such as pharmaceuticals, medical technology and infectious diseases, medical professionals have all the equipment and knowledge they require to carry out routine operations or procedures with a 100% success rate.

- Even in the age of modern healthcare, there are so many reasons for patients to die in relatively routine operations or procedures. Firstly, this statement is unclear in how long after the treatment it believes patients should survive. It likely means that 'every time a patient dies from a direct complication of a routine operation or procedure, it's a case of medical error', because it is inevitable for everyone to die at some point after a routine operation or procedure.

- There are so many possible complications that can arise from all operations or procedures, routine or non-routine. Medical advances can reduce the chance of these occurring, but cannot eliminate risk completely, even in the hands of an extremely competent physician who does not ever make any errors. For example, in the short term, there is risk of haemorrhage. Sometimes, partially because the anatomy of our vasculature varies between individuals, it is impossible to predict the source of bleeding or its extent. This can result in fatality even when a very competent doctor is acting with exceptional medical judgement. Later complications of invasive medical treatment include infection. Advances have been able to reduce both the chance of infection and the mortality of infections. However, even with our current understanding of different pathogens, use of antibiotics, disinfectants and sterilization, it is currently impossible to reduce fatality from infection to zero.

- More generally, for reasons that are either yet to be fully understood or applied to clinical practice, different patients respond to treatments in different ways, and it is therefore impossible to predict with 100% certainty how someone will respond to a treatment.

- Overall, there remain so many factors that contribute to death after routine operations and procedures, many of which are outside the hands of medical professionals. Death can occur for many reasons, such as haemorrhage and infection. The probability is reduced with medical advances, but even in the hands of a highly skilled healthcare team, routine treatment can still go wrong, particularly in vulnerable patient groups such as the elderly and those with comorbidities such as diabetes or obesity. **END OF PAPER**

2019

SECTION 1

Question 1: C
Last year:
The ticket costs $330 and the booking fee is $50. Therefore, the total of the basic cost and taxes is: £330 - $50 = $280. The booking fee and taxes are in a ratio of 3:2 so the cost is calculated first by $280/5 = 56. Then multiply 56 by either 3 to give the booking fee of $168 or by 2 to give taxes of $112.

This year:
Booking fee is halved to $25.
Basic cost of flight $168 x 1.2 = $201.6
Taxes: $112 x 1.1 = $123.2
Total cost = $123.2 + $201.6 + $25 = $349.80

Question 2: C
The main argument of the passage is that due to an increase in internet-connect smartphones there is a decrease in personal interactions which is impairing children's emotional development. Statement **C** provides an additional example of a feature of smartphones that prevents personal interactions. The other statements are irrelevant or weaken the argument.

Question 3: C
This easiest way to do this is by a process of elimination. Looking at the requirements for the car having both a sun-roof and air con we can quickly eliminate all the options except for Cronal and Elox. The car must be below $160 for 7 days so Elox is too expensive at $180. Therefore, Cronal is the only answer that fits this.

Question 4: B
The main argument is that the proposed internet sales tax will have negative implications for both individual companies and the government through fewer sales and so taxes. **B** is the correct answer as the assumption of the passage is that an online tax will stop people shopping completely as opposed to going to a store. If this is the case, the negative implications will not occur and so this is the flaw.

Question 5: B
Option **B** is the answer as on folding up the cube the two arrows point in the same direction as opposed to opposite directions like the others.

Question 6: C

The passage describes how the value of a product was historically calculated based on how much people were willing to pay for it, but this methodology is now flawed due to new services like social media being financially free but having other hidden costs such as a lack of privacy. This supports statement **C** of needing a new alternative method to value products.

Question 7: E

The question asks for the chance a child has a disease if they have a positive test result. This can be rephrased as the proportion of true positives out of the total positives so 72 out of 164.

Question 8: E

The first two paragraphs discuss the statistics and potential factors for the pay gap but ends with the statement that 64% of the pay gap is currently unexplained, supporting **E** as the correct answer and demonstrating that **A** and **D** are incorrect. **C** is irrelevant as the passage does not talk about this study and **B** is given as a suggestion in the paragraph but not proven.

Question 9: D

The average wage in 2017 was £550 with a gender pay gap of 9.1%. This means if we take the average female salary as x, the average male is $1.091x$.

We can then use this to make an algebraic equation to solve this: $x + 1.091x = 550 \times 2$

This is simplified to: $2.091x = 1100$

This is quite hard to calculate using mental maths so we can round this to approximately $2.1x = 1100$. This gives x approximately equal to £525.

Question 10: H

Statement 1: This is an incorrect statement as if women have longer tenures which pay more, they will earn more than men.

Statement 2: This is a feasible explanation as if women have shorter tenures which pay less, they will earn less than men.

Statement 3: This is an incorrect statement as if men have longer tenures that give them less pay, they will earn less than women.

Statement 4: This is a feasible explanation as if men have shorter tenures which pay more, they will earn more than women.

Question 11: E

In 1997 the pay gap was 17.4% whilst in 2016 it was 9.4% and the male increase is 75%. We can use algebra to help to solve this by assigning the women in 1997 as y.

Male pay in 1997 = 1.174y
Male pay in 2017 = 1.174y x 1.75
Women's pay in 2017 = 1.174y x 1.75/ 1.094

These numbers are too hard to work with using mental arithmetic so we will need to round them for the calculation
Women's pay in 2017 = 1.2y x 1.75/1.1 = 91% which is approximately 92%

Question 12: C

The first ferry from X will depart at 00:00 and so arrives at 2:30 and can leave Y again at 3:30 so can be the 3:45 ferry. In the meantime, 2 ferries are needed to depart from Y.

The first ferry from Y will depart at 00:45 and so arrives at 3:15 at X. It can then leave X again at 4:15 so can act as the 4:30 ferry. In the meantime, 3 ferries are needed to depart from X.

Therefore, 5 ferries are needed.

Question 13: B

The information from the passage can be summarised as: Italy has short café times, low wages and so low prices. England has medium café times and high wages and so medium prices. Greece has long café times and lower wages and the highest prices. This shows that it is the café times that have the greatest correlation to price, so **B** is the correct answer.

Question 14: C

The black section is slightly over a quarter of the total area whilst the pale grey is slightly below, and together, they create 50% of the pie chart. These must be Monday and Friday respectively as they are the two days from the graph with values that are the closest together. The combined total of the sales from these is 150. Therefore, as the two darker grey sections form the other half of the pie-chart they must add up to this. Looking at the graph, the smallest wedge must be Tuesday which has sales of 30. Therefore, the other section must equal 120 and so is Thursday. Therefore, Wednesday is missing.

Question 15: C

C is the correct statement as the passage describes the logic and rules regarding how several indefinite hyperbolic numbers are formed. There is no mention of **A** or **B,** and **D** is the opposite of **C**.

Question 16: A

	5 to 17 years Female	5 to 17 years Male	Adults Female	Adults Male	Total Na
Numbers tested	50	50	200	150	450
Percentage who could read the bottom line	28%	32%	12%	18%	
Number who can read the bottom line	14	16	24	27	81

Total = 81/450 = 18%

Question 17: D

The main conclusion of the passage is that it was a sensible decision not to review the cases and give legal pardons to the executed soldiers given the amount of time passed since WWI which would make it difficult to review the cases. Statement **D** is the correct answer as it is the underlying assumption necessary for the above to be true particularly regarding the quote 'too much time has elapsed for individual cases to be reviewed fairly'.

Question 18: D

In order to not need to go to the locker at break Jasper needs the books for both before and after break and this needs to be 5 or less.

Book numbers	Monday	Tuesday	Wednesday	Thursday	Friday
First lesson	Maths: 3	Music: 1	Maths: 3	Music: 1	Art: 1
Second lesson	English: 1	Science: 2	History: 2	English: 1	Geography: 3
Lesson after break	History: 2	Geography: 3		Maths: 3	Science: 2
Conclusion	6 so no	6 so no	No as 5 by break	**5 so correct**	6 so no

Question 19: D

To answer this, we need the value for daily Mirror sales in 2018, which is approximately 0.6 million, and the percentage decrease which is 80%. We do not need to use exact numbers as the question asks for the approximate answer.

We can use algebra to answer this question by assigning a as sales in 1992 and solving the equation.

(0.6 million − a)/ a = -0.8
0.6 million − a = -0.8a
0.6 million = 0.2a
3 million = a
Therefore, this is approximately 2.9 million.

Question 20: C

Line 1: This starts the highest so will be the Sun if line 2 is the Mail.
Line 2: The graph declines, but starts high and then declines slightly between 2005 and 2000 so cannot be the mirror but may be the Mail.
Line 3: This is the Mirror as it is the third highest paper and follows the correct trend.
Line 4: This line is flat between 2000 and 2005 so is not the Mirror.
Line 5: This line is too flat and so is likely to be a paper like the FT.

Question 21: D

The argument in the second and third paragraph is that the Sun and Mail both had large influences on public opinion and an example of this is supporting the Leave campaign. Statement **D** most weakens this as if people already have these opinions and so buy that paper, the paper is not influencing them but is a reflection of people's beliefs.

Question 22: B

1. False: (6m − 5m) / 6m = -17%
2. The pro-Leave papers are Express, Star, Sun, Mail and Telegraph. There total sales in 2016 = 0.4 m + 0.5m + 1.8m + 1.6m + 0.4m = 3.7m
   4.7m/ 6.3m = 75% so true

Question 23: D

Numbers capable of making the starting score are 2,3,7. We can ignore the first digit as from the answers it has to be a 2. They first tell us that the score didn't appear to change when they increased the score by 4. This means the starting digit score must be a 3 which will increase to 7. We therefore just need to find out what the tens digit is. After this score an increase in 6 the middle digit changes so that it can be an 8,9,0. This means the starting digit must be a 7. Therefore, the starting score is 273. The final score = 273 + 30 = 303

Question 24: B

The passage states that Cuckoos, cowbirds and honeyguides all use 'brood parasitism', however they have all developed the mechanisms of thicker eggs shells to enable them to do this separately providing evidence of convergent evolution, so **B** is the correct answer. Whilst statement **A** and **C** are true, they are not the main conclusions as they only address one aspect of the paragraph. **D** is false.

Question 25: E

She can have only Venus bars as 2.40 / 0.3 = 8. This means than 1, 2, 3 are all incorrect. Statement 4 is also incorrect as 30p + 40p + 50p = £1.20 so it is possible to have 2 of each bar. Therefore, none are correct.

Question 26: C

Statement 1: This is incorrect as there is no statement on whether these advancements will continue into the future.
Statement 2: There is no mention of this and is factually incorrect.
Statement 3: The final sentence shows this to be the case as computer software is used to analyse strokes.

Question 27: D

George has quite a few conditions that need to be met here:
1. He must have the Times
2. He must read 2 printed papers per week
3. He must read 3 different papers per week

The cheapest papers are the Express and the Dispatch so we can use these two with the Times. We can also therefore ignore the other two papers and the printed copy subscription as these are more expensive than buying the printed papers weekly.

Paper	Cost to buy the printed newspaper every week for the year	Online subscription	Difference between printed and online
Dispatch	£ 1 x 52 = £52	£30	£22
Times	£1.50 x 52 = £78	£60	£18
Express	£0.80 x 52 = £41.60	£20	£21.60

The difference between the printed and online Times and Express are smallest so we will use the printed version of that and online subscriptions for Dispatch.
Total cost = £30 + £78 + £41.60 = £149.60

Question 28: E

The conclusion of the passage is that drugs should not be legalised due to the increase in crime it caused after legalisation in Colorado. This is why statement 3 is correct as no other examples are given and the conclusion is reached only using this evidence. The example given also only shows a correlation but not necessarily causation as we do not have more information to conclude that there is causation so statement 1 is true. Statement 2 is incorrect as it is irrelevant to the passage as other substances are not mentioned.

Question 29: A

For this question it is useful to draw a Venn diagram.

As 70% use the swimming pool and 40% use the climbing wall then there must be a minimum of a 10% overlap as the total cannot be greater than 100%. This means that statement 1 is correct. It also means that all members must use one service so statement 2 is correct as well as statement 3 as the overlap in the venn diagram must be 10%.

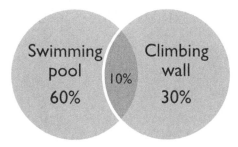

Question 30: C

The passage mentions that mothers use the self-help books to advise on feeding and sleeping routines, suggesting that they find difficulty in establishing them. There is no mention of **A** or **B. D** is incorrect as may be beneficial at low numbers for the mother and also be having a positive impact on the babies.

Question 31: F

The smallest amount occurs between days 4 to 5. On day 4, he can buy £6000 / 0.25 = 24,000 shares. On day 6, he will get 24,000 x 0.15 = £3,600.

The largest difference occurs if you buy on day 5 and sell on day 7. On day 7, he can buy £6000 / 0.15 = 40,000 shares. On day 6, he will get 40,000 x 0.3 = £12,000.

Difference = £12,000 - £3,600 = £8,400

Question 32: C

Using the graph, the lowest risk is 6 units per week.

Question 33: A

The conclusion in the first line is that not drinking increases the risk of dementia as much as drinking in middle age. This is based on a study of the drinking habits of middle-aged civil servants. The author therefore assumes that this group is proportional to all people and so relies on the assumption that occupation has no effect meaning **A** is the correct answer. **D** is incorrect as the conclusion is only for middle aged people so makes no statement on whether this is true for drinking at other ages.

Question 34: B

3 glasses of wine per day is 6 units which gives 42 units per week.
Proportion of adults who got dementia drinking 7 glasses of wine = 80/ 0.4 x 10,000 = 2%
1.5 = X/ 2% so X = 3%

Question 35: C

C provides an alternative explanation for why deaths from dementia are lower in drinkers and so undermines the conclusion the most. **A, B, D** all relate to changes in alcohol consumption in later life, which the study does not look into, and if consistent would have already been accounted for in death counts.

END OF SECTION

SECTION 2

Question 1: E
W is a gene and the larger structure it is found on is the chromosome.
X is used to remove the gene so is a restriction enzyme.
Z is used to insert the gene so is a ligase.
Y is a plasmid, recognised by the circular structure.

Question 2: C
This requires the use of the equation:

Acid + Calcium Carbonate → Calcium Salt + CO_2 + Water

A is therefore correct.
B is incorrect as the fizzing is due to carbon dioxide, shown above.
C is incorrect as the solution becomes neutral so increases.
D is correct as $CaCO_3$ has low water solubility.

Question 3: B
F = ma
F: force
m: mass
a: acceleration

600N – (300N + air resistance) = 50kg x 4 m s^{-2}
300N – air resistance = 200N
Air resistance = 100N

Question 4: B
Using the first equation:

p + q = 3(p – q)
p + q = 3p – 3q
2p = 4q
p = 2q

Then substitute this into the second equation: $\dfrac{pq}{p^2+q^2} = \dfrac{(2q)q}{(2q)^2+q^2} = \dfrac{2q^2}{4q^2+q^2} = \dfrac{2q^2}{5q^2} = \dfrac{2}{5}$

Question 5: A

Process 1 always requires ATP to occur so is active transport. From this we can rule out options C-G. Process 2 has to be diffusion. Process 3 is therefore osmosis as in secondary osmosis ATP is involved but normally does not require it.

Question 6: D

The rules for electroplating are:

Positive electrode: the coating metal, so in this scenario it is a silver rod
Negative Electrode: the metal to be electroplated, so in this scenario it is the copper rod
Electrolyte: solution of the coating metal (silver), so in this scenario it is a silver nitrate solution

Question 7: C

Volume of the object = 500 – 375 = 125 cm^3
mass = 200g

Density = mass/ volume
Density = 200g/ 125cm^3 = 1.6 g cm^{-3}

Question 8: E

$$\sqrt{\frac{6\times10^2+4\times10^1}{1.2\times10^{-2}+4\times10^{-3}}} = \sqrt{\frac{60+4}{0.012+0.004}} = \sqrt{\frac{64}{0.016}} = \sqrt{4000} = \sqrt{4\times100} = 2\times10 = 20$$

Question 9: D

Statement 1: false - it is proportional, the lower the temperature, the lower the CO_2 concentration observed.
Statement 2: this cannot be shown based on this experiment as it is likely that both are occurring.
Statement 3: this is true as limewater turns cloudy in the presence of CO_2.

Question 10: H

Statement 1 and 3: correct as Group 1 elements become more reactive down the group
Statement 2: correct all Group 1 elements have 1 electron in the outer shell

Question 11: C

For the object to have an overall negative charge there must be fewer protons than electrons.
The smaller sphere has an overall positive charge as it has lost electrons to electrons to the earth.

Question 12: H

Equation for a line is $y = mx + c$

m = gradient

c = intercept

$y = -3x + r$

First, we can use the gradient to find out p

Gradient = change y/ change x

$$-3 = \frac{(3p-1)-2}{6-(1-p)} = \frac{3p-3}{5+p}$$

$3p - 3 = -3p - 15$

$6p = -12$

$6p = -2$

Then we need to work out the coordinates of point N or M and substitute in to find r

Point N:

$(1 - -2, 2)$ so $(3,2)$

$2 = -3 (3) + r$

$2 = -9 + r$

$r = 11$

Question 13: H

Amino acids, cellulose and lipids all contain carbon so are taken up by organisms and incorporated into animals, plant cell walls and bacterial cell membranes respectively. These then return to the environment by decomposers, so all are involved in the carbon cycle.

Question 14: A

Y reacts faster so must be a powder (remember, powders have a larger surface area for collisions, hence they react much faster!)

X produces 0.1 mol dm^{-3} x 50 cm^3/1000 = 5×10^{-3} moles.

This would then produce 5×10^{-3} x 2 x 24 dm^3 = 0.24 dm^3 of O_2. This stage however is unnecessary as this step is the same for both equations.

From the graph we can see that there are 4×10^{-3} moles of Y produce.

We then can use the answers, along with the equation mol = concentration $\times$ volume to determine that the volume must be 20 cm^3 and the concentration 0.2 mol dm^{-3}

Question 15: F

F is true as ultrasonic, sound and P-waves are all longitudinal waves. The other options are either factually incorrect or confuse longitudinal and transverse waves.

Question 16: D

If we take x as the unknown side, we can write the following equation: $66 = x(7 - \sqrt{5})$

This can be rewritten as: $x = \dfrac{66}{(7 - \sqrt{5})}$

We can simplify this by multiplying by the inverse fraction:

$$\frac{66}{(7 - \sqrt{5})} = \frac{66(7 + \sqrt{5})}{(7 - \sqrt{5})(7 + \sqrt{5})} = \frac{66(7 + \sqrt{5})}{44} = \frac{3(7 + \sqrt{5})}{2}$$

To calculate the perimeter, we need to multiply the four sides together so

$$\text{Perimeter} = 2 \times \frac{3(7 + \sqrt{5})}{2} + 2(7 - \sqrt{5})$$
$$= 3(7 + \sqrt{5}) + 2(7 - \sqrt{5})$$
$$= 21 + 3\sqrt{5} + 14 - 2\sqrt{5}$$
$$= 35 + \sqrt{5}$$

Question 17: D

The pea plant is heterozygous for seed colour so must be Yy
The pea plant is homozygous for shape so could be either RR or rr.
The genotypes are therefore either Yy RR or Yy rr.

This means Statements 2, 4 and 7 are correct.

Question 18: E

For this question we need the equation:

Concentration = mol/volume

	Fe^{2+}	MnO_4^-
Volume (cm^3)	25	10
Concentration ($mol\ dm^{-3}$)	Unknown	0.05
Moles	25/1000 x volume = 0.025 x volume = 2.5×10^{-2} x volume	$0.05 \times 10/1000 = 0.0005$ = 5×10^{-4}
Mole ratio	5	1

$2.5 \times 10^{-2} \times v = 5 \times 10^{-4} \times 5$

$2.5 \times 10^{-2} \times v = 2.5 \times 10^{-3}$

volume $= 1 \times 10^{-1}$

Question 19: A

Statement 1 is false as microwaves have a shorter frequency than gamma rays.

Statement 2 is false as microwaves have a longer wavelength than gamma rays.

Statement 3 is false as the change is indirectly proportional.

Question 20: E

Probability that she catches the train if bus is on time: $0.6 \times 0.8 = 0.48$

Probability that she catches the train if bus is late: $0.4 \times 0.6 = 0.24$

Probability bus is on time given she caught the train = Probability that she catches the train if bus is on time/ probability she catches the train

Probability bus is on time given she caught the train = 0.48 / (0.24 + 0.48) = 0.48/0.72 = 2/3

Question 21: B

Statement 1 is false as 500/3 = 166.7 but one of these is a stop codon and another is a start codon, so the maximum is 164.

Statement 2 is true as all healthy human cells have 46 chromosomes.

Statement 3 is false as the adenine base pairs number equals the thymine base pair number, not the guanine.

Question 22: D

Concentration = moles / volume x 1000

$C_nH_{n+2} + xO_2 \rightarrow nCO_2 + (n+2)/2 . H_2O$

	C_nH_{n+2}	CO_2
Volume of gas (cm³)	35	105
Moles	35/ 24,.000	105/ 24,000
Mole ratio	1	n

We can ignore the 24dm³ for moles as it is a shared factor so:

35n =105

n = 3

C_3H_8

Question 23: A

Momentum = mass x velocity

Momentum = 400kg x 15 = 6000 kg m/s

Force = mass x velocity/ time

F = (400kg + 400kg) x 15m/s /12s

F = 800 x 1.25

F = 1000N

Question 24: H

The total angles in a pentagon = 540° and so each angle is 108°.

We are told that the bearing is 110° so we can fill this into the diagram as the internal angle is 108° the final angle must be 142°. Using Z angles, we can see that angle we are trying to work out is 142° − 108° = 34°. The question asks for the bearing, so the final answer is 360° − 34° = 326°.

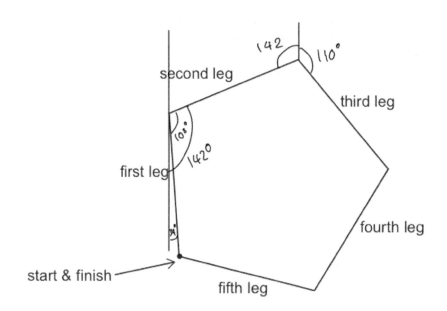

Question 25: B

Statement 1 is correct as damage to the Bowman's capsule will allow proteins in, as it affects ultrafiltration.
Statement 2 is incorrect as this affects reabsorption not filtration so proteins cannot enter the Bowman's capsule.
Statement 3 is incorrect as the effects on ADH receptors will reduce water reabsorption.

Question 26: E

There are three possible masses:
1. $(^{10}B)_2H_6$: 26 – the probability of this occurring is 0.2 x 0.2 = 0.04
2. $^{10}B^{11}BH_6$: 27 – The combination can be reversed so the probability of this occurring is 2x 0.2 x 0.8 = 0.32
3. $(^{11}B)_2H_6$: 28 – the probability of this occurring is 0.8 x 0.8 = 0.64

The ratio is 0.04 : 0.032 : 0.64
So, 1:8:16

Question 27: B

Np/ Ns = Vp/Vs

Np – number of turns on the primary transformer
Ns – number of turns on the secondary transformer
Vp – primary transformer voltage
Vs – secondary transformer voltage

400/100 = 240V/Vs

Vs = 240/4 = 60 V

VsIs = VpIp
60x2 = 240Ip

Ip = 0.5A

Ip - primary transformer current
Is – secondary transformer current

P = IV
P – power
I – current
V – voltage

P = 2 x 60V = 120 W

END OF SECTION

SECTION 3

People are often motivated to deny the existence of problems if they disagree with the solutions to those problems.

Explain what you think is meant by the statement. Present a counter-argument. To what extent do you agree with the statement?

Structure your answer such that you have one paragraph relating to each point in the question.

- The statement is saying that people who disagree with the result of something will pretend it doesn't exist which is particularly relevant in the case of politics. To demonstrate this, we can use the example of climate change. Many scientists have produced very strong proof of the fact that increases in carbon dioxide lead to increased global temperatures. As a result, committees like COP21 have recommended cutting activities that result in high CO2 levels. However, as some people disagree with this, they may deny that climate change is real. There are many other cases where this occurs such as evolution and the existence of HIV. Denial can also occur as one of the five stages of grieving, so may be seen in the medical setting.

- There are many examples where this is not the case and instead people may work to provide an alternative solution to the problem which they agree with more. This is why for some problems there are many theories which are used to explain the solution to the problem and people can choose which one to believe but all acknowledge the original problem.

In the final paragraph, introduce a more personal opinion.

- On one level, denial does not occur in all cases and its existence in others can sometimes undermine the ability to solve the problem. In these scenarios, it can be problematic, and useful to debunk the myths that people create to justify it.
  At the end of this essay, it is important to come to some sort of conclusion as to what extent you agree with the statement.

'In science, there are no universal truths, just views of the world that have yet to be shown to be false.' (Brian Cox and Jeff Forshaw)

Explain what you think is meant by the statement. Argue that scientists need to accept some things as 'truths' to advance their understanding. To what extent do you agree with the statement?

This statement makes reference to the ever-evolving nature of science where one theory constantly replaces the previous by using new evidence or technology. An example of this is the understanding of the creation of the universe which is currently believed to be caused by the Big Bang however some scientists do not agree with this, suggesting other theories.

- As scientists do not fully understand this, it may be replaced by a more concrete theory once more evidence is discovered or backed-up further but slightly modified with increased evidence.

- Whilst historically this statement may have been true, currently there is more technology to provide evidence for scientific theories. For example, our understanding of disease has evolved from theories of bad air to our current understanding of microorganisms. Currently scientists accept this as evidence of the truth, which is essential in order to learn more about microorganisms in order to treat and prevent diseases. The discovery of this and then antibiotics as a result has led to many lives being saved so this was an essential decision.

- Many scientists have also accepted the Big Bang Theory as truth in order to allow them to understand more about space and the next stages that occurred in the development of the universe after the Big Bang.

- I agree with Cox and Freshaw's statement to some extent in regards to some topics in science particularly our understanding of the universe. In other areas of science which are more tangible and therefore easier to experiment on, there is greater evidence for scientists' understanding of the truth and so these have not been contested or disproven for decades.

Teamwork is more important for surgical innovation than the skills of an individual surgeon.

Explain the reasoning behind this statement. Argue that the skills of individual surgeons are more important for surgical innovation or progress. To what extent do you agree with the statement?

- This statement suggests that it is the combined operating team who is responsible for surgical innovation and not the surgeon alone. An operation involves a team of many people including nurses, anaesthetists, radiographers and surgeons. It is therefore the collective improvement by all of them that leads to significant advancements in surgery. Evidence to support this is the introduction of the Surgical Checklist used at the start of an operation to check the patient and the operation being performed to prevent operating on the wrong body parts. This is a team led process. This has drastically reduced deaths from surgery. Other drastic reductions in deaths have been caused by the advancements in the understanding of anaesthetics to make them safer.

- It is the surgeon individually who operates on the patient so the advancement of the procedure itself could be argued to be mainly dependent on the surgeon. The first surgeries were performed by one surgeon experimenting and cutting open a patient to try to fix the problem. Currently many surgeons are involved in academic research into their chosen speciality and it is this that leads to improvements in the techniques a surgeon decides to perform which can drastically reduce the risk of complications. With the invention of the Da Vinci robot, it makes fewer staff necessary to help in the procedure as one doctor can sit behind the computer and complete the whole surgery.

In the final paragraph, introduce a more personal opinion. It may also be possible to bring in relevant work experience if done in surgery here.

- Overall, there are many factors that contribute to advances in surgery. Whilst the individual skills and techniques used by the surgeon are very important, I think based on my own work experience in surgery that it is the collective improvements by each department that will have the greatest impact. As surgery is integrating more technology, it could be argued that it is in fact this that will have the greatest impact in the future and so collaboration between computer scientists, engineers and healthcare professionals working to build a competent technological robot.

END OF PAPER

2020

Section I

Question 1: D

We can turn the statement into code, replacing 'practical driving test' with 'X' and 'passing the theory test' with Y: To have X, you need Y. You do not have Y, you cannot have X.

If we then do the same for statement D, the similarity is apparent:

If entering China is X and requiring a visa is Y, the statement becomes 'By doing X you need Y . If you do not have Y, you cannot do X.'

Question 2: B

$$30\% \ of \ £500 = £150$$
$$\frac{4700}{150} = 31.3$$

It will therefore take 32 quarters for the boiler to pay for itself.

Question 3: D

The passage states that 'the best evidence for the importance of community to our lives is summarised by our gravestones, which more commonly reference our relationships to others than financial achievements. In making this argument, there is an assumption that people's gravestones display what they consider to be the most important features of their lives.

Question 4: D

The main conclusion of this argument is that in order to have influence, innovators must campaign and embrace change. This is supported by the second sentence of the paragraph, which is then further expanded on in the third sentence.

Question 5: B

The argument makes the point that individuals cannot justify poor moral choices by arguing that those they are harming are also making poor moral choices. This is best illustrated by C, where Jorge argues that he can mistreat his ex-girlfriend because she mistreated him.

Question 6: E

Orange juice in Delicious Tropical:

$$\frac{1}{4} \times \frac{1}{2} = \frac{1}{8}$$

Orange juice in Vibrant Sunset:

$$\frac{3}{4} \times \frac{2}{3} = \frac{6}{12} = \frac{1}{2}$$

Orange juice total:

$$\frac{1}{8} + \frac{1}{2} = \frac{5}{8}$$

Question 7: C

8 small cards:

$$3 \times 8 = \$24$$

20 medium cards with their own image:

$$5 \times 20 = \$100$$

5 personalised greetings:

$$5 \times 2 = \$10$$

Postage charge:

$$3 \times 1 = \$3$$

Total:

$$24 + 100 + 10 + 3 = \$137$$

Question 8: E

Looking at the graph of change in the balance, we can disregard month 1 as this refers to the difference between the balance at the end of month 1 and the month preceding it. Months 2 and 3 show a fall in balance, and then month 4 shows that the balance stays the same. This pattern is best represented by E.

Question 9: A

Putting the argument into code, where high unemployment is X and weak economy is Y, we can code this argument as 'X results in Y. Malaysia has the opposite of X, so it must have the opposite of Y.' This is most closely paralleled by A:

If tanned skin is X and being out in the sun is Y, it can be coded as 'X results in Y. If you have the opposite of X, you must have the opposite of Y'.

Question 10: C

The argument states that 'Since almost all of the mongooses were removed from the island five years ago, it is unlikely that the frogs observed have ever seen one, and so this [hopping away faster] cannot have been learned from contact with mongooses'. It then goes on to conclude that 'The greater wariness in the frogs must have been caused by a genetic change.' However, the argument fails to address the point that genetic change is not the only alternative explanation to the frogs' greater wariness – they may instead have learnt this from another source, such as their experience with other predators.

Question 11: B

The paragraph states that 'Messages and other content posted on Facebook... tend to be edited somewhat to present a more attractive or successful image.' This suggests that if researchers are interested in how people wish to be perceived by others, they should look at their Facebook profiles. According to the statement, users tend to present an improved version of themselves, which presumably would be what they wish to present to others.

Question 12: B

100% blue costumes:

$$100\% \text{ of } 2m = 2m$$
$$2 \times 4 = 8m$$

50% blue costumes:

$$50\% \text{ of } 2m = 1m$$
$$1 \times 4 = 4m$$

25% blue costumes:

$$25\% \text{ of } 2m = 0.5m$$
$$0.5 \times 2 = 1m$$

20% blue costumes:

$$20\% \text{ of } 2m = 0.4m$$
$$0.4 \times 10 = 4m$$

Total length of blue material used:

$$8 + 4 + 1 + 4 = 17m$$
$$30 - 17 = 13m$$

Question 13: D

As direct flights are needed, A can be excluded. E can also be excluded as these tours are only partly escorted.

Beyond:

$$1200 + 10 = €1210 \; per \; person$$
$$1210 \times 3 = €3630$$

Cosmic:

$$1150 + 30 = €1180 \; per \; person$$
$$1180 \times 3 = €3540$$

Discreet:

$$1100 + 40 = €1140 \; per \; person$$
$$1140 \times 3 = €3420$$

Discreet is, therefore, the most affordable of the options fitting Tom's requirements.

Question 14: C

Letter	Morse code	Display morse code	Correctly or incorrectly labelled?
T	Dash	Dash	Correct
M	Dash dash	Dash dash	Correct
Q	Dash dash dot dash	Dash dash dash	Incorrect
G	Dash dash dot	Dash dash dot	Correct
W	Dot dash dash	Dash dot	Incorrect
K	Dash dot dash	Dash dot dash	Correct
D	Dash dot dot	Dash dot dot	Correct
E	Dot	Dot	Correct
A	Dot dash	Dot dash	Correct
N	Dash dot	Dot dash dash	Incorrect
R	Dot dash dot	Dot dash dot	Correct
I	Dot dot	Dot dot	Correct
U	Dot dot dash	Dot dot dash	Correct
S	Dot dot dot	Dot dot dot	Correct

Three letters have therefore been incorrectly labelled.

Question 15: A

The passage states that giving to 'large, well-known' charities is a mistake. The statement goes on to justify this by saying that for small charities, 'because administrative costs are low, nearly all of the money donated is used for the object of the charity.' Therefore, there is an assumption underlying the argument that people giving to charity wish to give as much money as possible to the object of charity, which is why they are making a mistake by giving to larger charities.

Question 16: C

The passage argues that the publishing industry is hypocritical as it preaches a message of environmental protection, but profits from their advertising space being bought by the airline industry. This argument would be <u>most</u> weakened by the finding that overall, newspapers and magazines are doing good as their message is stronger than that of their advertisers.

Question 17: A

The passage states that antiretrovirals, when used in HIV-positive individuals, resulted in a 62% reduction in the likelihood of developing MS. The conclusion that can then be drawn is that these drugs may prevent the onset of MS.

Question 18: D

A: Correct - There is no possible way to buy all three types of fruit and the sum to add to £1.50

B: Correct - This is possible, for example, 6 bananas and a nectarine:

$$(6 \times 20) + 30 = 150$$

C: Correct - 3 mangoes will cost £1.35, leaving 15p in change which cannot buy any other fruits, meaning it is not possible to buy three mangoes and the total cost to add to £1.50

D: Incorrect - The price of two mangoes and two nectarines will add to £1.50:

$$(2 \times 45) + (2 \times 30) = 150$$

E: Correct - Six bananas and a nectarine make 7 items, which is the maximum number of items that can be bought for £1.50.

Question 19: D
6 AAA batteries:
James could buy 3 x pack size 2:

$$3 \times 3.5 = \$11.50$$

Or he could buy 1 x pack size 4 and 1 x pack size 2:

$$6.5 + 3.5 = \$10$$

Or he could buy 1 x pack size 8:

$$1 \times 10.50 = \$10.50$$

The option of buying 1 x pack size 4 and 1 x pack size 2 is, therefore, the cheapest.

3 D batteries:
James could buy 3 x pack size 1:

$$3 \times 3.5 = \$11.50$$

Or he could buy 1 x pack size 2 and 1 x pack size 1:

$$3.5 + 5.5 = \$9$$

Or he could buy 1 x pack size 4:

$$1 \times 8.5 = \$8.50$$

The option of buying 1 x pack size 4 is, therefore, the cheapest.

Overall, the least James could spend is therefore: $8.5 + 10 = \$18.50$

Question 20: B
For each option, the arrow denotes which sphere would be pointing upwards to achieve the view.

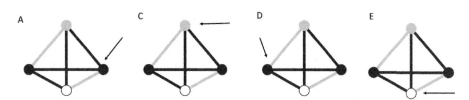

B is not a possible view as one of the black spheres will be pointing upwards with three black rods connected to it. The diagram in the question stem shows no black sphere with three black rods connected to it. Also, as part of the structure, the grey rods remain non-adjacent to each other, whereas, for B, these grey rods converge on the same black sphere.

Question 21: D

The passage states, 'Although many countries, including the US, allow the practice of positive discrimination, it remains illegal in the UK under the Equality Act 2010, and so it should.' The passage then goes on to justify this position. The main conclusion is, therefore, that positive discrimination should remain illegal in the UK.

Question 22: A

The passage states, 'Actively seeking the evidence that goes against your own beliefs allows you to make a proper assessment, and consequently a far more reliable judgment.' Therefore, it can be concluded that the passage argues that considering the reasons why beliefs may be false can serve to strengthen or weaken beliefs.

Question 23: B

If reducing carbon footprint is X, longer journeys is Y, and more expensive journeys is Z, then the argument is: If we want X, we must have Z and Y. I can have Z but not Y, therefore, I cannot have X.

In B, if going to her first choice university is X, achieving top grades is Y and doing well at interview is Z, the statement can be coded as Kristina cannot have X, as to have X she requires Z and Y. She has Z but not Y.

Question 24: B

If 10 saddles were built but only 7 cycles, we can deduce that 3 tandem bikes were built and 4 other bikes were built. 9 handlebars were used, and of these 6 will be for the tandem bikes. Therefore 3 bicycles or tricycles were built, and one unicycle was built. 14 wheels were built, and 6 of these were used for tandem bikes, and 1 was used for the unicycle. This leaves 7 to be used for the 3 bicycles or tricycles. Bicycles require two wheels, and tricycles require 3, so there must have been two bicycles and one tricycle built.

Question 25: D

The easiest way to start is to realise that we want to use as many of the smaller denominations as possible. Also, working backwards is a good idea!

Using all 8 of the 1p = 8p.
Use 3 of the 2p = 6p. Total is now 14p.
49p = 14p is 35p, which we can make up with the 5p and 10p coins.
Use all 5 of the 5p = 25p. Total is now 39p.
Use 1 of he 10p = 10p. Total is now 49p.

We have used 8(1p) + 3(2p) + 5(5p) + 1(10p) → 8 + 3 + 5 + 1 = 17 coins, equalling 49p.

Question 26: A

$$\frac{2}{3} of\ 80\% = 53.33$$
$$\frac{6}{10} of\ 53.33 = 32$$

Question 27: B

The passage argues that even though THC-containing vaping oils pose a greater risk, as this level of risk is not well understood, people should be allowed to make their own decision about whether to use them. B argues similarly; that because we do not have a complete understanding of the risk of passenger trips into space, the government should not ban them.

Question 28: C

The folded component of the arrangement consists of sides 3cm in length and a top that is 3cm in length. This means that when flat, the length of this portion of the paper will be 9cm in length. Therefore if x is taken to be the length of one rectangle:

$$75 = 2x - 5 - 6$$
$$86 = 2x$$
$$43 = x$$

Question 29: C

The passage states that 'At the time it [the finding that supercooled silk does not get brittle] was just an interesting curiosity, but plainly it has enormous practical significance'. Therefore, it can be concluded that the discovery is highly significant, to which the passage goes on to justify why.

Question 30: C

From the examples provided, we can see that the 'Y' in a paint's YMC is presented by the final two digits of the OCT, the 'M' is represented by the middle two digits, and the first two digits represent 'C'. 'Regal purple' has a C of 41, and the first two digits of its OCT is 50. 'Grizzly Grey' has an M of 34, and the middle two digits of its OCT is 39. None of the example colours has a Y of 50, but we can deduce that the lower the Y number is, the lower the final two digits a colour's OCT is. Therefore, the answer is C.

Question 31: C

The passage states that obesity has increased in the UK in recent years, then argues that obese people should not be blamed because, in Europe, there is more readily available unhealthy food. If it were true that levels of obesity in the UK are higher than in other European countries, this would weaken the argument that it is the European food environment that is the major contributor to rising obesity levels. This is because a similar trend would be expected in the rest of the continent.

Question 32: D

If spatial reasoning isn't your forte, try ruling out options!

For example, B can be ruled out quickly – this cannot be made without using 2 identical tiles. The same goes for E.

The four tiles may be arranged as shown below in order to achieve the pattern in D.

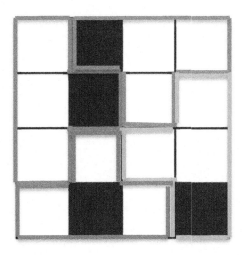

SECTION 2

Question 1: F
Statement 1: Asexual reproduction does not produce genetic variation.
Statement 2: Gamete production is one source of genetic variation, but there are other sources, such as crossing over.
Statement 3: This is true.
Statement 4: This is true.

Question 2: E
Isotope 1 has a mass of 69, and isotope 2 has a mass of 71.

When calculating the relative abundance of each isotope:

$$0.6 \times 69 = 41.4$$
$$0.4 \times 71 = 28.4$$
$$41.4 + 28.4 = 69.8$$

Question 3: D
In order for the products to thaw, hot air must travel down into the freezer. As hot air is less dense than cold air, the hot air will rise, meaning it will not travel down to thaw the product.

Question 4: A

$$2 - \frac{2x + 1}{4x^2 + 4x + 1} = 2 - \frac{2x + 1}{(2x + 1)(2x + 1)}$$

$$= 2 - \frac{1}{2x + 1}$$

$$= \frac{2(2x + 1)}{2x + 1} - \frac{1}{2x + 1}$$

$$= \frac{4x + 2}{2x + 1} - \frac{1}{2x + 1}$$

$$= \frac{4x + 1}{2x + 1}$$

Question 5: G
Statement 1: Both early embryo and bone marrow stem cells from an adult are diploid, producing diploid daughter cells.
Statement 2: This is true as adult bone marrow cells can only differentiate into blood cells.
Statement 3: This is true.

Question 6: E
Statement 1: **Correct.** Oxygen has an oxidation state of -2, and sodium has an oxidation state of +1. Therefore, to balance the oxidation states, chlorine must have an oxidation state of +5.
Statement 2: **Correct.** A disproportionation reaction is when an element gets oxidised and reduced, and here chlorine is oxidised and reduced.
Statement 3: **Incorrect.** Oxygen's oxidation state is -2 throughout the reaction.

Question 7: G
Statement 1: The current in the heating element will be 0.20A.
Statement 2: The voltage will stay the same, so it will still be 240V.
Statement 3: The resistance of a closed switch is infinity as no current will flow through the open switch.

Question 8: D
Take x to be the normal price of a camera. The equation to calculate the sale price is, therefore:
$$0.8x = 180$$

$$so \ x = \frac{180}{0.8}$$

Question 9: B
The homozygous rabbit in the first generation will have two alleles for the condition.
All the offspring in the second generation will be heterozygous for the condition, so these rabbits will have three alleles in total.
The offspring in the third generation will be heterozygous if they have the condition, so they contribute two alleles in total.
These sum to 7 alleles.

Question 10: A
The energy profile shows that if two moles of ammonia are decomposed into its constituent elements, 92kJ is absorbed. This means that if one mole is decomposed, 46kJ will be absorbed.

Question 11: D

Gold-silver volume:

$$\frac{24}{16} = 1.5 cm^3$$

Silver-nickel volume:

$$256 - 24 = 232 g$$

$$\frac{232}{10} = 23.2 cm^3$$

Total volume:

$$23.2 + 1.5 = 24.7 cm^3$$

Question 12: D

Length QO is the radius of the circle. If 'M' is taken as the midpoint of QR, we can therefore consider the right-angled triangle OQM:

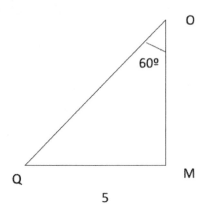

$$sin60 = \frac{5}{OQ}$$
$$OQ = \frac{5}{sin60}$$

Question 13: D

- Adds carbon dioxide: aerobic respiration in plants uses CO_2.
- Removes carbon dioxide: photosynthesis fixes CO_2 in plants.
- Adds oxygen: photosynthesis releases oxygen from plants.
- Removes oxygen: decomposers use oxygen for respiration.

Question 14: F

Since T is precipitated when aqueous T reacts with R, R must be more reactive as it has displaced T. If X occurs naturally in its elemental form, this suggests that it is very unreactive. R fizzes when it reacts with HCl, suggesting that it is somewhat reactive, but as it does not have an explosive reaction, it is not highly reactive. Z cannot be extracted by heating with carbon meaning it is more reactive than carbon. This suggests that the reactivity hierarchy is Z, R, T, X.

Question 15: C

Alpha particles consist of two protons and two neutrons. Beta particles consist of a single electron. Beta-decay does not affect the mass number but changes the atomic number by +1, whilst alpha decay changes the mass number by -4 and the atomic number by -2.

Four alpha decays mean that 8 protons and 8 neutrons will be lost. Two beta decays mean two electrons will be lost, causing the nucleus to gain 2 protons but lose 2 electrons. The resulting nuclide is, therefore, C, as the proton number will decrease by 6 and the mass number by 16.

Question 16: E

These can be plotted graphically:

We can see that lines 2 and 4 do not intersect.

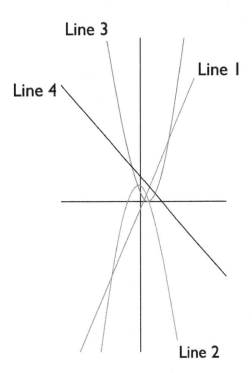

Question 17: F

Amino acids are the building blocks of proteins. Proteins can be split into amino acids by using a protease. As tubes 5 and 6 contain protease and protein, these will therefore contain amino acids. As enzymes are made of protein, adding protease to another enzyme such as lipase will also result in the production of amino acids.

Question 18: D

When an alkene is made into a polymer, the C=C double bond is broken, and reforms as a single bond, meaning the two carbons involved can bond with other carbons to make a polymer. This means that D will be the repeating unit in this polymer, as there will be a methyl group as part of the repeating unit.

Question 19: E

Energy = mass x gravity x change in height

$$45 \times 10 \times 10 = 4500J$$

Kinetic energy = ½ x mass x velocity²

$$4500 = \frac{1}{2} \times 45 \times v^2$$

$$4500 = \frac{1}{2} \times 45 \times v^2$$

$$200 = v^2$$

$$10\sqrt{2} = v^2$$

Question 20: F

A kite is a four-sided shape formed of two pairs of sides of equal length. Therefore $QP = QR$ and $PS = RS$. Point S lies at the intersection of the lines PS and QS. The line QS crosses the line PR at right-angles. The equation of the line PR is $y = x + 1$, so line QS is $y = -x + 5$. We can therefore solve these two equations to find the coordinates of point S, which is $(8, -3)$, meaning that $2l + m = 13$.

Question 21: A

The producer in this food chain is the rose bush, and the secondary consumers are ladybirds. The rose bush has $800AU$, and each of the ladybirds has $10/50 = 0.2AU$ in biomass.

$$\frac{0.2}{800} \times 100 = 0.025\%$$

Question 22: C

In order for these compounds to melt, the intermolecular forces between molecules must be overcome. If the intermolecular forces are strong, more energy will be required to overcome them, hence why BrI has a higher melting point.

Question 23: B

Speed = distance/time

$$\frac{1500}{1} = 1500\text{m/s}$$

Wavelength = speed/frequency

$$\frac{1500}{6000} = 0.25\text{m} = 25.0\text{cm}$$

Question 24: D

The radius of the new circle can be calculated by considering the triangle OXM, where M is the centre of circle XY. Length OX is the radius of the new circle, and using Pythagoras, we can calculate that this is $\frac{\sqrt{2}d}{2}$:

$$OX = \frac{d^2}{4} + \frac{d^2}{4} = \frac{\sqrt{2}d}{2}$$

The angle XOY is 90°. This means the length of the arc XPY is:

$$\frac{1}{4} \times 2\pi \times \frac{\sqrt{2}d}{2} = \frac{\sqrt{2}\pi d}{4}$$

Therefore, the perimeter of the grey shape is:

$$\frac{\sqrt{2}\pi d}{4} + d = \frac{\sqrt{2}\pi d}{4} + \frac{4d}{4} = \frac{\sqrt{2}\pi d + 4d}{4} = \frac{d(\sqrt{2}\pi + 4)}{4}$$

Question 25: E

Statement 1: **Correct.** Using the rule that A pairs with T and C pairs with G, we can work out that strand 2 will have 30% T, 28% A, 25% C and therefore 17% G.

Statement 2: **Correct.** The diagram shows that p is from strand 1 and r is part of the new strand. Therefore, if p is originally from strand 1, it should have paired with cytosine in the parent DNA molecule, meaning that r should be guanine if no mutation occurs.

Statement 3: **Incorrect.** DNA replication occurs during the S-phase of the cell cycle before the cell enters mitosis.

Question 26: B

As $moles = \frac{mass}{Mr}$, 3.40kg of ammonia is equal to 200 moles. 80% of 200 is 160 moles. Ammonia reacts to form diammonium hydrogen in the ratio 2:1, so 80 moles of diammonium hydrogen will be formed. 80 moles of diammonium hydrogen is equal to approximately 10.6kg.

Question 27: C

Statement 1: Adding more turns will increase the value of the voltage but not the frequency.

Statement 2: As the magnet is rotating at 120 revolutions per minute (i.e. one full revolution lasts 0.5 seconds), halfway through the revolution, and at the start of each new revolution, the direction of the induced voltage will change every 0.25 seconds.

Statement 3: The conductor must be part of a complete circuit for this to be true.

SECTION 3

'Power tends to corrupt, and absolute power corrupts absolutely.'

John Dalberg-Acton

When planning this response, separate the individual sub-questions/points to the question:

- Explain the reasoning behind this statement
- Argue that power does not necessarily degrade or weaken the morals of those who hold it.
- To what extent is it possible for someone to hold power without using it for their own personal gain?

To score a mark of 3 and higher, the candidate needs to cover all of the points above. To ensure that all points are covered above, it is recommended that the candidate 'mind dumps' points surrounding these points relating to the main quote. Once completed, these points should then be ordered in a manner that resembles a plan. The written section of the exam should only be 4 paragraphs maximum that fits within the exam sheet – approximately 3 quarters of a side of A4 (550 words on a word processor). As the exam does not permit handing out a new answer sheet, the candidate should plan what they write to ensure they cover the points clearly and concisely without the need to restart. This cannot be stressed enough.

Due to the specific sub-points, it makes sense to have a paragraph dedicated to each. It also makes sense to include a concluding statement at the end about morals and gain to round off the piece. A potential plan for this response is:

Paragraph 1 – explain the reasoning behind this statement.

- This quotation is rather specific and conveys the opinion that, as a person's individual power increases, their moral senses diminish. It highlights that those who also have 'absolute power' are destined to be corrupt in their absolute intentions.
- Alternatively, the statement also suggests that individuals who save themselves from power are humble people.
- Yet, this statement is not just for the individual, but is also applicable to institutions and governments as well.
  - An example of this includes the actions of financial institutions in 2007, where banks (who have such power in the market) acted with the sheer intentions of financial gain and facilitated one of the largest economic crashes known.

Paragraph 2 – argue that power does not necessarily degrade or weaken the morals of those who hold it.

- However, power is not definable solely in terms of moral corruption. As seen in clinicians, power could be through knowledge that may, in fact, help those in which Dalberg-Acton may define as humble. Hence, it could be argued that these members of power utilise it to better others – in this case, empower patients through treatment.
- Also, power may come abruptly to an individual or institution that requires immediate action to assist others. This is observed when a natural disaster occurs. Both people affected, governments and institutional charities utilise the power that they have to help others and rebuild. This demonstrates that the morals to support our peers remain intact along with their power.

Paragraph 3 – to what extent is it possible for someone to hold power without using it for their own personal gain?

- An individual's perception of their good doings may be perceived as 'corrupt' in another.
    - When we look at the language used to describe politicians (recently in the distribution of private contracts, for instance, PPE, to companies that have allies to Matt Hancock – Health secretary during COVID); although it may be seen as good that the PPE will be used by the NHS, others may perceive a corruption in power by allocating contracts to friends, especially when other companies that may do a better job were never given a fair chance for the contract.
- We could also question the term gain in regards to altruism. Although gain is often in reference to finance and profit, it may also be defined in terms of emotional gain. Could those who have power, such as clinicians, who are deemed to utilise their power to help for good, gain in terms of feeling better about themselves. If this were to be the case, then we may all have a predisposition for personal gain.
    - *If the above point were to be used and expanded on, there is the potential for it to tangent away from the thread of the essay. This is an example of making sure this point answers the question of the paragraph!*

Science and art once collaborated as equals to further human knowledge about the world. Today, science is far too advanced and specialised to work together with the arts for this purpose.

When planning this response, separate the individual sub-questions/points to the question:

- Explain what you think is meant by the statement.
- Argue that science and the arts can still work together to further understanding of the world.
- To what extent do you agree with the statement?

To score a mark of 3 and higher, the candidate needs to cover all of the points above. To ensure that all points are covered above, it is recommended that the candidate 'mind dumps' points surrounding these points relating to the main statement. Once completed, these points should then be ordered in a manner that resembles a plan. The written section of the exam should only be 4 paragraphs maximum that fits within the exam sheet – approximately 3 quarters of a side of A4 (550 words on a word processor). As the exam does not permit handing out a new answer sheet, the candidate should plan what they write to ensure they cover the points clearly and concisely without the need to restart. This cannot be stressed enough.

Once again, due to the points that are being made, It makes sense to make each point a paragraph. This should then be followed by a summary sentence on your opinion of the statement. Although this statement mentions both science and the arts, it must be reiterated that you should keep the thread of the two together throughout the piece. By differentiating the role of the arts separately from science, you run the risk of writing at a tangent, away from the point in the statement. You are also allowed to disagree, agree or partially agree with the statement above. Ultimately, as long as you can articulate your points in a manner that is demonstrable towards the question, the response will be valid. A potential plan for this response is:

Paragraph 1 - explain what you think is meant by the statement.
- The statement here argues that although art and science are human attempts to describe the world around us, science and art are now unable to operate in the same capacity due to the rate of progression in science overtaking and leaving behind that of the progression of art.
- It is innate for humans to develop our understanding of the world, and hence to the extent to which this can be explored in different ways is subject to debate.

Paragraph 2 – argue that science and the arts can still work together to further understanding of the world.
- Social sciences (especially psychology) still have strong influence and hypothesis generation from the world of art.
  - For example, we still use art such in vision studies, from Mondrian plots to inform our understanding of colour perception to optical illusions allowing us to comprehend whether our perception of scenes is determined innately or through nurture. This means that we utilise art and progressive art creativity to help provide evidence for scientific theorem.
- The arts are critical for communication.
  - Research papers require the development of digital models, formatting and figures, which could be argued as art. When communicating this to individuals with less knowledge of the subject area, animations and visual presentations are essential. It is hence the arts that are essential for furthering our peer's knowledge.

Paragraph 3 – to what extent do you agree with the statement?

- The subjects and methods of the two may have different audiences and methodological considerations, however, the motivations and goals are fundamentally the same. They allow us to explore the world around us and then share our understandings in a myriad of ways.
- The direction for the relationship between art and science appears to be more communication focused. As areas of science become more and more complex, the communication of it needs to be comprehensible through both written and visual communication. This is where art may be of most use and hence furthers human knowledge in the natural sciences in particular.
- Although there are advancements in digital technology to communicate science to the masses, this is at a slower rate than that of the progression of science and scientific explanation in many fields. Hence, one would argue that the furthering of other humans' knowledge is that of art via communicating what science has done.

There are now many different kinds of internet sites and apps offering medical advice, but they all share one thing in common: they do more harm than good.

When planning this response, separate the individual sub-questions/points to the question:

- Why might online sources of medical advice be said to 'do more harm than good'?
- Present a counter-argument.
- To what extent do you agree with the statement?

To score a mark of 3 and higher, the candidate needs to cover all of the points above. To ensure that all points are covered above, it is recommended that the candidate 'mind dumps' points surrounding these points relating to the main statement. Once completed, these points should then be ordered in a manner that resembles a plan. The written section of the exam should only be 4 paragraphs maximum that fits within the exam sheet – approximately 3 quarters of a side of A4 (550 words on a word processor). As the exam does not permit handing out a new answer sheet, the candidate should plan what they write to ensure they cover the points clearly and concisely without the need to restart. This cannot be stressed enough.

Unlike the previous essay questions, it may not be the best approach to allocate a paragraph to each of the sub-points. Instead, using them as a guide to build your response may be better. It may be recommended to make two points for and two points against, as this will give definition to the essay. A potential plan for this essay is:

Paragraph 1 – what are the sort of services being offered? What does the statement mean?

- From online prescription services and health apps to a simple search of a rash on the internet, technology has enabled us to be more independent with our health than ever before. But may this ever-developing area of the health and virtual sector in fact lead to 'more harm than good'.

Paragraph 2 – points why they 'may do more harm than good'.

- These sites are an inaccurate manner for self-diagnosis.
    - Patients may not be able to communicate or recognise the severity of their symptoms and not be able to discriminate between different descriptions of symptoms that a clinician may be then able to reword. This may lead to the wrong suggestive diagnosis that could be more severe than what it truly is, or much worse; a downplay of the severity of an individual's illness.
    - Both may result in increased anxiety, stress and even fear to see a clinician. Especially for certain mental health and even physiological conditions, anxiety can exacerbate symptoms and lead to the progression of an illness. Alternatively, it could be a hypochondriac's worst nightmare, which may lead to them seeking medical help when they simply do not need it.

Paragraph 3 – counter-arguments.

- If medical apps lead to individuals checking symptoms that they otherwise may not give a second look at, then they may be more likely to see a clinician.
    - For example, an odd symptom of a particular lung cancer involves sweating on one side of the face. This is a symptom that many may in fact breeze over, but an oncologist would take a medical intervention if presented to them.
- If these medical apps' accuracy is focused on seeing whether a person should seek clinical advice and where to get it, this may be advantageous.
    - If these sites are to filter individuals who do not require a GP or A and E appointments, it alleviates the demands on the NHS and gets the patient to where they need to be (for instance remaining at home or going to see a pharmacist).

Paragraph 4 – concluding statement.

- While it is difficult to accurately diagnose an individual through online platforms, if these websites and apps can filter individuals to see the relevant service, I believe it would benefit the healthcare service, patient, and clinicians time. This is the conservative state at which anything more may lead to 'more harm than good'.

2021

SECTION 1

Question 1: C

'Can be drawn as conclusion' – let's look for any indicating words or anything that the argument hinges upon.

"However, people should not put themselves in danger unnecessarily".

"In addition, it is wrong to benefit from the exploitation of workers"

These two sentences are the main points of the paragraph, as all other points act as evidence for them.

Therefore the answer would be C, as these points provide both personal and moral reasons.

The answer is not:

A – luxury cruises are liked by retired people but there is no indication of cost.

B – There is no mention of governments being responsible for fixing any piracy risk.

D – There is no mention of the business environment of cruise ships

E – The main points of this passage are not related to how employees should try to amend the working conditions.

Question 2: C

There must be at least 4 children – as child X must have a brother and two sisters.

However, if child X is a girl, then the 'brother' does not have a brother themselves. In this case we need one more boy.

If child X is a boy, each of the sisters would have two brothers and a sister, and we would need one more girl.

Therefore, the minimum number of children in the Ahmed family is 5 and the answer is C.

Question 3: C

'Most closely parallels the reasoning' – re-write the argument in shorthand.

If you do A, people will respond with B.

You cannot avoid the response B.

If you try to avoid B by increasing A, you will get response C, leading to B. If you try to avoid B by decreasing A, you will get response D, leading to B.

The answer to this question is C.

If teachers are strict, they alienate students, causing a breakdown of discipline.

If teachers are not strict at all, they will be seen as soft, causing a breakdown of discipline.

Therefore teachers cannot win.

This follows the same reasoning as the shorthand above.

Question 4: A

'Best illustrates the principle' – Find the foundation of the chain of reasoning

The passage is indicating that end-to-end encryption causes worry for some, but the majority of people and many different services are using it without worry. Therefore, there is little to be concerned about. A mirrors this reasoning, with many people skating, and therefore the assumption that it is safe. The other answers do not have the same underpinning to their arguments.

Question 5: D

The conclusion is the overarching argument and doesn't have to come at the end of the passage.

The main points in this passage are all indicating the drawbacks of the Royal Society investing in fossil fuels and why it is hypocritical to have so much invested. The concluding sentence is 'The Society should abandon its investments in fossil fuels'.

Question 6: B

Area of flower bed including path = $12 \times 8 = 96m^2$
Area of flower bed minus path = $11 \times 7 = 77m^2$
Area of path = $96 - 77 = 19m^2$
Each slab = $0.5 \times 0.5 = 0.25m^2$
Number of slabs needed = $19 / 0.25 = 76$ slabs $\rightarrow$ B

Question 7: D

Looking at salt:
Not chocolate sundae (0.85g)
Not Toffee and date pudding (0.64g)
Looking at fat:
Not Bakewell tart (56g)
Not chocolate sundae (66g)
Looking at protein:
Not crème brulee (4g)

Now compare carbohydrates in apple pie, chocolate brownie and rocky road sundae,

Rocky road sundae $\rightarrow$ 81g carbohydrates

Question 8: B

Price per day:

21st - $10

22nd- $10

23rd - $8

24th - $8

25th - $6

26th - $3

27th - $3

Money per day:

21st - $10 × 120 = $1200

22nd- $10 × 140 = $1400

23rd - $8 × 150 = $1200

24th - $8 × 80 = $640

25th - $6 × 50 = $300

26th - $3 × 70 = $210

27th - $3 × 90 = $270

We are looking for a graph that has the same revenue on days 1 and 3, which then continues to decline afterwards until the final day where it increases again. This best lines up with graph B.

Question 9: A

Assumption = unwritten link between reasons and conclusion.

If you are unsure, use the negation test. If the conclusion no longer follows, this is your conclusion.

Conclusion = If you want to reduce your chances of developing myopia, you should not go to university.

Reasons = Myopia is more common in those that went to university; those that do an unusually large amount of reading develop myopia.

Assumption = Those that go to university read a lot.

Negate = Those that go to university do not read much → conclusion no longer follows so our assumption is A.

Question 10: B

This passage is focussed on ways to keep the body functioning effectively. It mentions cardiovascular exercise for heart health; NEAT for effective calorie burning; strength training to keep muscles strong and reduce age related illnesses.

Overall, we can see that different types of exercise can have different benefits for the body. The conclusion is likely B.

Question 11: E

The passage argues that, although the hour-by-hour gig work is great for some, who are against government action, some people who are more vulnerable may be poorly affected and therefore we should help them with regulations.

i.e. Take action to protect those most at risk of being placed in a difficult situation.

The only answer that follows the same reasoning as the above is E.

Question 12: C

Jane = $(0.6 \times 240) - 20$

Tom = $(0.4 \times 240) - 20$

Jane needs to give Tom: $(0.4 \times 240) + 20 = 96 + 20 = £116$

Question 13: C

Total number of days medication taken for = $2 + 14 + 7 = 23$ days total

2 adults will need 1 100mg tablet each per day for 23 days
= 2×23 of the 100mg tablets = 46 tablets
28 tablets in each 100mg box, so 2 boxes needed

Child 1 weighs 25kg, so they need 1 25mg tablet daily for 23 days = 23 25mg tablets
Child 2 weighs 32kg, so they need 3 25mg tablets daily for 23 days = 69 25mg tablets
Between them, the children need 92 25mg tablets
30 tablets in each 25mg box, so 4 boxes needed.

The answer is therefore C.

Question 14: A

Work out the scale of the graphs and try to allocate which pupil's marks are expected to be.
The scale can be found using the highest and lowest marks for economics and statistics.
As you plot each students' marks, you find that Lin's statistics mark is mis-plotted. Therefore the answer is A. To speed things along, only look at the students whose names are in the answers.

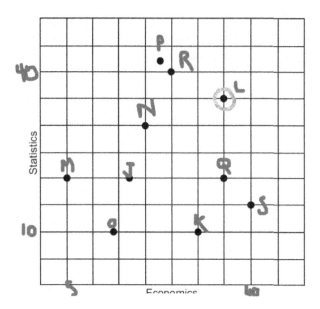

Question 15: E

If you want to control budget, consider land and buildings – can any be sold?

Wildflower has already sold some, so they wanted to keep their finances in good order.

If you want to do A, do B.
If B has already been done, then A must be the goal.

If you want to study English, consider learning to read quickly.
Vanessa has already done this, so she likely is wanting to study English.
The answer is E.

Question 16: B

We're looking for something that negates the assumption or makes the conclusion invalid. Conclusion = It is unreasonable to blame sports stars for their behaviour off the pitch when they are just enjoying rewards.

Reasons = They are selected based on their ability to help their team win matches and are rewarded accordingly.

Assumptions = Playing well leads to rewards that do not affect their ability to help their team win.

If B is true, the assumption is not correct and we **are** able to blame sports stars for their behaviour off the pitch. Therefore B is the answer.

Question 17: B

Conclusion = Faith schools restrict choice for parents.

Reasons = 21% of parents that opted for a non-faith secondary school for their children were allocated a faith school. 14% for primary school children.

Unwritten link/ assumption = Parents that apply for non-faith schools as a first choice for their children absolutely do not want their children going to a faith school.

Therefore B is correct

Question 18: D

2 jars of white flour = 2l

2 jars of wholemeal flour = 2O

1 jar of rye flour = R

2l > 2R

We could take a mathematical approach, but in this case it is likely quicker to eyeball the numbers and do some quick trial-and-error.

Double the weight of the rye is the same as the total of the white flour weights.

Let's say the rye is in the 900g jar → double this is 1800g

1050g + 750g = 1800g

Therefore the rye jar must be the 900g one. No other combinations work.

Question 19: D

Only look at the rows and columns relevant to the question i.e. only 'France' and 'UK' and only '2000' columns.

France:

14 deaths per 100,000 in 2000

Population of 60 million

60,000,000 / 100,000 = 600

600 x 14 = 8,400 deaths

UK:

8 deaths per 100,000 in 2000

Population of 58 million

58,000,000/100,000 = 580

580 x 8 = 4,640 deaths

Difference = 8,400 – 4,640 = 3760 → D

Question 20: C

This question will require you to visualise shapes in your head.

Work through the answer combinations to minimise the work and draw places where the shape would connect to help you visualise.

With this method, you will spot that 3 and 6 do not fit together, as the smaller semi-circular piece of 6 has flat sides, whereas in 3 we need a flat/curved surface. The answer is therefore C.

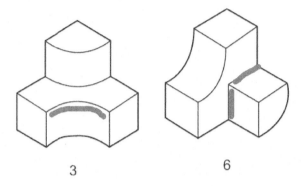

3   6

Question 21: A

A flaw question looks for a logic/reasoning issue in the passage.

Reading this passage, we can see that it wants to suggest that one cigarette can be responsible for a habit or addiction. However, all we know is:

People have tried cigarettes

They are now daily smokers

There is nothing to suggest that these two share a causative link or that the first cigarette tried is the one responsible for addiction in these people. It jumps to conclusions a bit.

Therefore the answer is A.

B is a similar answer, but it is incorrect as we are told only 2/3 people become daily smokers.

Question 22: E

The overarching idea of the argument in the passage is that publicly funded services are overstretched. If we have the choice to pay, and can afford it, we should do so to reduce stress on public services for those that cannot pay.

The only answer that follows the same idea of 'if you can pay for something that <u>could</u> be free, you should pay' is E.

Question 23: D

We want to look for an answer that takes away from our conclusion that 'there is no advantage in pursuing a policy of encouraging patients to consult their doctor over the phone'. In other words, which answer shows a clear benefit of getting patients to speak to doctors over the phone.

D is the answer as it shows that one benefit is reducing administrative costs.

B and E are in favour of the conclusion.

A mentions the emergency department, which is not relevant here.

C discusses banks, which is outside of the scope.

Question 24: A

At 10:30am on Monday, the clock shows 9:53am → 37 minutes late

22 hours go by

At 8:30am on Tuesday, the clock shows 7:42 am → 48 minutes late

Within 22 hours, the clock gets 11 minutes slower. It is therefore losing 30 seconds/0.5 minutes per hour.

4:30pm is 80 hours later, so 40 minutes extra delay will be seen on the clock.

88 minutes slower than 4:30pm is 3:02pm → 15:02 → The answer is A.

Question 25: A

This question requires careful thought, as actually there is more information given than needed. We only need to compare cost per day, so the number of grams of coffee is irrelevant.

Daughter: 255p/30 days = **8.5p per day**
Parent: 300p/40 days = **7.5p per day**

The answer is A – 1p difference.

Question 26: E

Small candle = $2
Medium candle = $5
Large candle = $10

Last month = 3 x as many medium candles as small and 5 x as many small candles as large.

S : M   = 1 : 3   = $2 : $15
S : L   = 5 : 1   = $10 : $10

S : M : L   = 5: 15 : 1   = $10 : $75 : $10

The only pie chart with small and large with the same proportion of the money, is E.

Question 27: E

This passage argues that regardless of **why** we pursue science, it must be for a good reason and as a result we should encourage scientific discoveries.

However, there is a flaw in reasoning here – just because the reasoning behind the science is positive, doesn't mean the discoveries will also be. It is excluding the alternative outcome – that some scientific discoveries will be negative.

The answer can be summarised as E.

Question 28: D

Let's start by eyeballing the answers.

Just by looking, in the top half of the shape, it is approximately half white and half grey. In the bottom half it is 2/3 white and 1/3 grey. This means we are expecting more white than grey. Therefore answers A, B and C are incorrect.

We now need to decide between D and E.

Question 29: C

'**Best** expresses the conclusion' – look for the overriding argument, and potentially any indicator words. The key here is the word 'therefore' – it indicates where the conclusion lies within the passage. 'University departments should therefore reject any research proposal that might cause offence'. As such, the answer must be C.

If you read around this sentence, you will find that the passage contains evidence/reasoning for this major conclusion e.g. avoiding fighting litigation.

Question 30: C

Pupils could have chosen cricket and basketball in one of two ways:

Group 1 - cricket (as opposed to rounders) in week 1 and then basketball (as opposed to hockey) in week 2
Group 2 - basketball (as opposed to hockey) in week 1 and then cricket (as opposed to rounders) in week 2

To work out the maximum number, look at the number of people doing cricket in week 1: 37. A maximum of 24 people (the number of people doing basketball in week 2) from Group 1 fulfil our criteria; since we want the maximum number we assume that all 24 people doing basketball in week 2 also chose cricket in week 1. Likewise, we must assume that the 15 people doing basketball in week 1 all then chose cricket in week 2.

In both cases, we choose the smaller number.

24 + 15 = 39 and hence the answer is C.

Question 31: C

'Most closely parallels the reasoning'.

Simplify the argument:

Democracy relies on consent.

People must freely vote based on evidence for results to be legitimate.

If not based on evidence, results are not legitimate.

A must have B to be done.

If no B, then no A.

Based on this logic, the answer must be C.

In order to get onto a PhD programme, you must have a Master's degree.

No Master's degree, no PhD programme.

Question 32: A

Hint: use a rubber or draw a net to help you visualise the shape in front of you.

Based on the net below, we can see that A is the correct answer.

These questions tend to take some time, so triage accordingly.

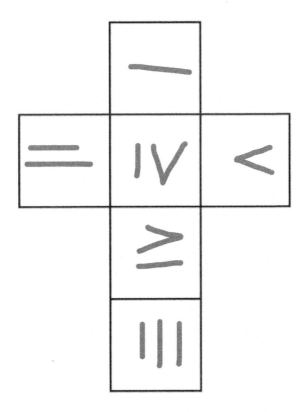

SECTION 2

Question 1: G
Statement 1 is incorrect. Cardiovascular disease is <u>not</u> a communicable disease, as it is not transmitted from person to person. Therefore, answers B, E, F and H are incorrect.
Statement 2 is correct, as we know that cardiac muscle cells get damaged when starved of oxygen. If stem cells can differentiate into new cardiac muscle cells, we can repair the damage and treat the issue. Therefore, answers A and D are incorrect.
Statement 3 is also correct. Cancer is a result of uncontrolled cell proliferation, often with some cells inappropriately differentiating. If we use stem cells, we run the risk of this happening. As both 2 and 3 are correct, our answer is G.

Question 2: D
Z^{3-} indicates that our element Z has gained 3 electrons to form the configuration 2, 8, 8.
Therefore, element Z must have the configuration 2, 8, 5 – as we lose electrons from the outermost shell.
5 electrons in the outer shell indicates that we are in group 15 – so we limit our answers to C and D.
3 shells indicates that we are in the third period. Therefore, our answer is D.

Question 3: G
F = ma
   = 5 x 10
   = 50N downwards force.
This counters the 50N upwards air resistance, so there is no resultant force.
No resultant force means that we are at a terminal velocity, and the object itself is not accelerating.
Therefore, the answer is G.

Question 4: **E**
$$3xy\left(\frac{x^2}{y^4}\right)^2 = 3xy\left(\frac{x^6}{y^8}\right) = \frac{3x^7 y}{y^8} = \frac{3x^7}{y^7}$$

Question 5: A
Statement 1 is incorrect. Denaturing indicates that the enzyme is no longer in the correct shape to be effective. As addition of alkaline increases the time, but not to the extent of a low (3.0) pH, it is unlikely that the enzyme is denatured. Statement 2 is incorrect as there is a lack of information. The time taken is the same for pH 7.0 and 9.0, but we do not have any information between these values. Statement 3 is incorrect as 'activated' indicates that the rate would increase and the time would decrease, but the time is increased compared to pH 7.0 alone.
None of the above are true therefore answer is A.

Question 6: B

Increasing pressure shifts the position of equilibrium to the right to resist the change as per Le Chatelier's principle (2 moles of gas on the left, one mole on the right).

The reaction is exothermic, as shown by the negative enthalpy. Therefore increasing the temperature will shift the position of equilibrium to the left, and decrease the number of moles of ethanol.

Adding more catalyst will not move the position of equilibrium, just increase the rate of reaction.

Thus, I is correct and B is the answer.

Question 7: D

24 counts per minute = background radiation.

Sample at T_0 = 248 − 24 = 224 counts per minute

Sample at T_{48} = 31 − 24 = 7 counts per minute

Number of half lives to decrease from 224 to 7:

224 → 112 → 56 → 28 → 14 → 7

5 half lives in 48 hours.

48/5 = 9.6 hours per half life.

Question 8: B

$$\sqrt{12} = \sqrt{4\sqrt{3}} = 2\sqrt{3}$$
$$\sqrt{27} = \sqrt{9\sqrt{3}} = 3\sqrt{3}$$
$$\sqrt{147} = \sqrt{49\sqrt{3}} = 7\sqrt{3}$$

$$\frac{2\sqrt{3}+3\sqrt{3}+ 7\sqrt{3}}{3} = \frac{12\sqrt{3}}{3} = 4\sqrt{3} = \sqrt{16\sqrt{3}} = \sqrt{48}$$

Question 9: E

The cell being described is a red blood cell.

Statement 1 is correct – red blood cells are carried in the blood.

Statement 2 is correct – red blood cells are responsible for carrying oxygen to respiring tissues.

Statement 3 is incorrect – red blood cells do not have mitochondria.

Therefore the answer is E.

Question 10: H

This question is looking at deducing ions based on inorganic composition tests.

A lilac flame is typically indicative of K^+ ions. This limits our answer to B, D, F, H.

In a metal hydroxide precipitate test, Fe^{2+} typically forms a green precipitate.

The addition of acidified barium chloride produces a white precipitate. This is barium sulphate that forms.

Our answer is now limited to F and H.

We need to consider how this all balances. Given that we have Fe^{2+} the ratio of ions aligns more with H. SO_4 is 2-, Fe(II) is 2+ and K is +, so we need two K^+, one Fe^{2+} and two SO_4^{2-} to have neutral charge.

Question 11: E

Dimensions of cuboid = 0.2 x 0.1 x 0.1 = 0.002 m³

Density, ρ = 2000 kgm⁻³

Surface are in contact = 0.2 x 0.1 = 0.02 m²

$$\text{Pressure} = \frac{Weight}{A} = \frac{mg}{A} = \frac{\rho Vg}{A}$$
$$= \frac{2000 \times (0.2 \times 0.1 \times 0.1) \times 10}{0.02} = 2000 \text{ Pa}$$

Question 12: A

Since the rectangles are similar, the ratio of short side to long side is the same. Hence x/y is equal to y (the short side of the long rectangle) over 5x (the long side of the long rectangle). Cross multiplying, we get that $\frac{x^2}{y^2} = \frac{1}{5}$ and hence x/y = 1/sqrt5 = A

Question 13: A

During Q, the uterus lining is no longer being maintained, therefore progesterone must be decreasing. Answers limited to A, C, E.

During R, uterus lining is increasing in thickness, so oestrogen must be increasing. C is no longer an option.

At point S, ovulation is occurring, so a mature egg must be released. Therefore LH must be increasing. Our only option is now A.

Question 14: H

Where the dotted line is the division between the initial carboxylic acid an alcohol.

The alcohol's longest carbon chain is 3 carbons long, so it must begin *prop-*.

The alcohol functional group *-OH* would be on the second carbon.

The alcohol's name is propan-2-ol.

Question 15: D

Work done is equal to force x distance moved. In this case, 800 N of force is moved, over a distance of 30 x 20 cm = 600 cm = 6 m. (Only the vertical distance matters since only work is only being done to move the individual from one floor to another, regardless of the gradient of that path. Thus the work done is 4800 J = D.

Question 16: **E**

The gradient of the reflected line will have the same magnitude but the opposite sign. Solving the two equations simultaneously, we can work out that they cross at the point (2,-3) and thus the reflected line will go through a point which has y coordinate -7:

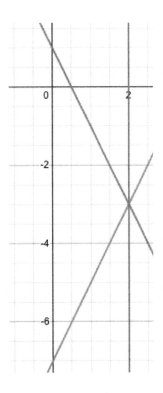

Thus the equation of the line is 2x -7 = **E**.

Question 17: C

Statement 1 is incorrect, as temperature is an abiotic (non-living) factor.

Statement 2 is correct, as mutualism indicates a relationship beneficial for both organisms, which this is. The octopus gets protection from predators using the toxin, and the bacteria get nutrient.

Statement 3 is incorrect, as that would be a community. A population is a group of organisms of the same species.

Therefore the answer is C.

Question 18: D

Since oxygen is approximately 21% by volume of the air, we can approximately multiply 25 by 5 to get 125 dm^3 and thus select D as our answer.

Question 19: E

One watt = one joule/second.

Thus the shower is transferring 8400 joules per second.

$q = mc\Delta T$

Thus, m = 8400/(4200*20) = 0.1 kg = E (100g)

Question 20: H

Since theta is acute, sine is positive. Thus, if 4 is the hypotenuse and 3 is the adjacent, then the opposite must be sqrt7.

$$\text{Area} = 42 = \frac{1}{2} * 8 * QR * \sin \emptyset$$

Hence $QR = \dfrac{42}{\left(\frac{1}{2}\right)*8*\frac{sqrt7}{4}}$

= 42/sqrt7 = 6 sqrt 7 = E

Question 21: C

600 subunits indicates 200 nucleotides.

28 of 200 bases are adenine.

28 of 200 bases are therefore thymine.

200 − 28 − 28 = 144 bases are cytosine and guanine.

72 bases are guanine.

(72/600) *100 = 12%.

Question 22: D

The key information in this question is that there is **one mole** of each element, despite those in group 17 (halogens) existing as molecular pairs.

Reactivity increases down group one, so our most reactive element here is potassium.

Boiling point increases down group 17, so our element here is bromine.

$K + \frac{1}{2} Br_2 \rightarrow KBr$

$39 + 80 \rightarrow 119$

The answer is D.

Question 23: D

$F = IL \times B$

$\quad = 2000 \times 20 \times 5 \times 10^{-5}$

$\quad = 2N$

$F = 120 \times 10$

$\quad = 1200 \ N$

Overall $= 1200 - 2 = 1198 \ N$

So the answer is D.

Question 24: B

The gradient of the tangent will be the negative inverse of the gradient of the radius. The radius is the line connecting the centre and the tangent point: i.e. the line between (3,5) and (6,7) = 2/3. Therefore the answer is B (-3/2).

Question 25: E

This is a recessive condition, so if an individual has the condition, they need 2 copies of the recessive allele.

Statement 1 is correct. 1 and 2 must be carriers (heterozygous) if one of their offspring is affected by the condition even though they are not.

Statement 2 is correct, as one of their parents is homozygous recessive and therefore will always pass on the recessive allele for the condition. As neither 3 nor 4 is affected by the condition, they must have inherited the dominant allele from their other parent, They must therefore both be heterozygous.

Statement 3 is incorrect.

Question 26: C

$Ca_3N_2 + 6H_2O \rightarrow 3(Ca(OH)_2) + zN_xH_y$

Using the information in the question, we can create an equation with some unknowns as above. Any balancing done at this point is done with reference to hydrogen, oxygen and calcium.

On the left of this equation:

3 x Ca

2 x N

12 x H

6 x O

On the right of the equation:

3 x Ca

? x N

6 x O

6 x H

We need 6 more H and 2 N to balance this.

If x = 1, y = 3 and z = 2, we get our equation balanced.

$Ca_3N_2 + 6H_2O \rightarrow 3(Ca(OH)_2) + 2NH_3$

However, we need the empirical (simplest) formula i.e. NH_3

Our answer is C.

Question 27: F

In this question, the person on the platform will initially hear the train as it accelerates into the station. The sound will be at a lower amplitude but higher frequency as it is 'blue shifted' (see Doppler effect).

It will then slow down and decelerate, such that eventually the person on the platform hears the frequency and amplitude that the driver heard throughout.

The amplitude of the wave, will therefore increase as the frequency decreases to f_0. This is because frequency x wavelength = 1 and one must increase as the other decreases.

The answer is therefore F.

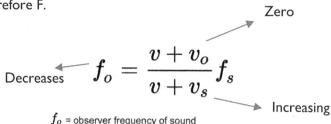

Zero

Decreases

$$f_o = \frac{v + v_o}{v + v_s} f_s$$

Increasing

f_o = observer frequency of sound

v = speed of sound waves

v_o = observer velocity

v_s = source velocity

f_s = actual frequency of sound waves

SECTION 3

1. The real aim of education is to teach moral values.

Explain the reasoning behind this statement. Present a counter-argument. To what extent do you agree with the statement?

Education has the transformative ability to advance societal outcomes but exactly how it does this is complex and multifaceted. Some would argue that, despite mainstream education being dominated by fact-learning and information, its real aim to inculcate a sense of morality. To allow pupils to operate in a civil society, they must be able to effectively differentiate between what is right and wrong to make their decisions. Only some of the facts that we learn in school will be used in our later lives, but the principles and values of humility, kindness and hard work, for example, can be generalised and applied to any situation.

Yet, having said that, fact-learning and knowledge acquisition are fundamental components of education, leading to scientific and technological advancement. This aspect of education cannot be ignored and is surely part of its real aim. Furthermore, it is difficult to say which moral values are the correct moral values. Prescribing in too much detail a set curriculum would devalue the concept of a moral education.

Perhaps the true aim of education is a balance of the two. If educators are able to cultivate a burnishing intellectual curiosity in their pupils, in its true Socratic sense, then parts of both aims would be achieved. Ultimately, members of society must have a rounded education so that they can contribute purposefully to a collective aim.

2. Science today gives people a false sense of certainty in an unpredictable world

Explain what you think this statement means. Argue to the contrary. To what extent can we regard science as a source of certainty.

At the heart of this statement is the struggle we face when seeking the truth in an increasingly complex, erratic world. Science, underpinned by the scientific method, is often viewed as the paragon of objective fact, largely due to its rigour and reliability. Theories and laws are backed up by swathes of compelling evidence as a canon of information is gradually built up.

Nonetheless, that is not say that science has never been wrong, and that it should never be doubted. If something is backed up by scientific evidence, we are often inclined to believe, but this can sometimes have detrimental consequences, as was the case with the thalidomide tragedy. Millions of mothers in the 1950s and 60s trusted the advice of scientific experts and used this drug as a cure for symptoms of nausea during pregnancy. Sadly, scientists can missed the fact that this drug could also have disastrous effects in certain forms. Blindly trusting scientific claims is certainly a recipe for disasters. Likewise, exceptions always exist to the trend which means that there are aspects of nature which we cannot yet explain. We can only assume that the scientific method is leading us down the right track but unfortunately there is no way to verify this. Our scientific understanding can change at any time, and we can never guarantee that we have obtained 'objective truth'.

However, the recent trend of scientific denialism, casting doubt on established fact and reason, is equally worrying. Dismissing one of our most powerful tools to uncover the truth and advance frontiers is self-limiting. It is not science that gives people a false sense of certainty, but our unwillingness to scrutinise the evidence for ourselves and hold a certain degree of scepticism.

Ultimately, scientists do not proclaim themselves as providers of certainty; indeed, the scientific method is built upon the principle of refutation, questioning established hypotheses in an effort to prove it wrong. We can, and should, trust science but we must learn to mediate our trust. This trust cannot be blind since nothing is ever certain, but science can be used to get closer to the truth.

3. 'Before you examine the body of a patient, be patient to learn his story. For once you learn his story, you will also come to know his body.' (Suzy Kassem)

Explain the argument behind this statement. Offer a counter-argument. To what extent is Kasseem's view relevant to the practice of modern medicine.

In recent decades, overmedicalisation has led to a change in the delivery of healthcare; medicine today sometimes seems to be less about the patient as a person, and more about checklists, procedures and technological advancements. Whilst these are positive developments which have improved the lives of millions, there is a case to be made that perhaps we need to turn back the clock slightly and rethink our position.

The statement is arguing for a more holistic approach to modern medicine. Kassem is arguing that rather than jumping straight into the physical symptoms of a presenting complaint, it can often be more beneficial to let the patient talk and share their story. At its most basic level, new facts can be uncovered as a result that wouldn't have been uncovered otherwise. But, more fundamentally, taking this approach can help to build up a rapport and encourage openness and trust in the doctor-patient relationship. Putting the patient at the centre of everything and understanding their needs and wants can allow us to better serve them as doctors.

Having said that, there are certain cases where Kassem's advice cannot apply. Clearly, doctors working in the emergency room cannot think about anything except the treatment they must provide in the moment. On a separate note, professional boundaries must be maintained and the focus must be on the patient's medical concern, rather than vagaries which seem nice on paper. Taking a holistic approach to everything would place a lot of strain on doctors, and would therefore require a lot of funding to make it a viable option.

Ultimately, Kassem's thoughts echo an existing shift in mentality in the medical world, but they do seem somewhat idealistic. While it would be fantastic if we could adopt her advice in every single scenario (meaning that patients receive the best possible care), we cannot let the perfect get in the way of the good. A more holistic approach is definitely required, looking at all aspects of health, but this must be managed and funded in the correct way.

BMAT PRACTICE PAPERS

PREPARING FOR THE BMAT

Before going any further, it's important that you understand the optimal way to prepare for the BMAT. Rather than jumping straight into doing further mock papers, it's essential that you start by understanding the components and the theory behind the BMAT by using a BMAT textbook. Once you've finished the non-timed practice questions, you can progress to past BMAT papers. These are freely available online at www.uniadmissions.co.uk/bmat-past-papers and serve as excellent practice. You're strongly advised to use these in combination with the BMAT Past Worked Solutions book so that you can improve your weaknesses. Finally, once you've exhausted past papers, move onto the mock papers in this book.

ALREADY SEEN THEM ALL?

So, you've run out of past papers? Well, that is where this book comes in. It contains eight unique mock papers; each compiled by expert BMAT tutors at *UniAdmissions* who scored in the top 10% nationally. Having successfully gained a place on their course of choice, our tutors are intimately familiar with the BMAT and its associated admission procedures. So, the novel questions presented to you here are of the correct style and difficulty to continue your revision and stretch you to meet the demands of the BMAT.

GENERAL ADVICE

Practice

This is the best way of familiarising yourself with the style of questions and the timing for this section. Although the exam will essentially only test GCSE level knowledge, you are unlikely to be familiar with the style of questions in all sections when you first encounter them. Therefore, you want to be comfortable with this before you sit the test.

Practising questions will put you at ease and make you more confident for the exam. The more comfortable you are, the less you will panic on the test day and the more likely you are to score highly. Initially, work through the questions at your own pace, and spend time carefully reading the questions and looking at any additional data. When it becomes closer to the test, **make sure you practice the questions under exam conditions**.

Repeat Questions

When checking through answers, pay particular attention to questions you have got wrong. If there is a worked answer, look through that carefully until you feel confident that you understand the reasoning, and then repeat the question without help to check that you can do it. If only the answer is given, have another look at the question and try to work out why that answer is correct. This is the best way to learn from your mistakes, and means you are less likely to make similar mistakes when it comes to the test. The same applies for questions which you were unsure of and made an educated guess which was correct, even if you got it right. When working through this book, **make sure you highlight any questions you are unsure of**, this means you know to spend more time looking over the solutions to them once marked – even if you guessed correctly.

Manage your Time:

It is highly likely that you will be juggling your revision alongside your normal school studies. Although it is tempting to put your A-levels on the back burner, falling behind in your school subjects is not a good idea. Don't forget that to meet the conditions of your offer should you get one you will need at least one A*. So, time management is key!

Make sure you set aside a dedicated **90 minutes** (and much more closer to the exam) to commit to your revision each day. The key here is not to sacrifice too many of your extracurricular activities as everybody needs some down time, but instead to be efficient. Take a look at our list of top tips for increasing revision efficiency below:

1. Create a comfortable workstation
2. Declutter and stay tidy
3. Treat yourself to some nice stationery
4. See if listening to music works for you → if not, find somewhere peaceful and quiet to work
5. Turn off your mobile or at least put it into silent mode
6. Silence social media alerts
7. Keep the TV off and out of sight
8. Stay organised with to do lists and revision timetables – more importantly, stick to them!
9. Keep to your set study times and don't bite off more than you can chew
10. Study while you're commuting
11. Adopt a positive mental attitude
12. Get into a routine
13. Consider forming a study group to focus on the harder exam concepts
14. Plan rest and reward days into your timetable – these are an excellent incentive for you to stay on track with your study plans!

Keep Fit & Eat Well:

'A car won't work if you fill it with the wrong fuel' - your body is exactly the same. You cannot hope to perform unless you remain fit and well. The best way to do this is not underestimate the importance of healthy eating. Beige, starchy foods will make you sluggish; instead start the day with a hearty breakfast like porridge. Aim for the recommended 'five a day' intake of fruit/veg and stock up on oily fish or blueberries – the so-called "super foods".

When hitting the books, it's essential to keep your brain hydrated. If you get dehydrated, you'll find yourself lethargic and possibly developing a headache, neither of which will do any favours for your revision. Invest in a good water bottle that you know the total volume of and keep sipping through the day. Don't forget that the amount of water you should be aiming to drink varies depending on your body mass, so calculate your own personal recommended intake as follows: 30 ml per kg per day.

It is well known that exercise boosts your wellbeing and instils a sense of discipline - all of which will reflect well in your revision. It's well worth devoting half an hour a day to some exercise, get your heart rate up, break a sweat, and get those endorphins flowing.

Sleep

It's no secret that when revising you need to keep well rested. Don't be tempted to stay up late revising as sleep actually plays an important part in consolidating long-term memory. Instead aim for a minimum of 7 hours good sleep each night, in a dark room without any glow from electronic appliances. Install flux (https://justgetflux.com) on your laptop to prevent your computer from disrupting your circadian rhythm. Aim to go to bed the same time each night and no hitting snooze on the alarm clock in the morning!

Revision Timetable

Still struggling to get organised? Try filling in the example revision timetable below. Remember to factor in enough time for short breaks, and stick to it! Schedule in several breaks throughout the day and actually use them to do something you enjoy e.g. light exercise, TV, reading, YouTube, listening to music etc.

	8AM	10AM	12PM	2PM	4PM	6PM	8PM
MONDAY							
TUESDAY							
WEDNESDAY							
THURSDAY							
FRIDAY							
SATURDAY							
SUNDAY							
EXAMPLE DAY		School			Biology	Critical Thinking	Physics

Top tip! Ensure that you take a watch that can show you the time in seconds into the exam. This will allow you have a much more accurate idea of the time you're spending on a question. In general, if you've spent >150 seconds on a section 1 question or >90 seconds on a section 2 questions – move on regardless of how close you think you are to solving it.

GETTING THE MOST OUT OF MOCK PAPERS

Mock exams can prove invaluable if tackled correctly. Not only do they encourage you to start revision earlier, they also allow you to **practice and perfect your revision technique**. They are often the best way of improving your knowledge base or reinforcing what you have learned. Probably the best reason for attempting mock papers is to familiarise yourself with the exam conditions of the BMAT as the time pressure is particularly tough.

Start Revision Earlier

Thirty five percent of students agree that they procrastinate to a degree that is detrimental to their exam performance. This is partly explained by the fact that exams often seem a long way in the future. In the scientific literature this is well recognised. Dr. Piers Steel, an expert in the field of motivation, states that *'the further away an event is, the less impact it has on your decisions'*.

Mock exams are therefore a way of giving you a target to work towards and motivate you in the run up to the real thing – every time you do one treat it as the real deal! If you do well then it's a reassuring sign; if you do poorly then it will motivate you to work harder (and earlier!).

Practice and perfect revision techniques

In case you haven't realised already, revision is a skill in itself, and it can take some time to learn how to revise effectively. For example, the most common revision techniques including **highlighting and/or re-reading can be ineffective** ways of committing things to memory. Unless you are thinking critically about something you are much less likely to remember it or indeed understand it.

Mock exams therefore allow you to test your revision strategies as you go along. Try spacing out your revision sessions so you have time to forget what you have learned in-between. This may sound counterintuitive but the second time you remember it for longer. Try teaching another student what you have learned as this forces you to structure the information in a logical way that may aid memorisation. Always try to question what you have learnt and appraise its validity. Not only does this aid memory but it is also a useful skill for BMAT section 3, Oxbridge interviews, and your studies beyond school.

Improve your knowledge

The act of applying what you have learned reinforces that piece of knowledge. A question may ask you to think about a relatively basic concept in a novel way (not cited in textbooks), which deepens your understanding. Exams rarely test word for word what is in the syllabus, so when running through mock papers try to understand how the basic facts are applied and tested in the exam. As you go through the mocks or past papers take note of your performance and see if you consistently under-perform in specific areas, thus highlighting areas for future study.

Get familiar with exam conditions

Pressure can cause all sorts of trouble for even the most brilliant students. The BMAT is a particularly time pressured exam with high stakes for competitive university admission. The real key to the BMAT is actually overcoming this pressure and remaining calm to allow you to think efficiently.

Mock exams are therefore an excellent opportunity to devise and perfect your own exam techniques to beat the pressure and meet the demands of the exam. **Don't treat mock exams like practice questions – it's imperative you do them under time conditions.**

Remember! It's better that you make all the mistakes you possibly can now in mock papers and then learn from them so as not to repeat them in the real exam.

BEFORE USING THIS BOOK

Do the ground work

- Read in detail: the background, methods, and aims of the BMAT as well as logistical considerations such as how to take the BMAT in practice. A good place to start is a BMAT textbook like *The Ultimate BMAT Guide* (flick to the back to get a free copy!) which covers all the groundwork but it's also worth looking through the official BMAT site for an overview of the syllabus too (www.admissionstesting.org/bmat).
- It is generally a good idea to start re-capping all your GCSE maths and science.
- Practice substituting formulas together to reach a more useful one expressing known variables e.g. $P = IV$ and $V = IR$ can be combined to give $P = V^2/R$ and $P = I^2R$.
- Remember that calculators are not permitted in the exam, so get comfortable doing more complex long addition, multiplication, division, and subtraction.
- Get comfortable rapidly converting between percentages, decimals, and fractions.
- These are all things which are easiest to do alongside your revision for exams before the summer break. Not only gaining a head start on your BMAT revision but also complimenting your year 12 studies well.
- Discuss scientific problems with others - propose experiments and state what you think the result would be. Be ready to defend your argument. This will rapidly build your scientific understanding for section 2 but also prepare you well for an Oxbridge interview.
- Read through the BMAT syllabus before you start tackling whole papers. This is absolutely essential. It contains several stated formulae, constants, and facts that you are expected to apply - or may just be an answer in their own right. Familiarising yourself with the syllabus is also a quick way of teaching yourself the additional information other exam boards may learn which you do not. Sifting through the whole BMAT syllabus is a time-consuming process so we have done it for you. **Be sure to flick through the syllabus checklist** later on, which also doubles up as a great revision aid for the night before!

Ease in gently

With the groundwork laid, there's still no point in adopting exam conditions straight away. Instead invest in a beginner's guide to the BMAT, which will not only describe in detail the background and theory of the exam, but take you through section by section what is expected. *The Ultimate BMAT Guide: 800 Practice Questions* is the most popular BMAT textbook – you can get a free copy by flicking to the back of this book.

When you are ready to move on to past papers, take your time and puzzle your way through all the questions. Really try to understand solutions. A past paper question won't be repeated in your real exam, so don't rote learn methods or facts. Instead, focus on applying prior knowledge to formulate your own approach.

If you're really struggling and have to take a sneak peek at the answers, then practice thinking of alternative solutions, or arguments for essays. It is unlikely that your answer will be more elegant or succinct than the model answer, but it is still a good task for encouraging creativity with your thinking. Get used to thinking outside the box!

Accelerate and Intensify

Start adopting exam conditions after you've done two past papers. Don't forget that **it's the time pressure that makes the BMAT hard** – if you had as long as you wanted to sit the exam you would probably get 100%. If you're struggling to find comprehensive answers to past papers then *BMAT Past Papers Worked Solutions* contains detailed explained answers to every BMAT past paper question and essay (flick to the back to get a free copy).

Doing all the past papers from 2009 – present is a good target for your revision. Note that the BMAT syllabus changed in 2009 so questions before this date may no longer be relevant. In any case, choose a paper and proceed with strict exam conditions. Take a short break and then mark your answers before reviewing your progress. For revision purposes, as you go along, keep track of those questions that you guess – these are equally as important to review as those you get wrong.

Once you've exhausted all the past papers, move on to tackling the unique mock papers in this book. In general, you should aim to complete one to two mock papers every night in the ten days preceding your exam.

HOW TO USE PRACTICE PAPERS

If you have done everything this book has described so far then you should be well equipped to meet the demands of the BMAT, and therefore **the mock papers in the rest of this book should ONLY be completed under exam conditions**.

This means:

- Absolute silence – no TV or music
- Absolute focus – no distractions such as eating your dinner
- Strict time constraints – no pausing half way through
- No checking the answers as you go
- Give yourself a maximum of three minutes between sections – keep the pressure up
- Complete the entire paper before marking
- Mark harshly

In practice this means setting aside two hours in an evening to find a quiet spot without interruptions and tackle the paper. Completing one mock paper every evening in the week running up to the exam would be an ideal target.

- Tackle the paper as you would in the exam.
- Return to mark your answers, but mark harshly if there's any ambiguity.
- Highlight any areas of concern.
- If warranted read up on the areas you felt you underperformed to reinforce your knowledge.
- If you inadvertently learnt anything new by muddling through a question, go and tell somebody about it to reinforce what you've discovered.

Finally relax… the BMAT is an exhausting exam, concentrating so hard continually for two hours will take its toll. So, being able to relax and switch off is essential to keep yourself sharp for exam day! Make sure you reward yourself after you finish marking your exam.

SCORING TABLES

SECTION 1	1st Attempt	2nd Attempt	3rd Attempt
Mock A			
Mock B			
Mock C			
Mock D			
Mock E			
Mock F			
Mock G			
Mock H			

SECTION 2	1st Attempt	2nd Attempt	3rd Attempt
Mock A			
Mock B			
Mock C			
Mock D			
Mock E			
Mock F			
Mock G			
Mock H			

Fortunately for our mock papers our tutors have compiled model answers for you to compare your essays against! If you're repeating a mock paper, its best to attempt a different essay title to give yourself maximum experience with the various styles of BMAT essays.

SECTION 3	Essay 1	Essay 2	Essay 3	Essay 4
Mock A				
Mock B				
Mock C				
Mock D				
Mock E				
Mock F				
Mock G				
Mock H				

MOCK PAPER A

SECTION 1

Question 1:

A square sheet of paper is 20cm long. How many times must it be folded in half before it covers an area of 12.5cm^2?

A. 3　　　　B. 4　　　　C. 5　　　　D. 6　　　　E. 7

Question 2:

Mountain climbing is viewed by some as an extreme sport, while for others it is simply an exhilarating pastime that offers the ultimate challenge of strength, endurance, and sacrifice. It can be highly dangerous, even fatal, especially when the climber is out of his or her depth, or simply gets overwhelmed by weather, terrain, ice, or other dangers of the mountain. Inexperience, poor planning, and inadequate equipment can all contribute to injury or death, so knowing what to do right matters.

Despite all the negatives, when done right, mountain climbing is an exciting, exhilarating, and rewarding experience. This article is an overview beginner's guide and outlines the initial basics to learn. Each step is deserving of an article in its own right, and entire tomes have been written on climbing mountains, so you're advised to spend a good deal of your beginner's learning immersed in reading widely. This basic overview will give you an idea of what is involved in a climb.

Which statement best summarises this paragraph?
A. Mountain climbing is an extreme sport fraught with dangers.
B. Without extensive experience embarking on a mountain climb is fatal.
C. A comprehensive literature search is the key to enjoying mountain climbing.
D. Mountain climbing is difficult and is a skill that matures with age if pursued.
   The terrain is the biggest unknown when climbing a mountain and therefore presents the biggest danger.

Question 3:

50% of an isolated population contract a new strain of resistant Malaria. Only 20% are symptomatic, of which 10% are female. What percentage of the total population do symptomatic males represent?

A. 1%　　　　　B. 9%　　　　　C. 10%　　　　　D. 80%

Question 4:

John is a UK citizen who is looking to buy a holiday home in the South of France. He is purchasing his new home through an agency. Unlike a normal estate agent, they offer monthly discount sales of up to 30%. As a French company, the agency sells in Euros. John decides to hold off on his purchase until the sale in the interest of saving money. What is the major assumption made in doing this?

A. The house he likes will not be bought in the meantime.

B. The agency will not be declared bankrupt.

C. The value of the pound will fall more than 30%.

D. The value of the pound will fall less than 30%.

E. The value of the euro may increase by up to 35% in the coming weeks.

Question 5:

In childcare professions, by law, there must be an adult to child ratio of no more than 1:4. Child minders are hired on a salary of £8.50 an hour. What is the maximum number of children that can be continually supervised for a period of 24 hours on a budget of £1,000?

A. 1   B. 8   C. 12   D. 16   E. 468

Question 6:

A table of admission prices for the local cinema is shown below:

	Peak	Off-peak
Adult	£11	£9.50
Child	£7	£5.50
Concession	£7	£5.50
Student	£5	£5

How much would a group of 3 adults, 5 children, a concession and 4 students save by visiting at an off-peak time rather than a peak time?

A. £11.50   B. £13.50   C. £15.50   D. £17.50   E. £18.50

Question 7:

All musicians play instruments. All oboe players are musicians. Oboes and pianos are instruments. Karen is a musician. Which statement is true?

A. Karen plays two instruments.

B. All musicians are oboe players.

C. All instruments are pianos or oboes.

D. Karen is an oboe player.

E. None of the above.

Question 8:

Flow mediated dilatation is a method used to assess vascular function within the body. It essentially adopts the use of an ultrasound scan to measure the percentage increase in the width of an artery before and after occlusion with a blood pressure cuff. Ultrasound scans are taken by one sonographer, and the average lumen diameter is then measured by an analyst. What is a potential flaw in the methodology of this technique?

A. Results will not be comparable within an individual if different arteries start at different diameters.

B. Results will not be comparable between individuals if they have different baseline arterial diameters.

C. Ultrasound is an outdated technique with no use in modern medicine.

D. This methodology is subject to human error.

E. This methodology is not repeatable.

Question 9:

If it takes 20 minutes to board an aeroplane, 15 minutes to disembark and the flight lasts two and a half hours. In the event of a delay, it is not uncommon to add 20 minutes to the flight time. Megan is catching the flight in question as she needs to attend a meeting at 5pm. The location of the meeting is 15 minutes from the airport without traffic; or 25 minutes with traffic. Which of the following statements is valid considering this information?

A. If Megan wants to be on time for her meeting, given all possibilities described, the latest she can begin boarding at the departure airport is 1.30pm.

B. If Megan starts boarding at 1.40pm she will certainly be late.

C. If Megan aims to start boarding at 1.10pm she will arrive in time whether the plane is delayed or not.

D. If Megan wishes to be on time, she doesn't have to worry about the plane being delayed as she can make up the time during the transport time from the arrival airport to the meeting.

Question 10:

A cask of whiskey holds a total volume of 500L. Every two and a half minutes half of the total volume is collected and discarded. How many minutes will it take for the entire cask to be emptied?

A. 80   B. 160   C. 200   D. 240   E. ∞

Question 11:

The keypad to a safe comprises the digits 1 - 9. The code itself can be of indeterminate length. The code is therefore set by choosing a reference number so that when a code is entered the average of all the different numbers entered must equal the chosen reference number.

Which of the following is true?

A. If the reference number was set greater than 9, the safe would be locked forever.

B. This safe is extremely insecure as if random digits were pressed for long enough it would average out at the correct reference number.

C. More than one number is always required to achieve the reference number.

D. All of the above are true.

E. None of the above are true.

Question 12:

The use of antibiotics is one of the major paradoxes in modern medicine. Antibiotics themselves provide a selection pressure to drive the evolution of antibiotic resistant strains of bacteria. This is largely due to the rapid growth rate of bacterial colonies and asexual cell division. As such, a widespread initiative is in place to limit the prescription of antibiotics.

Which of the following is a fair assumption?

A. Antibiotic resistance is impossible to avoid as it is driven by evolution.

B. If bacteria reproduced at a slower rate antibiotic resistance would not be such an issue.

C. Medicine always creates more problems than it solves.

D. In the past antibiotics were used frivolously.

E. All of the above could be possible.

Question 13:

At a society meeting, 1000 people are entitled to vote in the elections for Chairperson with a one-person-one-vote system. The election rules state if no candidate obtains more than 50% of the votes cast in the first ballot, a second ballot must be held between the top two candidates. 350 votes were cast for a particular candidate in the first ballot. Then a second ballot took place.

Under these circumstances which one of the following is possible?

A. The candidate won the election, came second, or came third.

B. The candidate either won the election or came second.

C. The candidate came second or third, but did not win.

D. The candidate came third.

E. The candidate definitely won the election.

Question 14:

Ever since Uranus was discovered, astronomers have believed there may be more planets in the Solar System. Small deviations in the orbits of Uranus and Neptune suggest another planet might exist. Astronomers have named this hypothetical undiscovered planet 'Planet X'. These deviations in orbit can also be explained by incorrect predictions. Since Uranus and Neptune take many decades to circle the sun, astronomers rely on old data to calculate their orbits. As this data is likely to be inaccurate, the calculated orbits are probably wrong, and so Uranus and Neptune will deviate from them even if there is no 'Planet X'.

Which one of the following best expresses the main conclusion of the above argument?

A. The use of old and inaccurate data indicates that Planet X cannot exist.

B. Astronomers are right to think that there must be an undiscovered planet.

C. The deviations in the orbits of Uranus and Neptune cannot tell us whether Planet X exists.

D. The calculations of the orbits of Uranus and Neptune are probably wrong.

E. Uranus and Neptune will deviate from the predicted orbits whether or not Planet X exists.

Question 15:

One in four deaths caused by road accidents involving commercial vehicles is caused by the driver falling asleep at the wheel. The problem even affects police officers, who are now more likely to die while driving when tired than from physical attacks. Evidence at the scene (such as tyre marks) can tell investigators how quickly the car driver braked. Late breaking indicates a lack of concentration which might be caused by tiredness. The problem with this evidence is that it is not conclusive, whereas conclusive evidence can be offered for other offences such as drink driving.

Which of the following can be drawn as a conclusion of the passage above?

A. Accidents caused by drivers falling asleep at the wheel are a greater problem than drink driving.

B. Commercial vehicle drivers and the police are more prone to falling asleep at the wheel because of the long hours they work.

C. The number of hours per day that commercial drivers should be allowed to drive should be reduced.

D. It will not be as easy to prosecute drivers for falling asleep at the wheel as it is for drink driving.

E. It would be unfair to prosecute people for falling asleep at the wheel.

Question 16:

A study involving a brain-training exercise was carried out on more than a thousand adults aged 65 and over, some of whom later developed dementia. Results showed that the benefits of the five-week mental agility course undertaken by some of the adults lasted for at least five years. This led to an improvement in everyday activities such as money management and the ability to do housework. If those with trained brains developed dementia, they did so later than those in the control group. The results also showed that, for those people in the study who developed dementia, following their diagnosis, their mental decline occurred faster than for those who had undertaken the training.

Which one of the following can be drawn as a conclusion from the above passage?

A. People do a decreasing amount of housework as they grow older.

B. It is preferable to have swift mental decline once dementia develops.

C. Older people do not perform mentally challenging tasks unless forced to do so.

D. Keeping the mind active delays the onset of dementia.

E. All over-65s who undertake brain training live for at least five years afterwards.

Question 17:

According to the current mainstream scientific view, Near Death Experiences (NDEs) are explicable in purely physiological terms. Specifically, they are caused by cerebral anoxia (oxygen deficiency in brain tissue), which occurs in a dying brain. On the other hand, recent research on hundreds of successfully resuscitated cardiac patients found that only twenty per cent reported NDEs. If NDEs had purely medical causes then most of the patients should have experienced them, since they had all been clinically dead and experienced cerebral anoxia. NDEs therefore do not have purely physiological causes.

Which one of the following best expresses the main conclusion of the above passage?

A. Not all successfully resuscitated cardiac patients have NDEs.

B. Not all clinically dead patients have NDEs.

C. NDEs are caused by oxygen deficiency in the brain.

D. NDEs are not necessarily caused by physical events alone.

E. NDEs are a physical property of the human brain

Question 18:

A study on identical twins concluded that genetics contribute roughly half of the attributes we need to be happy. People often find such studies scary, seeing something sinister about us being mere puppets of our biology. However, put in non-scientific terms, it sounds like common sense. Parents frequently notice their children have different personality traits from a very young age. Perhaps it is nicer to think this is caused by something 'fluffy' like a soul. Even if this were true, why is it more reassuring than the thought that genes are responsible? Either way, you are born as you are.

Which one of the following statements is best supported as the conclusion of the passage above?

A. Roughly half of what we need to be happy is decided by our genetic make-up.

B. We may as well accept the idea that our potential for happiness in life is to some extent decided at birth.

C. Whether or not you are happy in life is either determined by your soul or your genes.

D. Whether or not you are happy in life is not something over which you yourself have any control.

E. The person you are at birth is the person you will be throughout your life.

Question 19:

Horrific images of the earthquake in Haiti were seen immediately all over the world, and by the next day the full extent of the damage was seen by the entire world. Clearly, the main problem was moving aid from the airport to distant areas, and with the roads largely blocked, the only practical method was to use helicopters. The great nations of the world should be ashamed that food was not getting to the people who needed it, and that even a week later, their relief still depended on the ability of courageous and skilful drivers to reach them in trucks.

Which one of the following is an underlying assumption of the argument above?

A. The relief agencies were able to import trucks to Haiti but not helicopters.
B. The great nations of the world had helicopters at their disposal which could reach Haiti within a week.
C. There was enough food in Haiti to supply all the people in the weeks after the earthquake.
D. The images failed to prompt the great nations of the world into relief operations after the earthquake.
E. The people of Haiti were able to clear their roads within a week of the earthquake.

Question 20:

A company sells custom design t-shirts. A breakdown of their costs is shown below:

Number of Items	Cost per Item	
	Black and white	Colour
0 – 99	£3.00	£5.00
100 - 499	£2.50	£4.50
500 - 999	£2.00	£4.00
1000+	£1.00	£3.00

Customers with a never-before-printed design must also pay a surcharge of £50 to cover the cost of building a jig. What is the total cost for an order of unique stag-do t-shirts: 50 in colour, and 200 in black and white?

A. £650   B. £700   C. £750   D. £800   E. £850

Question 21:

The Scouts is a movement for young people first established by Lord Baden Powell. As the founder he was the first chief scout of the association. Since his initial appointment there have been a number of notable chief scouts including Peter Duncan and Bear Grylls. Some of the first camping trips conducted by Lord Powell's scout troop were on Brown Sea Island.

Now the Scout movement is a worldwide global phenomenon giving children from all backgrounds the opportunity not only to embark upon adventure but also to engage in the understanding and teaching of foreign culture. Traditionally religion formed the backbone of the scouting movement which was reflected in the scouts promise: "I promise to do my duty to God and to the Queen".

Which of the following applies to the scout movement?

A. Scouts work for the Queen.
B. The scout network is aimed at adventurous individuals.
C. Chief scout is appointed by the Queen.
D. You have to be religious to be a scout.
E. None of the above.

Question 22:

Three rats are placed in a maze that is in the shape of an equilateral triangle. They pick a direction at random and walk along the side of a triangle. Sophie thinks they are less likely to collide than not. Is she correct?

A. Yes, because mice naturally keep away from each other.
B. No, they are more likely to collide than not.
C. No, they are equally likely to collide than not collide.
D. Yes, because the probability they collide is 0.25.
E. None of the above.

Question 23:

The use of human cadavers in the teaching of anatomy is hotly debated. Whilst many argue that it is an invaluable teaching resource, demonstrating far more than a textbook can, others argue that it is an outdated method, which puts unfair stress on an already bereaved family. One of the biggest pros for using human tissue in anatomical teaching is the variation that it displays. Whilst textbooks demonstrate a standard model averaged over many 100s of specimens, many argue that it is the variation between cadavers that really reinforces anatomical knowledge.

The opposition argues that it is a cruel process that damages the grieving process of the affected family since the use of the cadaver often occupies a period of up to 12 months. As such the relative in question is returned to the bereaved family for burial around the time it would be expected that they were recovering as described in the grieving model.

Does the article support or reject the use of cadavers in anatomical teaching?
A. Supports the use
B. Rejects the use
C. Impartial
D. Can't tell
E. None of the above

Question 24:

A ferry is carrying its full capacity. At the time of departure (7am) the travel time to the nearest hour is announced as 13 hours. What is the latest that the ferry could arrive at its destination?

A. 08.29   B. 20.00   C. 20.29   D. 20.30   E. 20.59

Question 25:

A game is played using a circle of 55 stepping-stones. A die is rolled showing the numbers 1 - 6. The number on the die tells you how many steps you may take during your go. The only rule is that during your go you must take your steps in the routine two steps forward, 1 step back. The winner is the first person to move around the full circle of stepping-stones.

What is the minimum number of rolls required to win?

A. 17   B. 18   C. 19   D. 32   E. 55

Question 26:

On a racetrack there are 3 cars recording lap times of 40 seconds, 60 seconds, and 70 seconds. They all started simultaneously 4 minutes ago. How much longer will the race need to continue for them to all cross the start line again at the same time?

A. 4 minutes   C. 14 minutes   E. 1 hour 12 minutes
B. 10 minutes   D. 32 minutes

Question 27:

A class of 60 2nd year medical students are conducting an experiment to measure the velocity of nerve conduction along their radial arteries. This work builds on a previous result obtained demonstrating the effects of how right-handed men have faster nerve conduction velocities than gender matched left-handed individuals. 60% of the class are female of which 3% were unable to take part due to underlying heart conditions. 2 of the male members of the class were also unable to take part. On average the female cohort had faster nerve conduction velocities than men in their dominant arm.

Right-handed women have the fastest nerve conduction velocities.

A. True   B. False   C. Can't tell

Question 28:

Mark is making a double tetrahedron dice by joining two square based pyramids together at their bases. Each square based pyramid is 5cm wide and 8cm tall. What area of card would have been required to produce the nets for the whole die?

A. 150cm^2 B. 180 cm^2 C. 210 cm^2 D. 240 cm^2 E. 270 cm^2

Question 29:

A serial dilution is performed by lining up 10 wells and filling each one with 9ml of distilled water. 1 ml of a concentrated solvent is then added to the first well and mixed. 1 ml of this new solution is drawn from the first well and added to the second and mixed. The process is repeated until all 10 wells have been used.

If the solvent starts off at concentration x, what will its final concentration be after 10 wells of serial dilution?

A. $x/10^9$ B. $x/10^{10}$ C. $x/10^{11}$ D. $x/10^{12}$ E. $x/10^{13}$

Question 30:

A student decides to measure the volume of all the blood in his body. He does this by injecting a known quantity of substrate into his arm, waiting a period of 20 minutes, then drawing a blood sample and measuring the concentration of the substrate in his blood. What assumption has he made here?

A. The substrate is only soluble in blood.   D. The substrate is not degraded.
B. The substrate is not bioavailable.   E. All of the above.
C. The substrate is not excreted.

Question 31:

Jason is ordering a buffet for a party. The buffet company can provide a basic spread at £10 per head. However more luxurious items carry a surcharge. Jason is particularly interested in cupcakes and shell fish. With these items included the buffet company provides a new quote of £10 per head. In addition to simply ordering the food Jason must also purchase cutlery and plates. Plates come in packs of 20 for £8 whilst cutlery is sold in bundles of 60 sets for £10.

With a budget of £2,300 (to the nearest 10 people) what is the maximum number of people Jason can provide food on a plate for?

A. 180   B. 190   C. 209   D. 210   E. 220

Question 32:

What were once methods of hunting have now become popular sports. Examples include archery, the javelin throw, the discus throw and even throwing a boomerang. Why such dangerous hobbies have begun to thrive is now being investigated by social scientists. One such explanation is that it is because they are dangerous that we find them appealing in the first place. Others argue that it is a 'throwback' to our ancestral heritage, where, as a hunter gatherer, being a proficient hunter was something to show off and flaunt. Whilst this may be the case, it is well observed that many find the chase of a hunt exciting if not controversial.

Sports like archery provide excitement analogous to that of the chase during a hunter gatherer hunt.
A. True
B. False
C. Can't tell

END OF SECTION

SECTION 2

Question 1:

A crocodile's tail weighs 30kg. Its head weighs as much as the tail and one half of the body and legs. The body and legs together weigh as much as the tail and head combined.

What is the total weight of the crocodile?

A. 220kg   B. 240kg   C. 260kg   D. 280kg   E. 300kg

Question 2:

A body is travelling at x ms^{-1} with y J of kinetic energy. After a period of retardation the kinetic energy of the body is $1/16y$. Assuming that the mass of the body has remained constant what is its new velocity?

A. $1/196x$ B. $1/16x$ C. $1/8x$ D. $1/4x$ E. $4x$

Question 3:

Which of the following cannot be classified as an organ?

1. Blood   3. Larynx   5. Prostate   7. Skin
2. Bone   4. Pituitary Gland   6. Skeletal Muscle

A. 1 and 6   B. 2 and 3   C. 5 and 7   D. 1 and 5   E. 1,4, 5 and 6

Question 4:

An increase in aerobic respiratory rate could be associated with which of the following physiological changes?

1. A larger percentage of water vapour in expired air
2. Increased expired CO_2
3. Increased inspired O_2
4. Perspiration
5. Vasodilatation

A. 3 only   C. 1, 2 and 3 only   E. All of the above
B. 1 and 2 only   D. 2, 3 and 5

Question 5:

The nephron is to the kidney, as the _____ is to striated muscle:

A. Actin filament   C. Myofibril   E. Vein
B. Artery   D. Sarcomere

Question 6:

A diabetic patient's glucagon and insulin levels are measured over 12 hours.

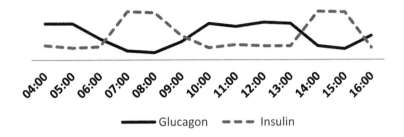

During this time the patient is given two large boli of glucose. A graphical representation of this is shown above.

At which times would you expect the patients' blood glucose to be greatest?

A. 05:00 and 12:00
B. 07:00 and 14.00

C. 08:00 and 15:00
D. 10:00 and 13:00

E. 06:00, 10:00 and 16:00

Question 7:

In addition to the A, B or O classification, blood groups can also be distinguished by the presence of Rhesus antigen (Rh). Care must be taken in blood transfusion as once blood types are mixed a Rh -ve individual will mount an immune response against Rh +ve blood. This is particularly well exemplified in haemolytic disease of the newborn – where a Rh-ve mother carries a Rh+ve foetus.

Applying what is written here and your knowledge of the human immune system, explain why the mother's first child would be relatively safe and unaffected, yet further offspring would be at high risk.

A. The first pregnancy is always such a shock to the body it compromises the immune system.
B. Antibodies take longer than 9 months to produce and mature to an active state.
C. First born children are immunologically privileged.
D. There is a high risk of haemorrhage to both mother and child during birth.
E. Plasma T cells require time to multiply to lethal levels.

Question 8:

Which of the following is NOT present in the Bowman's capsule?

A. Urea
B. Glucose

C. Sodium
D. Water

E. Haemoglobin

Question 9:

At present a large effort is being made to produce tailored patient care. One of the ultimate goals of this is to be able to grow personal, genetically identical organs for those with end stage organ failure. This process will first require the harbouring of what cell type?

A. Cells from the organ that is failing

B. Haematopoietic stem cells

C. Embryonic stem cells

D. Adult stem cells

E. All of the above

Question 10:

Below are three statements about electromagnetic radiation. Which is / are correct?

1. For identical amplitude, waves with the smallest wavelength transfer the most energy.
2. The speed of electromagnetic waves is directly proportional to their wavelength.
3. Microwaves can be dangerous because they can be easily absorbed by water molecules.

A. 1 only

B. 2 only

C. 3 only

D. 1 and 2

E. 1 and 3

F. None of the above

Question 11:

From which of the following elemental groups are you most likely to find a catalyst?

A. Alkali Metals

B. Transition metals

C. Alkaline Earth Metals

D. Noble Gases

E. Halogens

Question 12:

1.338kg of francium is mixed in a reaction vessel with an excess of distilled water. What volume will the hydrogen produced occupy at room temperature and pressure? Mr of Francium = 223

A. 20.4dm³   B. 36dm³   C. 40.8dm³   D. 60.12dm³   E. 72dm³

Question 13:

The composition by mass of a compound is Carbon 53%, Hydrogen 11%, and Oxygen 36%, to the nearest percentage.
What is the empirical formula of this compound?

A. CH_2O

B. $C_2H_6O_2$

C. C_2H_6O

D. CHO

E. $C_3H_5O_2$

Question 14:
What is the actual molecular formula of the compound in question 13 if the M_r is 45?

A. C_2H_6O
B. $C_4H_{12}O_2$
C. C_3H_8O
D. $C_4H_8O_2$
E. More information needed

Question 15:
1.2×10^{10} kg of sugar is dissolved in 4×10^{12} L of distilled water. What is the concentration?

A. 3×10^{-2} g/dL
B. 3×10^{-1} g/dL
C. 3×10^1 g/dL
D. 3×10^2 g/dL
E. 3×10^3 g/dL

Question 16:
Which of the following is not essential for the progression of an exothermic chemical reaction?

A. Presence of a catalyst
B. Increase in entropy
C. Achieving activation energy
D. Attaining an electron configuration more closely resembling that of a noble gas
E. None of the above

Question 17:
Which of the following combinations are commonly used in the treatment of drinking water?

A. F_2 and Cl_2
B. H_2 and Cl_2
C. F^- and Cl^-
D. F^- and H^+
E. H^+ and Cl^-

Question 18:
Which of the following is a unit equivalent to the Volt?

A. $A.\Omega^{-1}$
B. $J.C^{-1}$
C. $W.s^{-1}$
D. $C.s$
E. $W.C.\Omega$

Question 19:
Complete the sentence below:
A voltmeter is connected in _____ and therefore has _____ resistance; whereas an ammeter is connected in _____ and has _____ resistance.

A. Parallel, zero, parallel, infinite
B. Parallel, zero, series, infinite
C. Parallel, infinite, series, zero
D. Series, zero, parallel, infinite
E. Series, infinite, parallel, zero

Question 20:
A body "A" of mass 12kg travelling at 15m/s undergoes inelastic collision with a fixed, stationary object "B" of mass 20kg over a period of 0.5 seconds. After the collision body A has a new velocity of 3m/s. What force must have been dissipated during the collision?

A. 288N   B. 298N   C. 308N   D. 318N   E. 328N

Question 21:
What process is illustrated here: $^{14}_{6}C \rightarrow\ ^{14}_{7}N + x$

A. Thermal decomposition   C. Beta decay
B. Alpha decay   D. Gamma decay

Question 22:
A radio dish is broadcasting messages into deep space on a 20 Hz radio frequency of wavelength 3km. With every hour how much further does the signal travel into deep space?

A. 200,000 km   C. 232,000 km   E. 264,000 km
B. 216,000 km   D. 248,000 km

Question 23:
A formula: $\sqrt[3]{\dfrac{z(x+y)(l+m-n)}{3}}$ is given. Which of the following options would you expect this formula to calculate?

A. A length   C. A volume   E. A geometric average
B. An area   D. A volume of rotation

Question 24:
Evaluate the following: $(4.2 \times 10^{10}) - (4.2 \times 10^{6})$

A. 415,800,000   C. 41,995,800,000   E. 4,242,000,000
B. 415,800   D. 419,958,000

Question 25:

Calculate a – b

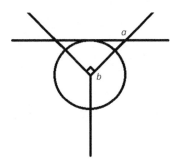

A. 0°
B. 5°
C. 10°
D. 15°
E. 20°

Question 26:

Jack has a bag with a complete set of snooker balls (15 red, 1 yellow, 1 green, 1 brown, 1 blue, 1 pink and 1 black ball) within it. Blindfolded, Jack draws two balls from the bag without replacing them. What is the probability that he draws a blue and a black ball in any order?

A. 2/41   B. 2/210   C. 1/210   D. 1/105   E. 2/441

Question 27:

An experiment is repeated using an identical methodology and upon further review it is proven to demonstrate identical scientific practice. If the result obtained is different to the first, this would be due to:

A. Calibration Bias
B. Systematic Bias
C. Random Chance
D. Serial dilution
E. Inaccuracies in the methodology

END OF SECTION

SECTION 3

1) *Doctors should wear white coats, as it helps to create a placebo effect, rendering the treatment more effective.*

Explain what is meant by this statement. Argue to the contrary. To what extent do you agree with the statement? What points can you see that contradict this statement?

2) *"Medicine is a science of uncertainty and an art of probability."*

William Osler

Explain what this statement means. Argue to the contrary. To what extent do you agree with the statement?

3) *"The New England Journal of Medicine reports that 9 out of 10 doctors agree that 1 out of 10 doctors is an idiot."*

Jay Leno

What do you understand by this statement? Explain why the assumption above may be inaccurate and argue to the contrary.

4) *"My father was a research scientist in tropical medicine, so I always assumed I would be a scientist, too. I felt that medicine was too vague and inexact, so I chose physics."*

Stephen Hawking

Explain what this statement means. Argue to the contrary. To what extent do you agree with the statement?

END OF PAPER

MOCK PAPER B

SECTION 1

Question 1:
"If vaccinations are now compulsory because society has decided that they should be forced, then society should pay for them." Which of the following statements would weaken this argument the most statement?

A. Many people disagree that vaccinations should be compulsory.
B. The cost of vaccinations is too high to be funded locally.
C. Vaccinations are supported by many local communities and GPs.
D. Healthcare workers do not want vaccinations.
E. None of the above

Question 2:
Josh is painting the outside walls of his house. The paint he has chosen is sold only in 10L tins. Each tin costs £4.99. Assuming a litre of paint covers an area of 5m², and the total surface area of Josh's outside walls is 1050m²; what is the total cost of the paint required if Josh wants to apply 3 coats?

A. £104.79   B. £209.58   C. £314.37   D. £419.16   E. £523.95

Question 3:
The stars of the night sky have remained unchanged for many hundreds of years, which allows sailors to navigate using the North Star. However, this only applies within the Northern Hemisphere as the populations of the Southern Hemisphere are subject to an alternative night sky.

An asterism can be used to locate the North Star, which comes by many names including the plough, the saucepan, and the big dipper. Whilst the North Star's position remains fixed in the sky (allowing it to point north reliably always) the rest of the stars traverse around the North Star in a singular motion. In a very long time, the North Star will one day move from its location due to the movement of the Earth.

Which of the following is **NOT** an assumption made in this argument?
A. The Earth is rotating on its axis.
B. Sailors still need to navigate using the stars.
C. An analogous southern star is used to navigate in the Southern hemisphere.
D. The plough is not the only method of locating the North Star.
E. None of the above.

Question 4:
John wishes to deposit a cheque. The bank's opening times are 9am until 5pm Monday to Friday, 10am until 4pm on Saturdays, and the bank is closed on Sundays. It takes on average 42 bank hours for the money from a cheque to become available.

If John wishes to have the money by 8pm on Tuesday, what is the latest he can cash the cheque?

A. 5pm the Saturday before
B. 5pm the Friday before
C. 1pm the Thursday before

D. 1pm the Wednesday before
E. 9am the Tuesday before

Question 5:
How many different squares can be visualised in the image shown to the right?

A. 25   C. 48   E. 63
B. 32   D. 58

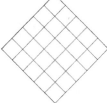

Question 6:
In 4 years, I will be one third of the age that my brother will be next year. In 20 years' time he will be double my age. How old am I?

A. 4   B. 9   C. 15   D. 17   E. 23

Question 7:
Aneurysmal disease has been proven to induce systemic inflammatory effects, reaching far beyond the site of the aneurysm. The inflammatory mediator responsible for these processes remains unknown, however the effects of systemic inflammation have been well categorised and observed experimentally in pig models.

This inflammation induces an aberration of endothelial function within the innermost layer of blood vessel walls. The endothelium not only represents the lining of blood vessels but also acts as a transducer converting the haemodynamic forces of blood into a biological response. An example of this is the NO pathway, which uses the shear stress induced by increased blood flow to drive the formation of NO. NO diffuses from the endothelium into the smooth muscle surrounding blood vessels to promote vasodilatation and therefore acts to reduce blood flow.
Failure of this process induces high risk of vascular damage and therefore cardiovascular diseases such as thrombosis and atherosclerosis.

What is a valid conclusion from the text above?

A. Aneurysmal disease does not affect the NO pathway.

B. Aneurysms directly increase the likelihood of cardiovascular disease.

C. Aneurysms are the opposite of transducers.

D. Observations of this kind should be made in humans to see if the results can be replicated.

E. Aneurysms induce high blood flow.

Question 8:

A traffic surveyor is stood at a T-junction between a main road and a side street. He is only interested in traffic leaving the side street. He logs the class of vehicle, the colour and the direction of travel once on the main road. During an 8-hour period he observes a total of 346 vehicles including bikes, of which 200 were travelling west whilst the rest travelled east. The overwhelming majority of vehicles seen were cars, at 90%, with bikes, vans and articulated lorries together comprising the remaining 10%. Red was the most common colour observed whilst green was the least. Black and white vehicles were seen in equal quantities.

Which of the following is an accurate inference based on his survey?

A. Global sales are highest for those vehicles which are coloured red.

B. Cars are the most popular vehicle on all roads.

C. Green vehicles are less popular than red vehicles in the area that the surveyor was based.

D. The daily average rate of traffic out of a T junction in Britain is 346 vehicles over 8 hours.

E. To the east of the junction is a dead end.

Question 9:

William, Xavier, and Yolanda race in a 100m race. All of them run at a constant speed during the race. William beats Xavier by 20m. Xavier beats Yolanda by 20m. By how many metres does William beat Yolanda?

A. 30m   B. 36m   C. 40m   D. 60m   E. 64m

Question 10:

A television is delivered in a box that has volume 60% larger than that of the television. The television is 150cm x 100cm x 10cm. How much surplus volume is there?

A. 0.09m³   B. 0.9 m³   C. 9 m³   D. 90 m³   E. 900 m³

Question 11:

Matthew and David are deciding where they would like to go camping from Friday to Sunday. Upon completing their research, they discover the following:

- Whitmore Bay charges £5.50 per night and does not require a booking. The site provides showers, washing up facilities and easy access to a beach

- Port Eynon charges £5 per night and a booking is compulsory. However, the site does not provide showers but does have 240V sockets free of charge

- Jackson Bay charges £7 per night and is billed as a luxury site with compulsory booking, private showers, toilets, mobile phone charging facilities and kitchens.

David presents the following suggestion:

As Port Eynon is the farthest distance to travel the benefit of its cheap nightly rate is negated by the cost of petrol. Instead, he recommends they visit Jackson Bay as it is the shortest distance to travel and will therefore be the cheapest.

Which of the following best illustrates a flaw in this argument?

A. Whitmore bay may be only a few miles further which means the total cost would be less than visiting Jackson Bay.

B. With kitchen facilities available they will be tempted to buy more food, increasing the cost.

C. The campsite may be fully booked.

D. There may be a booking fee driving the cost up above that of the other campsites.

E. All of the above.

Question 12:

The manufacture of any new pharmaceutical is not permitted without scrupulous testing and analysis. This has led to the widespread, and controversial use of animal models in science. Whilst it is possible to test cytotoxicity on simple cell cultures, to truly predict the effect of a drug within a physiological system it must be trialled in a whole organism. With animals being cheap to maintain, readily available, rapidly reproducing, and not subject to the same strict ethical laws, they have become an invaluable component of modern scientific practice.

Which of the following best illustrates the main conclusion of this argument?

A. New pharmaceuticals cannot be approved without animal experimentation.

B. Cell culture experiments are unhelpful.

C. Modern medicine would not have achieved its current standard without animal experimentation.

D. Logistically animals are easier to keep than humans for mandatory experiments.

E. All of the above.

Question 13:

After looking at interviews conducted with a number of adult learners, our research suggested that the learners who felt they were most successful were all highly motivated. We noticed that early success had heightened motivation in some cases and saw that both success and motivation may be due to a special aptitude for learning. We also noticed that many of those who felt they were most motivated were also learning in favourable conditions or for fun, which meant they may have become motivated since starting their classes. Though these conditions seem persuasive, the results led us to the same conclusion. It's impossible to learn anything without motivation.

Which one of the following is **NOT** a flaw in the above argument?

A. It assumes that those who felt they were successful actually were.
B. It assumes that those who felt they were motivated actually were.
C. The research does not establish that there are no successful learners who lacked motivation.
D. The research is only concerned with adult learners.
E. It assumes that in order to be motivated you may have to have a special aptitude for learning.

Question 14:

A nationwide survey showed that the majority of people would not be willing to give up their car in favour of public transport. However, in a recent survey of people living in an area with heavy traffic problems, 76% stated that they would prefer to travel to work by public transport if the system was made more reliable. This shows that the previous findings were wrong. We should therefore restrict car use and start a programme to improve the nation's public transport network as soon as possible.

Which one of the following is the best statement of the flaw in the argument above?
A. It fails to specify which types of public transport are to be improved.
B. The counter arguments are not explained in detail.
C. The statistic presented may not be representative of the whole population.
D. It does not consider the 24% who would not prefer to use public transport.
E. It fails to explain how the public transport system can be improved.

Question 15:

A restaurant owner makes 100 burgers and 50 hotdogs at the start of the day, to sell that day. The burgers are priced at £8.00, and the hotdogs are priced at £6.00.

By the end of lunchtime, there are 20 burgers and 15 hotdogs left, and the prices for these are halved for sale in the afternoon.

At the end of the day, there are still 2 burgers and 3 hotdogs left, which are disposed of.

Each burger costs £2.50 to make, and each hotdog costs £1.50 to make.

How much profit does the restaurant make from the sales of burgers and hotdogs on this day?

A. £958   B. £633   C. £1,283   D. £1,066   E. £741

Question 16:

A train driver runs a service between Cardiff and Merthyr. On average a one-way trip takes 40 minutes to drive but he requires 5 minutes to unload passengers and a further 5 minutes to pick up new ones. The distance between Cardiff and Merthyr is 22 miles.

Assuming he works an 8-hour shift with two 20-minute breaks, and when he arrives to work the first train is already loaded with passengers how far does he travel (in miles)?

A. 132   B. 143   C. 154   D. 176   E. 198

Question 17:

The massive volume of traffic that travels down the M4 corridor regularly leads to congestion at peak times. A case is being made by local councils in congested areas to introduce relief lanes thus widening the motorway in an attempt to relieve the congestion. This would involve introducing either a new 2 or 4 lanes to the motorway on average costing 1 million pounds per lane per 10 miles.

Many conservationist groups are concerned as this will involve the destruction of large areas of countryside either side of the motorway. They argue that the side of a motorway is a unique habitat with many rare species residing there.

The local councils argue that with many hundreds if not thousands of cars sitting idle on the motorway pumping pollutants out into the surrounding areas, it is better for the wildlife if the congestion is eased and traffic can flow through. The councils have also remarked that if congestion is eased there would be less money needed to repair the roads from car incidents with could in theory be given to the conservationist groups as a grant.

Which of the following is assumed in this passage?

A. Wildlife living on the side of the motorway cannot be re-homed.

B. Congestion causes car incidents.

C. Relief lanes have been proven to improve traffic jams.

D. A and B.

E. B and C.

F. All of the above.

G. None of the above.

Question 18:

Apples and oranges are sold in packs of 5 for the price of £1 and £1.25 respectively. Alternatively, apples can be purchased individually for 30p, and oranges can be purchased individually for 50p. Helen is making a fruit salad, and she remarks that her order would have cost her an extra £6.25 if she had purchased the fruit individually.

Which of the following could have been her order?

A. 15 apples 10 oranges

B. 15 apples 15 oranges

C. 25 apples 10 oranges

D. 25 apples, 15 oranges

E. 30 apples, 30 oranges

Question 19:

Laura is blowing up balloons for a birthday party. The average volume of a balloon is 300cm^3 and Laura's maximum forced expiratory rate in a single breath is 4.5L/min. What is the fastest Laura could inflate 25 balloons assuming it takes her 0.5 secs to breathe in per balloon, and somebody else ties the balloons for her?

A. 112.5 seconds

B. 122.5 seconds

C. 132.5 seconds

D. 142.5 seconds

E. 152.5 seconds

Question 20:

George reasons that A is equal to B which is not equal to C. In which case C is equal to D which is equal to E.

Which of the following, if true, would most *weaken* George's argument?

A. A does not equal D.

B. B is equal to E.

C. A and C are not equal.

D. C is equal to 0.

E. None of the above

Question 21:

In a single day how many times do the hour, minute and second hands of an analogue clock all point to the same number?

A. 12   B. 24   C. 36   D. 48   E. 72

Question 22:

"People who practice extreme sports should have to buy private health insurance."

Which of the following statements most strongly supports this argument?

A. Exercise is healthy and private insurance offers better reward schemes.

B. Extreme sports have a higher likelihood of injury.

C. Healthcare should be free for all.

D. People that practice extreme sports are more likely to be wealthy.

Question 23:

Explorers in the US in the 18th Century had to contest with a great variety of obstacles ranging from natural to man-made. Natural obstacles included the very nature and set up of the land, presenting explorers with the sheer size of the land mass, the lack of reliable mapping as well as the lack of paths and bridges. On a human level, challenges included the threat from outlaws and other hostile groups. Due to the nature of the settling situation, availability of medical assistance was sparse and there was a constant threat of diseases and fatal results of injuries.

Which of the following statements is correct with regards to the above text?

A. Medical supply was good in the US in the 18th Century.

B. The land was easy to navigate.

C. There were few outlaws threatening the individual.

D. Crossing rivers could be difficult.

E. All the above.

Question 24:

The statement "The human race is not dependent on electricity" assumes what?

A. We have no other energy resource.

B. Electricity is cheap.

C. Electrical appliances dominate our lives.

D. Electricity is now the accepted energy source and is therefore the only one available.

E. All of the above.

Question 25:
Wine is sold in cases of 6 bottles. A bottle of wine holds 70cl of fluid whereas a wine glass holds 175ml. Cases of wine are currently on offer for £42 a case buy one get one free. If Elin is hosting a 3-course dinner party for 27 of her friends, and she would like to provide everyone with a glass of wine per course, how much will the wine cost her?

A. £42   B. £84   C. £126   D. £168   E. £210

Question 26:
Hannah buys a television series in boxset. It contains a full 7 series with each series comprising 12 episodes. Rounded to the nearest 10 each episode lasts 40 minutes.

What is the shortest amount of time it could possibly take to watch all the episodes back-to-back?
A. 49 hours
B. 51 hours
C. 53 hours
D. 56 hours
E. 60 hours

Question 27:

Many are familiar with the story that aided in the discovery of the "germ". Semmelweis worked in a hospital where maternal death rates during labour were astronomically high. He noticed that medical students often went straight from dissection of cadavers to the maternity wards. As an experiment Semmelweis split the student cohort in half. Half did their maternity rotation instead before dissection whereas the other half maintained their traditional routine. In the new routine, maternity ward before dissection, Semmelweis recorded an enormous reduction in maternal deaths and thus the concept of the pathogen was born.

What is best exemplified by this passage?

A. Science is a process of trial and error.
B. Great discoveries come from pattern recognition.
C. Provision of healthcare is closely associated with technological advancements.
D. Experiments always require a control.
E. All of the above.

Question 28:

Jack sits at a table opposite a stranger. The stranger says here I have 3 precious jewels: a diamond, a sapphire, and an emerald. He tells Jack that if he makes a truthful statement Jack will get one of the stones, if he lies he will get nothing.

What must Jack say to ensure he gets the sapphire?

A. Tell the stranger his name.
B. Tell the stranger he must give him the sapphire.
C. Tell the stranger he wants the emerald.
D. Tell the stranger he does not want the emerald or the diamond.
E. Tell the stranger he will not give him the emerald or the diamond.

Question 29:

Simon invests 100 pounds in a savings account that awards compound interest on a 6-monthly basis at 50%. Simon's current account awards compound interest on a yearly basis at 90%.
After 2 years will Simon's investment in the savings account yield more money than it would have in the current account?

A. Yes
B. No
C. Can't tell

Question 30:

My mobile phone has a 4-number pin code using the values 1 – 9. To determine this, I use a standard algorithm of multiplying the first two numbers, subtracting the third and then dividing by the fourth. I change the code by changing the answer to this algorithm – I call this the key. What is the largest possible key?

A. 42   B. 55   C. 70   D. 80   E. 81

Question 31:

A group of scientists is investigating the role of different nutrients after exercise. They set up two groups of averagely fit individuals, consisting of the same number of both males and females aged 20 – 25, and weighing between 70 and 85 kilos. Each group will conduct the same 1hr exercise routine of resistance training, consisting of various weighted movements. After the workout they will receive a shake with vanilla flavour that has identical consistency and colour in all cases. Group A will receive a shake containing 50 g of protein and 50g of carbohydrates. Group B will receive a shake containing 100 g of protein and 50 g of carbohydrates. All participants have their lean body mass measured before starting the experiment.

Which of the following statements is correct?

A. The experiment compares the response of men and women to endurance training.
B. The experiment is flawed as it does not take into consideration that men and women respond differently to exercise.
C. The experiment does not consider age.
D. The experiment mainly looks at the role of protein after exercise.
E. None of the above.

Question 32:

A child weighs 35kg and is 120cm tall. Using the equation, $BMI = \dfrac{\text{weight/kg}}{(\text{height/m})^2}$, what is the BMI of the child to the nearest two decimal places?

A. 0.0024   B. 0.29   C. 24.31   D. 29.17   E. 1020

END OF SECTION

SECTION 2

Question 1:

GLUT2 is an essential and ATP independent mediator in the liver's uptake of plasma glucose. This is an example of:

A. Active transport   C. Exocytosis   E. Osmosis
B. Diffusion   D. Facilitated Diffusion

Question 2:

The molecular weight of glucose is 180 g/mol. 5.76Kg of glucose is split evenly between two cell cultures under anaerobic conditions. One cell culture is taken from human cardiac muscle, whilst the other is a yeast culture. What will be the difference (in moles) between the amount of CO_2 produced between the two cultures?

A. 0 mol   B. 4 mol   C. 8 mol   D. 12 mol   E. 16 mol

Question 3:

Which of the following cell types does not contain DNA?

A. Kidney cells   C. Nerve cells   E. None of the above
B. Liver cells   D. Red blood cells

Question 4:

Which of the following is a function of the cardiovascular system?

A. Distribution of heat
B. Oxygenation of blood
C. Removal of waste products from the body
D. All of the above
E. None of the above

Question 5:

Pepsin and trypsin are both digestive enzymes. Pepsin acts in the stomach whereas trypsin is secreted by the pancreas. Which graph below (trypsin in black and pepsin in grey) would most accurately demonstrate their relative activity against pH?

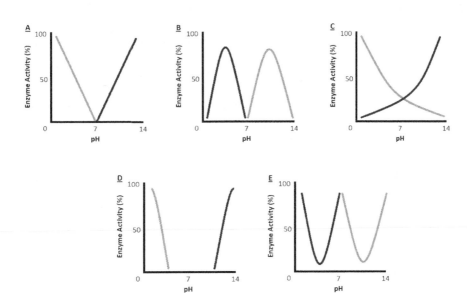

Question 6:

MRSA is a strain of Staphylococcus aureus that is resistant to an antibiotic called methicillin. It is responsible for several difficult-to-treat infections in humans. The fact that MRSA exists is a direct consequence of which of the following processes?

A. Natural selection   C. Sexual reproduction   E. Co-dominance
B. Genetic engineering   D. Lamarckism

Question 7:

What is the electron configuration of magnesium in $MgCl_2$?

A. 2,8   B. 2,8,2   C. 2,8,4   D. 2,8,8   E. None of the above

Question 8:

A calcium sample is run in a mass spectrometer. It is later discovered that the sample was contaminated with the most abundant isotope of chromium. A section of the trace is shown below. What was the actual abundance of the most common calcium isotope?

A. 1/9   B. 6/17   C. 1/2   D. 11/19   E. 17/19

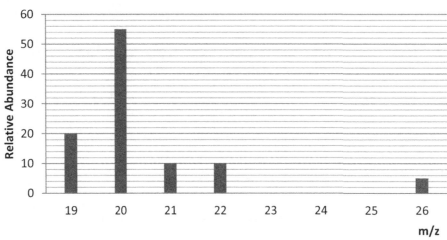

Question 9:

A warehouse receives 15 tonnes of arsenic in bulk. Assuming that the sample is at least 80% pure, what is the minimum amount, in moles, of arsenic that they have obtained? (Mr of arsenic = 75).

A. 1.6×10^5 B. 2×10^5 C. 1.6×10^6 D. 2×10^6 E. 1.6×10^7

Question 10:

A sample of silicon is run in a mass spectrometer. The resultant trace shows m/z peaks at 26 and 30 with relative abundance 60% and 30% respectively. What other isotope of silicon must have been in the sample to give an average atomic mass of 28?

A. 28   B. 30   C. 32   D. 34   E. 36

Question 11:

72.9g of pure magnesium ribbon is mixed in a reaction vessel with the equivalent of 54g of steam. The ensuing reaction produces $72dm^3$ of hydrogen. Which of the following statements is true?

A. This is a complete reaction
B. This is a partial reaction
C. There is an excess of steam
D. There is an excess of magnesium
E. Magnesium hydroxide is a product

Question 12:
Which species acts as the reducing agent in the following equation?:
$3Cu^{2+} + 3S^{2-} + 8H^+ + 8NO_3^- \rightarrow 3Cu^{2+} + 3SO_4^{2-} + 8NO + 4H_2O$

A. Cu^{2+} B. S^{2-} C. H^+ D. NO_3^- E. H_2O

Question 13:
Which of the following is **not** true of alkanes?

A. They have the homologous formula C_nH2_{n+2}
B. They are saturated
C. They are reactive
D. They produce only CO_2 and water when burnt in an excess of oxygen
E. None of the above

Question 14:
A rubber balloon is inflated and rubbed against a sample of animal fur for a period of 15 seconds. At the end of this process the balloon is carrying a charge of -5 coulombs. What magnitude of current must have been induced during the process of rubbing the balloon against the animal fur; and in which direction was it flowing?

A. 0.33A into the balloon   C. 0.33A in no net direction   E. 75A into the fur
B. 0.33A into the fur   D. 75A into the balloon

Question 15:
Which of the following is a unit equivalent to the Amp?

A. $V.\Omega$ B. $(W.V)/s$ C. $C.\Omega$ D. $(J.s^{-1})/V$ E. $C.s$

Question 16:
The output of a step-down transformer is measured at 24V and 10A. Given that the transformer is 80% efficient what must the initial power input have been?

A. 240W   B. 260W   C. 280W   D. 300W   E. 320W

Question 17:
An electric winch system hoists a mass of 20kg 30 metres into the air over a period of 20 seconds. What is the power output of the winch assuming the system is 100% efficient?

A. 100W   B. 200W   C. 300W   D. 400W   E. 500W

Question 18:

On day 1 of an experiment, a sample is tested and has a count rate of 130Bq.

The same sample is tested on day 7, and the count rate is now 40Bq.

Given that the background radiation count rate is 10Bq, on what day of the experiment will the count rate of the sample be 25Bq?

A. Day 8
B. Day 9
C. Day 10
D. Day 11
E. More information needed

Question 19:

An 80W filament bulb draws 0.5A of household electricity. Using the information that household electricity is available in the UK at 240V, determine the efficiency of the bulb.

A. 25%   B. 33%   C. 50%   D. 66%   E. 75%

Question 20:

Rearrange the following equation in terms of t: $x = \frac{\sqrt{b^3 - 9st}}{13j} + \int_{-z}^{z} 9a - 7$

A. $t = \frac{(13jx - \int_{-z}^{z} 9a-7)^2 - b^3}{9s}$

B. $t = \frac{13jx^2}{b^3 - 9s} - \int_{-z}^{z} 9a - 7$

C. $t = x - \frac{\sqrt{b^3 - 9s}}{13j} - \int_{-z}^{z} 9a - 7$

D. $t = \frac{x^2}{\frac{b^3 - 9s}{13j} + \int_{-z}^{z} 9a - 7}$

E. $t = \frac{[13j(x - \int_{-z}^{z} 9a - 7)]^2 - b^3}{-9s}$

Question 21:

An investment of £500 is made in a compound interest account. At the end of 2 years the balance reads £1125. What is the interest rate?

A. 20%   B. 35%   C. 50%   D. 65%   E. 80%

Question 22:

What is the equation of the line of best fit for the scatter graph below?

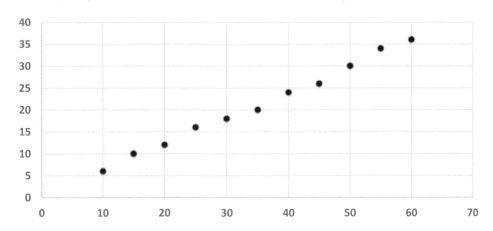

A. y = 0.2x + 0.35

B. y = 0.2x − 0.35

C. y = 0.4x + 0.35

D. y = 0.4x − 0.35

E. y = 0.6x + 0.35

Question 23:

Simplify: $m = \sqrt{\dfrac{9xy^3z^5}{3x^9yz^4}} - m$

A. $m = \sqrt{\dfrac{3y^2z}{x^8}} - m$

B. $m^2 = \dfrac{3y^2z}{x^8} - m$

C. $2m = \sqrt{\dfrac{3y^2z}{x^8}}$

D. $2m^2 = 3x^{-8}y^2z$

E. $4m^2 = 3x^{-8}y^2z$

Question 24:

Which of the following is a suitable descriptive statistic for non-normally distributed data?

A. Mean

B. Normal range

C. Confidence interval

D. Interquartile range

E. Mode

Question 25:

Which of the following best describes the purpose of statistics?

A. Evaluate acceptable scientific practice.

B. Reduce the ability of others to criticise the data.

C. To quickly analyse data.

D. Calculate values representative of the population from a subset sample.

E. To allow for universal comparison of scientific methods.

Question 26:

A rotating disc has two wells, in which bacteria are cultured. The first well is 10 cm from the centre whereas the second well is 20 cm from the centre. If the inner well completes a revolution in 1 second, how much faster is the outer well travelling?

A. 0.314m/s

B. 0.628m/s

C. 0.942m/s

D. 1.256m/s

E. 1.590m/s

Question 27:

Which is the equivalent function to: $y = 9x^{-\frac{1}{3}}$?

A. $y = \frac{1}{x}$

B. $y = \sqrt[3]{9x}$

C. $y = \frac{1}{\sqrt[3]{9x}}$

D. $y = \frac{9}{\sqrt[3]{x}}$

E. $y = \frac{3}{\sqrt[3]{x}}$

END OF SECTION

SECTION 3

1) *"Progress is made by trial and failure; the failures are generally a hundred times more numerous than the successes; yet they are usually left unchronicled."*

Williams Ramsey

Explain what this statement means. Argue to the contrary. To what extent do you agree with the statement?

2) *"He who studies medicine without books sails an uncharted sea, but he who studies medicine without patients does not go to sea at all."*

William Osler

Explain what this statement means. Argue to the contrary. To what extent do you agree with the statement?

3) *"'Medicine is the restoration of discordant elements; sickness is the discord of the elements infused into the living body"*

Leonardo da Vinci

Explain what this statement means. Argue to the contrary. To what extent do you think this simplification holds true within modern medicine?

4) *"Modern medicine is a negation of health. It isn't organized to serve human health, but only itself, as an institution. It makes more people sick than it heals."*

Ivan Illich

What does this statement mean? Argue to the contrary, that the primary duty of a doctor is not to prolong life. To what extent do you agree with this statement?

END OF PAPER

MOCK PAPER C

SECTION 1

Question 1:
Adam, Beth and Charlie are going on holiday together. A single room costs £60 per night, a double room costs £105 per night and a four-person room costs £215 per night. It is possible to opt out from the cleaning service and to pay £12 less each night per room.

What is the minimum amount the three friends could pay for their holiday for a three-night stay at the hotel?
A. £122   B. £144   C. £203   D. £423   E. £432

Question 2:
I have two 96ml glasses of squash. The first is comprised of $\frac{1}{6}$ squash and $\frac{5}{6}$ water. The second is comprised of $\frac{1}{4}$ water and $\frac{3}{4}$ squash. The contents of both glasses are fully mixed. I take 48ml from the first glass and add it to glass two. I then take 72ml from glass two and add it to glass one.

How much squash is now in each glass?
A. 16ml squash in glass one and 72ml squash in glass two.
B. 40ml squash in glass one and 32ml squash in glass two.
C. 48ml squash in glass one and 32ml squash in glass two.
D. 48ml squash in glass one and 40ml squash in glass two.
E. 80ml squash in glass one and 40ml squash in glass two.

Question 3:
It may amount to millions of pounds each year of taxpayers' money; however, it is strongly advisable for the HPV vaccination in schools to continue. The vaccine, given to teenage girls, has the potential to significantly reduce cervical cancer deaths and furthermore, the vaccines will decrease the requirement for biopsies and invasive procedures related to the follow-up tests. Extensive clinical trials and continued monitoring suggest that both Gardasil and Cervarix are safe and tolerated well by recipients. Moreover, studies demonstrate that a large majority of teenage girls and their parents are in support of the vaccine.

Which of the following is the conclusion of the above argument?
A. HPV vaccines are safe and well tolerated.
B. It is strongly advisable for the HPV vaccination in schools to remain.
C. The HPV vaccine amounts to millions of pounds each year of taxpayers' money.
D. The vaccine has the potential to significantly reduce cervical cancer deaths.
E. Vaccinations are vital to disease prevention across the population.

Question 4:

Anna cycles to school, which takes 30 minutes. James takes the bus, which leaves from the same place as Anna, but 6 minutes later and gets to school at the same time as Anna. It takes the bus 12 minutes to get to the post office, which is 3km away. The speed of the bus is $\frac{5}{4}$ the speed of the bike. One day Anna leaves 4 minutes late.

How far does she get before she is overtaken by the bus?

A. 1.5km   B. 2km   C. 3km   D. 4km   E. 6k

Question 5:

In a school year, there are 2 separate maths sets, and each student is assigned to one of them. Set 1 is the 'top set', where students tackled more difficult questions than in Set 2, which is the lower set.
The maths teacher is trying to work out who needs to be moved up from Set 2 to Set 1, and who to award a certificate at the end of term. The students must fulfil certain criteria:

Reward	Criteria
Move to Set 1	Attendance over 95%
	Average test mark over 92
	Less than 5% homework handed in late
Awarded a Certificate	Absences below 4%
	Average test mark over 89
	At least 98% homework handed in on time

	Terry	Alex	Bahara	Lucy	Shiv
Attendance %	97	92	97	100	98
Average test mark %	89	93	94	95	86
Homework handed in on time %	96	92	100	96	98

Who would move from Set 2 to Set 1, and who would receive a certificate?

A. Bahara would move up from Set 2 to Set 1 and receive a certificate.
B. Bahara and Lucy would move up from Set 2 to Set 1 and Bahara would receive a certificate.
C. Bahara, Terry and Lucy would move up from Set 2 to Set 1 and Bahara and Shiv would receive a certificate.
D. Lucy would move up from Set 2 to Set 1 and Bahara would receive a certificate.
E. Lucy would move up from Set 2 to Set 1 and Bahara and Terry would receive a certificate.

Question 6:

18 years ago, A was 25 years younger than B is now. In 21 years time, A will be 28 years older than B was 14 years ago. How old is A now if A is $\frac{5}{6}$B?

A. 27   B. 28   C. 35   D. 42   E. 46

Question 7:

The time now is 10.45am. I am preparing a meal for 16 guests who will arrive tomorrow for afternoon tea. I want to make 3 scones for each guest, which can be baked in batches of 6. Each batch takes 35 minutes to prepare and 25 minutes to cook in the oven and I can start the next batch while the previous batch is in the oven. I also want to make 2 cupcakes for each guest, which can be baked in batches of 8. It takes 15 minutes to prepare the mixture for each batch and 20 minutes to cook them in the oven. I will also make 3 cucumber sandwiches for each guest. 6 cucumber sandwiches take 5 minutes to prepare.

Assuming I can only work on one component of the meal at a time, what will the time be when I finish making all the food for tomorrow?

A. 4:35pm   B. 5.55pm   C. 6:00pm   D. 6:05pm   E. 7:20pm

Question 8:

Pyramid	Base edge (m)	Volume (m³)
1	3	33
2	4	64
3	2	8
4	6	120
5	2	8
6	6	120
7	4	64

What is the difference between the height of the smallest and tallest pyramids?

A. 1m   B. 5m   C. 4m   D. 6m   E. 8m

Question 9:

The wage of Employees at Star Bakery is calculated as: £210 + (Age x 1.2) – 0.8 (100 - % attendance). Jessica is 35 and her attendance is 96%. Samira is 65 and her attendance is 89%.

What is the difference between their wages?

A. £30.40   C. £248.80   E. £279.20

B. £60.50   D. £263.20

Question 10:

It is important that research universities demonstrate convincing support of teaching. Undergraduates comprise an overwhelming proportion of all students and universities should make an effort to cater to the requirements of the majority of their student body. After all, many of these students may choose to pursue a path involving research and a strong education would provide students with skills equipped towards a career in research.

What is the conclusion of the above argument?

A. Undergraduates comprise an overwhelming proportion of all students.
B. A strong education would provide a strong foundation and skills equipped towards a career in research.
C. Research universities should strongly support teaching.
D. Institutions should provide undergraduates with a high-quality learning experience.
E. Research has a greater impact than teaching and limited funds should mainly be invested in research.

Question 11:

American football has reached a level of violence that puts its players at too high a level of risk. It has been suggested that the NFL, the governing body for American football, should dispose of the use of the iconic helmets. The hard-plastic helmets all must meet minimum impact-resistance standards intended to enhance safety, however in reality they give players a false sense of security that only results in harder collisions. Some players now suffer from early onset dementia, mood swings and depression. The proposal to ban helmets for good should be supported. Moreover, it would prevent costly legal settlements involving the NFL and ex-players suffering from head trauma.

What is the conclusion of the above argument?

A. Sports players should not be exposed to unnecessary danger.
B. Helmets give players a false sense of security.
C. Players can suffer from early onset dementia, mood swings and depression.
D. The proposal to ban helmets should be supported.
E. American football is too violent and puts its players at risk.

Question 12:

At the final stop (stop 6), 10 people get off the tube. At the previous stop (stop 5) $\frac{1}{2}$ of the passengers got off. At stop 4, $\frac{3}{5}$ of the passengers got off. At stop 3, $\frac{1}{3}$ of the passengers got off and at stops 1 and 2, $\frac{1}{6}$ of the passengers got off.

How many passengers got on at the first stop?

A. 10   B. 36   C. 90   D. 108   E. 3600

Question 13:

Everyone likes English. Some students born in spring like maths and some like biology. All students born in winter like music and some like art. Of those born in autumn, no one likes biology, and everyone likes art.

Which of the following is true?

A. Some students born in spring like both biology and maths.
B. Students born in spring, winter, and autumn all like art.
C. No one born in winter or autumn likes biology.
D. No one who likes biology also likes art.
E. Some students born in winter like 3 subjects.

Question 14:

Until the twentieth century, the whole purpose of art was to create beautiful, flawless works. Artists attained a level of skill and craft that took decades to perfect and could not be mirrored by those who had not taken great pains to master it. The serenity and beauty produced from movements such as impressionism has however culminated in repulsive and horrific displays of rotting carcasses designed to provoke an emotional response rather than admiration. These works cannot be described as beautiful by either the public or art critics. While these works may be engaging on an intellectual or academic level, they no longer constitute art.

Which of the following is an assumption of the above argument?

A. Beauty is a defining property of art.
B. All modern art is ugly.
C. Twenty first century artists do not study for decades.
D. The impressionist movement created beautiful works of art.
E. Some modern art provokes an emotional response.

Question 15:

The cost of sunglasses is reduced over the bank holiday weekend. On Saturday, the price of the sunglasses is reduced by 10%, compared to the price on Friday. On Sunday the price of the sunglasses is reduced again by 10%, compared to the price on Saturday. On Monday, the price of the sunglasses is reduced by a further 10%, compared to the price on Sunday. What percentage of the price on Friday is the price of the sunglasses on Monday?

A. 55.12%   B. 59.10%   C. 63.80%   D. 70.34%   E. 72.9%

Question 16:

Putting the digit 7 on the right-hand side of a two-digit number causes the number to increase by 565. What is the value of the two-digit number?

A. 27   B. 52   C. 62   D. 66   E. 627

Question 17:

When folded, which box can be made from the net shown below?

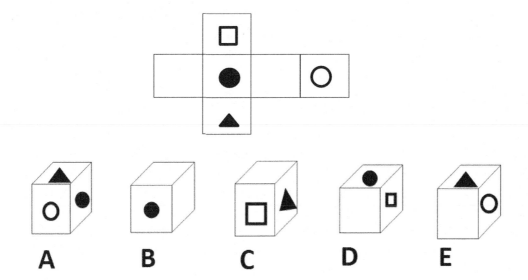

Question 18:

The grid below is comprised of 49 squares. The shaded area is 588cm². What is its perimeter in cm?

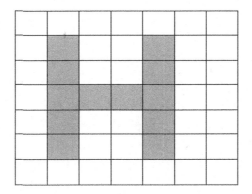

A. 26   B. 49   C. 84   D. 126   E. 182

Question 19:

The UK energy market is highly competitive. In an effort to attract more business and increase revenue, the company EnergyFirst has invested significant funds into its publicity. Last month, they doubled their advertising expenditures, becoming the energy company to invest the largest proportion of investment into advertising. As a result, it is expected that EnergyFirst will expand its customer base at a rate exceeding its competitors in the ensuing months. Other energy companies are likely to follow by example.

Which of the following, if true, is most likely to weaken the above argument?

A. Other companies invest more money into good customer service.

B. Research into the energy industry demonstrates a low correlation between advertising investment and new customers.

C. The UK energy market is not highly competitive.

D. EnergyFirst currently has the smallest customer base.

E. Visual advertising heavily influences customers.

Question 20:

The consumption of large quantities of red meat is suggested to have negative health ramifications. Carnitine is a compound present in red meat and a link has been discovered between carnitine and the development of atherosclerosis, involving the hardening and narrowing of arteries. Intestinal bacteria convert carnitine to trimethylamine-N-oxide, which has properties that are damaging to the heart. Moreover, red meat consumption has been associated with a reduced life expectancy. It may be that charring meat generates toxins that elevate the chance of developing stomach cancer. If people want to be healthy, a vegetarian diet is preferable to a diet including meat. Vegetarians often have lower cholesterol and blood pressure and a reduced risk of heart disease.

Which of the following is an assumption of the above argument?

A. Diet is essential to health, and we should all want to be healthy.

B. Vegetarians do the same amount of exercise as meat eaters.

C. Meat has no health benefits.

D. People who eat red meat die earlier.

E. Red meat is the best source of iron.

Question 21:

Auckland is 11 hours ahead of London. Calgary is 7 hours behind London. Boston is 5 hours behind London. The flight from Auckland to London is 22 hours, but the plane must stop for 2 hours in Hong Kong. The flight from London to Calgary is 8 hours 30 minutes. The flight from Calgary to Boston is 6 hours 30 minutes. Sam leaves Auckland at 10am for London. On arrival to London, he waits 3 hours then gets the plane to Calgary. Once in Calgary, he waits 1.5 hours and gets the plane to Boston. What time is it when Sam arrives in Boston?

A. 13:30pm   B. 22.30pm   C. 01:00am   D. 01:30am   E. 03:30am

Question 22:

Light A flashes every 18 seconds, light B flashes every 33 seconds and light C flashes every 27 seconds. The three lights all flashed at the same time 5 minutes ago.

How long will it be until they next all flash simultaneously?

A. 33 seconds
B. 294 seconds
C. 300 seconds
D. 333 seconds
E. 594 seconds

Question 23:

Drivers in the age group 17-19 comprise 1.5% of all drivers; however, 12% of all collisions involve young drivers in this age category. The RAC Foundation wants a graduated licensing system with a 1-year probationary period with restrictions on what new drivers can do on roads. Additionally, driving instructors need to emphasise the dangers of driving too fast and driving tests should be designed to make new drivers more focused on noticing potential hazards. These changes are essential and could stop 4,500 injuries on an annual basis.

What is the assumption of the above argument?

A. Young drivers are more likely to have more passengers than other age groups.
B. Young drivers spend more hours driving than older drivers.
C. Young drivers are responsible for the collisions.
D. The cars that young people drive are unsafe.
E. Most young drivers involved in accidents are male.

Question 24:

Many countries spent billions on vaccines in response to advice that a virus had the potential to kill millions. These countries are now trying to sell the stockpiles of vaccines which they do not need. There is concern that advice given by officials may have been influenced by pharmaceutical companies. Clearly such companies would have an interest in making sure that governments spend large sums of money on vaccines. It is essential that an investigation into this matter takes place as soon as possible so that those responsible can be held to account.

Which one of the following is an assumption on which this argument depends?

A. The pharmaceutical companies influenced the advice given by officials.
B. The advice given by officials was not appropriate.
C. It will not be possible for the stockpiles of vaccines to be sold.
D. The pharmaceutical companies misjudged the dangers of the virus.
E. Groups with financial interests do not advise officials in other areas of decision making.

Question 25:

Consider the following statements:

1 There are fewer rats than people. **2** There are not more people than rats. **3** There are at least as many rats as people. **4** There are not more rats than people.

Which two of the above statements are equivalent?

A. 1 and 3   B. 1 and 4   C. 2 and 3   D. 2 and 4   E. 3 and 4

Question 26:

A significant social trend in the 20th century was that people moved away from their place of birth in order to access education and work. This gave individuals more opportunities and helped the economy by producing mobility within the workforce. The negative side of this is now being felt as more and more elderly people face the problems of old age without family members nearby to care for them. This has negative effects on the economy as well as on the individual, as more and more state funding for care is needed.

Which one of the following could be drawn as a conclusion of the above passage?

A. The benefits of a mobile workforce have to be compared with the costs to elderly people and the economy.

B. Elderly people are expecting the state to provide care for them rather than relying on their children.

C. People should try to find education and work close to their place of birth.

D. The state should provide care for elderly people to make mobility of the workforce possible.

E. People should make caring for their elderly parents a priority over choice of work opportunities.

Question 27:

Any company that wishes to sell a new drug must provide the government with details of research about its safety and possible side effects. At present, this information is confidential, but there are plans to make it available to the public. While patients are surely entitled to more information about the drugs they are prescribed, this will also inevitably make public vital details about the ingredients of certain drugs and how they are manufactured. Drug companies are naturally reluctant to release this information to their competitors. Therefore, through fear of imitators, drug companies will no longer introduce new and important drugs into the country.

Which one of the following, if true, would most weaken the above argument?

A. There are sufficient drugs already on the market and so there is no need to introduce new ones.

B. The drug industry is a very competitive business and secrecy is vital if companies are to survive.

C. People may be reluctant to use certain drugs when they have fuller information about them.

D. People are better informed about the side effects of drugs abroad than they are in this country.

E. Strong patent laws prevent companies from using the information to create rival drugs.

Question 28:

There are an increasing number of historical and significant buildings in the UK which are said to be 'At Risk'. Without a change in the law most of these buildings are doomed to crumble to the ground. This is because these buildings are no longer structurally sound. The existing strict renovation laws mean that they are too expensive or impractical for private individuals or developers to renovate and repair. There are certainly people out there who would be willing to maintain these buildings if they could use more modern and less expensive techniques and materials. Surely it is better to sacrifice some of the original building's character than lose the entire structure?

Which one of the following best expresses the main conclusion of the above argument?

A. There is nothing wrong with changing the character of historic buildings.

B. 'At Risk' buildings need to be renovated according to strict rules.

C. A change in the law is needed if we hope to preserve more 'At Risk' buildings.

D. Existing laws make 'At Risk' buildings too expensive for most developers.

E. Historians can learn more from buildings which have not been modernised by modern developers.

Question 29:

Many people believe that foreign travel broadens the mind and that there is an inherent benefit in spending some time in a culture different from your own. Many students are taking 'gap' years where they spend time in another country. Whilst this may offer some benefits in terms of confidence and independence, it is wrong to assume that foreign travel alone can provide this. Global travel can have negative impacts on local cultures and the environment. Home country based 'gap' year projects are often seen as unglamorous, but the benefit of working with different groups of people and different cultures within our own society can be equally rewarding.

Which one of the following is the main conclusion of the above passage?

A) Foreign gap year projects must have an element of community work for them to be worthwhile.

B) Foreign travel is not the only way to gain confidence and independence.

C) Projects within our own society can be as rewarding as foreign travel.

D) There is inherent benefit in spending some time abroad.

E) It is important that gap year students consider the impact of their travel on the communities they work in.

Question 30:

"Sugar should be taxed like alcohol and cigarettes."

Which of the following statements, if true, most supports this claim?

A. Sugar can cause diabetes.

B. Sugar has high addictive potential and is associated with various health concerns.

C. High sugar diets increase obesity.

D. People that eat a lot of sugar are more likely to start abusing alcohol.

E. None of the above.

Question 31:

There is no empirical evidence that human activities directly result in global warming, and this is used as a reason against decreasing carbon emissions. However, many scientists believe that human activity is highly likely to cause global warming since higher levels of greenhouse gases cause the atmosphere to thicken, retaining heat. It therefore seems sensible that we should not wait for proof considering the catastrophic effects of climate change, regardless of subsequent findings. Similarly, if a tree branch had a significant chance of falling on you, it would be sensible to move away immediately.

What is the main conclusion?

A. Many scientists believe that human activity is highly likely to cause global warming.

B. We should not wait for proof of climate change.

C. If a tree branch had a significant chance of falling on you, it would be sensible to move away immediately.

D. The effects of climate change are catastrophic.

E. There is no empirical evidence that human activities directly result in global warming, so we should not reduce carbon emissions.

Question 32:

"Unpaid national service is a good way for young people to prepare themselves to become productive members of a democratic society"

Which of the following most closely parallels the reasoning of the above argument?

A. Young people should undertake work experience to prepare themselves for adulthood.

B. Voting should only be extended to those who contribute to society.

C. Internships are a good way for employers to learn which graduates are worth hiring.

D. Unpaid internships are a good way to learn how to become a productive employee.

E. Unemployed people should contribute to society through work schemes if they can't find jobs.

END OF SECTION

SECTION 2

Question 1:
Which of the following is / are **not** involved in the carbon cycle?

1. Lipid molecules in an animal cell
2. Plasmids in a bacterial species
3. Proteins made by a plant cell

A. 1 and 2   C. 1 only   E.   None of the above
B. 1 and 3   D. 2 only

Question 2:
Which of the following statements regarding enzymes are correct?

1. Enzymes are denatured at high temperatures or extreme pH values.
2. Amylase is produced in the salivary glands only and converts starch to sugars.
3. Lipases catalyse the breakdown of oils and fats into glycerol and fatty acids. This takes place in the small intestine.
4. Bile is stored in the pancreas and travels down the bile duct to neutralise stomach acid.

A. 1 and 3 only   C. 1, 2 and 3 only   E.   3 and 4 only
B. 1, 3 and 4 only   D. 2 and 4 only

Question 3:
Which of the following describes the role of the colon?

A. Food is combined with bile and digestive enzymes.
B. Storage of faeces.
C. Reabsorption of water.
D. Faeces leave the alimentary canal.
E. Any digested food is absorbed into the lymph and blood.

Question 4:

Which of the following statements regarding transmission of signals in the nervous system are true?

1. The signal is transmitted across the synapse by diffusion.
2. Transmitter molecules are stored in the pre-synaptic neuron.
3. Transmitter molecules bind to specific receptors on the post-synaptic membrane.

A. 1 and 3 only   C. 1, 2 and 3 only   E. 2 only
B. 1 and 2 only   D. 2 and 3 only

Question 5:

Which of the following statements are true regarding the transition elements?

1. Iron (II) compounds are light green.
2. Transition elements are neither malleable nor ductile.
3. Transition metal carbonates may undergo thermal decomposition.
4. Transition metal hydroxides are soluble in water.
5. When Cu^{2+} ions are mixed with sodium hydroxide solution, a blue precipitate is formed.

A. 1 and 2   B. 1 and 3   C. 1, 3 and 5   D. 3 and 5   E. 5 only

Question 6:

What is the value of C when the equation is balanced?

$\underline{5}$ PhCH$_3$ + $\underline{A}$ KMnO$_4$ + $\underline{9}$ H$_2$SO$_4$ = $\underline{5}$ PhCOOH + $\underline{B}$ K$_2$SO$_4$ + $\underline{C}$ MnSO$_4$ + $\underline{14}$ H$_2$O

A. 3   B. 4   C. 5   D. 7   E. 9

Question 7:

Tongue-rolling is controlled by the dominant allele T, while non-rolling is controlled by the recessive allele, t.

Red-green colour blindness is controlled by a sex-linked gene on the X chromosome. Normal colour vision is controlled by dominant allele B, while red-green colour blindness is controlled by the recessive allele, b.

The mother of a family is colour blind and heterozygous for tongue-rolling, while the father has normal colour vision and is a non-roller.

Which of the following statement(s) is / are correct?

1. More males than females in a population are red-green colour blind.
2. 50% of children will be non-rollers.
3. All the male children will be colour-blind.

A. 1 and 2 only   C. 2 only   E. 3 only

B. 1, 2 and 3   D. 2 and 3 only

Question 8:

Make y the subject of the formula: $\frac{y+x}{x} = \frac{x}{a} + \frac{a}{x}$

A. $y = \frac{x^2}{a} + a$ C. $y = \frac{-ax}{x^2+a^2}$ E. $y = a^2 - ax$

B. $y = \frac{x^2+a^2-ax}{a}$ D. $y = \frac{x^2}{ax} + a - x$

Question 9:

What is the mass in grams of calcium chloride, $CaCl_2$, in $25cm^3$ of a solution with a concentration of 0.1 mol.l⁻¹? (Ar of Ca is 40 and Ar of Cl is 35)

A. 0.28g   B. 0.46g   C. 0.48g   D. 0.72g   E. 1.28g

Question 10:

Consider the equations: A: $y = 3x$ and B: $y = \frac{6}{x} - 7$. At what values of x do the two equations intersect?

A. x=2 and x=9   C. x=6 and x=27   E. x=18

B. x=3 and x=6   D. x=6

Question 11:
Which of the following statement(s) regarding the circulatory system is / are correct?

1. The pulmonary artery carries oxygenated blood from the right ventricle to the lungs.
2. The aorta has a high content of elastic tissue and carries oxygenated blood from the left ventricle around the body.
3. The mitral valve is between the pulmonary vein and the left atrium.
4. The vena cava carries deoxygenated blood from the body to the right atrium.

A. 1 and 3   B. 1 and 2   C. 2 only   D. 2 and 4   E. 3 only

Question 12:
A compound with a molar mass of 120 g.mol^{-1} contains 12g of carbon, 2g of hydrogen and 16g oxygen. What is the molecular formula of the compound? (Ar C = 12, Ar H = 1, Ar O = 16).

A. CH_2O B. $C_2H_4O_2$ C. C_4H_2O D. $C_4H_8O_4$ E. $C_8H_{16}O_8$

Question 13:
Rupert plays one game of tennis and one game of squash.

The probability that he will win the tennis game is $\frac{3}{4}$

The probability that he will win the squash game is $\frac{1}{3}$

What is the probability that he will win one game only?

A. $\frac{3}{12}$ B. $\frac{7}{12}$ C. $\frac{4}{5}$ D. $\frac{13}{12}$ E. $\frac{7}{6}$

Question 14:
What is the median of the following numbers:

$\frac{7}{36}$; $0.\dot{3}$; $\frac{11}{18}$; 0.25; 0.75; $\frac{62}{72}$; $\frac{7}{7}$

A. $\frac{7}{36}$ B. $0.\dot{3}$ C. $\frac{11}{18}$ D. $\frac{62}{72}$ E. 0.75

Question 15:

16.4g of nitrobenzene is produced from 13g of benzene in excess nitric acid: $C_6H_6 + HNO_3 \rightarrow C_6H_5NO_2 + H_2O$

What is the percentage yield of nitrobenzene ($C_6H_5NO_2$)? (Ar C = 12, Ar N = 14, Ar H = 1, Ar O = 16)

A. 65%   B. 67%   C. 72%   D. 78%   E. 80%

Question 16:

Which of the following points regarding electromagnetic waves are correct?

1. Radiowaves have the longest wavelength and the lowest frequency.
2. Infrared radiation has a shorter wavelength than visible light and is used in optical fibre communication, and heater and night vision equipment.
3. All of the waves from gamma to radio waves travel at the speed of light (about 300,000,000 m/s).
4. Infrared radiation is used to sterilise food and to kill cancer cells.
5. Darker skins absorb more UV light, so less ultraviolet radiation reaches the deeper tissues.

A. 1 and 2   B. 1 and 3   C. 1, 3 and 5   D. 2 and 3   E. 2 and 4

Question 17:

Two carriages of a train collide and then start moving together in the same direction. Carriage 1 has mass 12,000 kg and moves at 5ms^{-1} before the collision. Carriage 2 has mass 8,000 kg and is stationary before the collision.

What is the velocity of the two carriages after the collision?

A. 2 ms^{-1} B. 3 ms^{-1} C. 4 ms^{-1} D. 4.5 ms^{-1} E. 5 ms^{-1}

Question 18:

Which of the following statements are true?

1. Control rods are used to absorb electrons in a nuclear reactor to control the chain reaction.
2. Nuclear fusion is commonly used as an energy source.
3. An alpha particle is comprised of two protons and two neutrons and is the same as a helium nucleus.
4. When $^{14}_{6}C$ undergoes beta decay, an electron and $^{14}_{7}N$ are produced.
5. Beta particles are less ionising than gamma rays and more ionising than alpha particles.

A. 1 and 2   C. 3 and 4   E. None of the statements are true
B. 1 and 3   D. 3, 4 and 5

Question 19:

Simplify fully: $\dfrac{(3x^{\frac{1}{2}})^3}{3x^2}$

A. $\dfrac{3x}{\sqrt{x}}$

B. $\dfrac{9}{x}$

C. $3x^{\frac{1}{2}}$

D. $3x\sqrt{x}$

E. $\dfrac{9}{\sqrt{x}}$

Question 20:

Which of the following are true?

1. Lightning, as well as nitrogen-fixing bacteria, converts nitrogen gas to nitrate compounds.
2. Decomposers return nitrogen to the soil as ammonia.
3. The shells of marine animals contain calcium carbonate, which is derived from dietary carbon.
4. Nitrogen is used to make the amino acids found in proteins.

A. 1 only

B. 1 and 2

C. 2 and 3

D. 2, 3 and 4

E. They are all true

Question 21:

Write $\dfrac{\sqrt{20}-2}{\sqrt{5}+3}$ in the form: $p\sqrt{5} + q$

A. $2\sqrt{5} - 4$

B. $3\sqrt{5} - 4$

C. $3\sqrt{5} - 5$

D. $4\sqrt{5} - 6$

E. $5\sqrt{5} + 4$

Question 22:

Which of the following statements is / are false?

1. Simple molecules do not conduct electricity because there are no free electrons and there is no overall charge.
2. The carbon and silicon atoms in silica are arranged in a giant lattice structure and it has a very high melting point.
3. Ionic compounds do not conduct electricity when dissolved in water or when melted because the ions are too far apart.
4. Alloys are harder than pure metals.

A. 1 and 2

B. 1, 2 and 4

C. 1, 2, 3 and 4

D. 2 and 4

E. 3 only

Question 23:

The graph below shows a circle with radius 5 and centre (0,0).

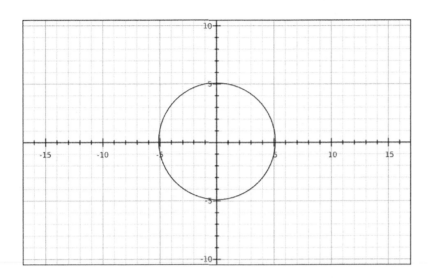

What are the values of x when the line $y = 3x - 5$ meets the circle?

A. $x = 0$ or $x = 3$
B. $x = 0$ or $x = 3.5$

C. $x = 1$ or $x = 3.5$
D. $x = 1.5$ or $x = -3$

E. $x = 1.5$ or $x = -2$

Question 24:

Which if the following statements regarding heat transfer is / are correct?

1. In liquids and gases, heat energy is transferred from hotter to colder places by conduction because particles in liquid and gases move more quickly when heated.
2. Liquid and gas particles in hot areas are less dense than in cold areas.
3. Heat transfer via radiation does not need particles to travel.
4. Dull surfaces are good at absorbing and poor at reflecting infrared radiation, whereas shiny surfaces are poor at absorbing, but good at reflecting infrared radiation.

A. 1 and 2
B. 1, 2 and 4

C. 2 and 3
D. 2 and 4

E. 4 only

Question 25:

The following points refer to the halogens:

1. Iodine is a grey solid and can be used to sterilise wounds. It forms a purple vapour when warmed.
2. The melting and boiling points increase as you go up the group.
3. Fluorine is very dangerous and reacts instantly with iron wool, whereas iodine must be strongly heated as well as the iron wool for a reaction to occur and the reaction is slow.
4. When bromine is added to sodium chloride, the bromine displaces chlorine from sodium chloride.
5. The hydrogen atom and chlorine atom in hydrogen chloride are joined by a covalent bond.

Which of the above statements is / are false?

A. 1, 3 and 5   C. 2 and 4   E. 3, 4 and 5
B. 1, 2 and 3   D. 3 only

Question 26:

Consider the triangle to the right where BE=4cm, EC=2cm and AC=9cm.

What is the length of side DE?

A. 4cm
B. 5.5cm
C. 6cm
D. 7.5cm
E. 8cm

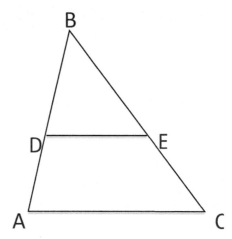

Question 27:

A ball is projected vertically upwards with an initial speed of 40 ms^{-1}. What is the maximum height reached? (Take gravity to be 10 ms^{-2} and assume negligible air resistance).

A. 25m   B. 45m   C. 60m   D. 75m   E. 80m

END OF SECTION

SECTION 3

1) *'The NHS should not treat obese patients'*

Explain what this statement means. Argue to the contrary, that we **should** treat obese patients. To what extent do you agree with this statement?

2) *'We should all become vegetarian'*

Explain what this statement means. Argue to the contrary, that we **should not** all become vegetarian. To what extent do you agree with this statement?

3) *'Certain vaccines should be mandatory'*

Explain what this statement means. Argue to the contrary, that vaccines **should not** be mandatory. To what extent do you agree with this statement?

4) *'Compassion is the most important quality of a healthcare professional'*

Explain what this statement means. Argue to the contrary, that there are more important qualities than compassion for health professionals. To what extent do you agree with this statement?

END OF PAPER

MOCK PAPER D

SECTION 1

Question 1:
"Competitors need to be able to run 200 metres in under 25 seconds to qualify for a tournament. James, Steven and Joe are attempting to qualify. Steven and Joe run faster than James. James' best time over 200 metres is 26.2 seconds." Which response is definitely true?

A. Only Joe qualifies.

B. James does not qualify.

C. Joe and Steven both qualify.

D. Joe qualifies.

E. No one qualifies.

Question 2:
You spend £5.60 in total on a sandwich, a packet of crisps and a watermelon. The watermelon cost twice as much as the sandwich, and the sandwich cost twice as much as the crisps.

How much did the watermelon cost?

A. £1.20   B. £2.60   C. £2.80   D. £3.20   E. £3.60

Question 3:
Jane, Chloe and Sam are all going by train to a football match. Chloe gets the 2:15pm train. Sam's journey takes twice as long Jane's. Sam catches the 3:00pm train. Jane leaves 20 minutes after Chloe and arrives at 3:25pm.

When will Sam arrive?

A. 3:50pm   B. 4:10pm   C. 4:15pm   D. 4:30pm   E. 4:40pm

Question 4:
Michael has eleven sweets. He gives three sweets to Hannah. Hannah now has twice the number of sweets Michael has remaining.

How many sweets did Hannah have before the transaction?

A. 11   B. 12   C. 13   D. 14   E. 15

Question 5:
Alex gets a pay rise of 10% plus an extra £5 per week. The flat rate of income tax on his salary is decreased from 20% to 15% at the same time. Alex's old weekly take-home pay after tax is £200 per week.

What will his new weekly take-home pay be, to the nearest whole pound?

A. £220   B. £232   C. £238   D. £245   E. £250

Question 6:
You have four boxes, each containing two cubes. Box A contains two white cubes, Box B contains two black cubes, and Boxes C and D both contain one white cube and one black cube. You pick a box at random and take out one cube. It is a white cube. You then draw another cube from the same box. What is the probability that this cube is not white?

A. ½   B. ⅓   C. ⅔   D. ¼   E. ¾

Question 7:
Anderson & Co. hire out heavy plant machinery at a cost of £500 per day. There is a surcharge for heavy usage, at a rate of £10 per minute of usage over 80 minutes. Concordia & Co. charge £600 per day for similar machinery, plus £5 for every minute of usage. For what duration of usage are the costs the same for both companies?

A. 100 minutes   C. 140 minutes   E. 180 minutes
B. 130 minutes   D. 170 minutes

Question 8:
Simon is discussing with Seth whether or not a candidate is suitable for a job. When pressed for a weakness at interview, the candidate told Simon that he is a slow eater. Simon argues that this will reduce the candidate's productivity, since he will be inclined to take longer lunch breaks.

Which statement **best** supports Simon's argument?
A) Slow eaters will take longer to eat lunch.
B) Longer lunch breaks are a distraction.
C) Eating more slowly will reduce the time available to work.
D) Eating slowly is a weakness.
E) Eating slowly will lead to less time to work efficiently.

Question 9:

Three pieces of music are on repeat in different rooms of a house. One piece of music is three minutes long, one is four minutes long and the final one is 100 seconds long. All pieces of music start playing at exactly the same time. How long is it until they are next starting together again?

A. 12 minutes   C. 20 minutes   E. 300 minutes

B. 15 minutes   D. 60 minutes

Question 10:

A car leaves Salisbury at 8:22am and travels 180 miles to Lincoln, arriving at 12:07pm. Near Warwick, the driver stopped for a 14-minute break.

What was its average speed, whilst travelling, in kilometres per hour? It should be assumed that the conversion from miles to kilometres is 1:1.6.

A. 51kph   B. 67kph   C. 77kph   D. 82kph   E. 86kph

Question 11:

"Recently in Kansas, a number of farm animals have been found killed in the fields. The nature of the injuries is mysterious, but consistent with tales of alien activity. Local people talk of a number of UFO sightings, and claim extra terrestrial responsibility. Official investigations into these claims have dismissed them, offering rational explanations for the reported phenomena. However, these official investigations have failed to deal with the point that, even if the UFO sightings can be explained in rational terms, the injuries on the carcasses of the farm animals cannot be. Extra terrestrial beings must therefore be responsible for these attacks."

Which of the following best expresses the main conclusion of this argument?

A. Sightings of UFOs cannot be explained by rational means.

B. Recent attacks must have been carried out by extraterrestrial beings.

C. The injuries on the carcasses are not due to normal predators.

D. UFO sightings are common in Kansas.

E. Official investigations were a cover-up.

Question 12:

"To make a cake you must prepare the ingredients and then bake it in the oven. You purchase the required ingredients from the shop, however, your oven is broken. Therefore, you cannot make a cake."

Which of the following arguments has the same structure as the passage above?

A. To get a good job, you must have a strong CV then impress the recruiter at interview. Your CV was not as good as other applicants; therefore, you didn't get the job.

B. To get to Paris, you must either fly or take the Eurostar. There are flight delays due to dense fog, therefore you must take the Eurostar.

C. To borrow a library book, you must go to the library and show your library card. At the library, you realise you have forgotten your library card. Therefore, you cannot borrow a book.

D. To clean a bedroom window, you need a ladder and a hosepipe. Since you don't have the right equipment, you cannot clean the window.

E. Bears eat both fruit and fish. The river is frozen, so the bear cannot eat fish.

Question 13:

"Making model ships requires patience, skill and experience. Patience and skill without experience is common – but often such people give up prematurely, since skill without experience is insufficient to make model ships, and patience can quickly be exhausted."

Which of the following summarises the main argument?

A. Most people lack the skill needed to make model ships.

B. Making model ships requires experience.

C. The most important thing is to get experience.

D. Most people make model ships for a short time but give up due to a lack of skill.

E. Successful model ship makers need to have several positive traits.

Question 14:

"Joseph has a bag of building blocks of various shapes and colours. Some of the cubic ones are black. Some of the black ones are pyramid shaped. All blue ones are cylindrical. There is a green one of each shape. There are some pink shapes."

Which of the following is definitely **NOT** true?

A. Joseph has pink cylindrical blocks.

B. Joseph doesn't have pink cylindrical blocks.

C. Joseph has blue cubic blocks.

D. Joseph has a green pyramid.

E. Joseph doesn't have a black sphere.

Question 15:

Sam notes that the time on a normal analogue clock is 1540hrs.

What is the smaller angle between the hands on the clock?

A. 110°   B. 120°   C. 130°   D. 140°   E. 150°

Question 16:

A fair 6-faced die has 2 sides painted red. The die is rolled 3 times.
What is the probability that at least one red side has been rolled?

A. $8/27$ B. $19/27$ C. $21/27$ D. $24/27$ E. 1

Question 17:

"In a particular furniture warehouse, all chairs have four legs. No tables have five legs, nor do any have three. Beds have no less than four legs, but one bed has eight as they must have a multiple of four legs. Sofas have four or six legs. Wardrobes have an even number of legs, and sideboards have and odd number. No other furniture has legs. Brian picks a piece of furniture out, and it has six legs."

What can be deduced about this piece of furniture?

A. It is a table.
B. It could be either a wardrobe or a sideboard.
C. It must be either a table or a sofa.
D. It must be either a table, a sofa or a wardrobe.
E. It could be either a bed, a table or a sofa.

Question 18:

Two friends live 42 miles away from each other. They walk at 3mph towards each other. One of them has a pet falcon which starts to fly at 18mph as soon as the friends set off. The falcon flies back and forth between the two friends until the friends meet.

How many miles does the falcon travel in total?

A. 63   B. 84   C. 114   D. 126   E. 252

Question 19:

"Antibiotic resistance is on the increase. As a result, many antibiotics in our vast armoury are becoming ineffective against common infections. Probably the most significant contributor to this is the use of preventative antibiotics in farming, as this exposes bacteria to antibiotics for no good reason, giving the opportunity for resistance to develop. If this worrying trend continues, we might, in 30 years time, be back in the Victorian situation, where people die from skin or chest infections, we consider mild and eminently treatable today."

Which of the following best represents the overall conclusion of the passage?

A. Antibiotic resistance is a serious issue.

B. Antibiotics use in farming is essential.

C. The use of antibiotics in farming could cause us serious harm.

D. Victorians used to die from diseases we can treat today.

E. Antibiotics can treat skin infections.

Question 20:

A complete set of maths equipment includes a pen, a pencil, a geometry set and a pad of paper. Pens cost £1.50, pencils cost 50p, paper pads cost £1, and geometry sets cost £3. Sam, Dave and George each want complete sets, but Mr Browett persuades them to share some items. Sam and Dave agree to share a paper pad and a geometry set. George must have his own pen but agrees that he and Sam can share a pencil.

What is the total amount spent?

A. £12.00   B. £13.50   C. £16.50   D. £17.50   E. £18.00

Question 21:

The figure below shows 12 individual planks arranged such that 5 squares are made with them.

To make 7 squares in total, which two planks need to be moved?

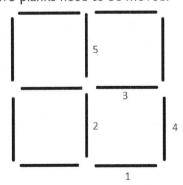

A. 1 and 2   B. 1 and 3   C. 1 and 4   D. 3 and 5   E. 4 and 5

Question 22:

A cube has six sides of different colours. The red side is opposite to black. The blue side is adjacent to white. The purple side is adjacent to blue. The final side is yellow.

Which colour is opposite the purple side?

A. Red   B. Black   C. Blue   D. White   E. Yellow

Question 23:

"Some people with stomach pains and diarrhoea have Giardiasis."

Using the information above, which of the following statements can be accurately concluded?

A. Some people have stomach pains, but do not have Giardiasis.
B. Some people with stomach pains and diarrhoea do not have Giardiasis.
C. Kate has Giardiasis. Therefore, she has stomach pains.
D. Giardiasis is defined as stomach pains and diarrhoea together.
E. None of the above

Question 24:

"Insect pests such as aphids and weevils can be a problem for farmers, as they cause destruction by feeding on crops. Thus, many farmers spray their crops with pesticides to kill these insects, increasing their crop yield. However, there are also predatory insects such as wasps and beetles that naturally prey on these pests – which are also killed by pesticides. Therefore, it would be better to let these natural predators control the pests, rather than by spraying needless chemicals."

Which of the following best describes the flaw in this logic?

A. Many pesticides are expensive, so should not be used unless necessary.
B. It fails to consider other problems the pesticides may cause.
C. It does not explain why weevils are a problem.
D. It fails to assess the effectiveness of natural predators compared to pesticides.
E. It does not consider the benefits of using fewer pesticides.

Question 25:

In regions of a comparatively low altitude, many birds fly to the far North in order to find the proper climatic conditions in which to rear their broods and spend their summer vacation. Some of them go to the subarctic provinces and others go beyond. How different among the sublime heights of the Rockies! Here, they are required to make a journey of only a few miles, say from five to one hundred, or slightly more, according to the locality selected. They travel in order to find the best conditions in terms of temperature, food, nesting sites, etc., that are precisely to their taste.

Which of the following statements can be reliably concluded from the above passage?

A. A journey of 100 miles is too far for these birds to travel for food.
B. Rearing their young is the most important part of these birds' lives.
C. These birds fly north as all their needs are in a localised area.
D. Nesting in the Rockies keeps the birds away from predators.
E. The birds struggle to survive in the harsh cold temperatures further north.

Question 26:

Three ladies X, Y and Z marry three men A, B and C. X is married to A, Y is not married to an engineer, Z is not married to a doctor, C is not a doctor and A is a lawyer.

Using the information above, which of the following statements can be inferred?

A. Y is married to C who is an engineer.
B. Z is married to C who is a doctor.
C. X is married to a doctor.
D. All of the above.
E. None of these.

Question 27:

The medical scientific establishment has a long-established system for naming body parts and medical phenomena. This system is based upon ease of understanding, such that a body part, or a process of the body, is named based on its clinical relevance. This means that features are named in a way which will help doctors understand and explain to patients what the body part is, or what is wrong with it in the case of a disease. However, this poses significant problems for scientific medical research. Often, the most important features of a body part from a scientific point of view are not the most clinically important features, leading to confusion within the scientific literature, as medical researchers misunderstand the purpose of a discussion, due to confusing nomenclature. Whilst it is important for doctors to be able to explain things clearly to patients, it is relatively easy for this to happen in spite of confusing nomenclature, whereas confusing names cause serious problems in the scientific world. Thus, the naming system for medical features should be edited, to reflect the scientifically important features of body parts, rather than the clinically important ones.

Which of the following best illustrates the main conclusion of this passage?

A. The naming system based on clinically important features causes problems in scientific literature.

B. Changing the naming system would allow faster progress to be made in scientific medical research.

C. The naming system should be changed to reflect the features of body parts which are most important scientifically.

D. The current naming system is sufficient and should not be changed to help lazy scientists who cannot be bothered to do fact-checking.

E. It is more important to have good doctor-patient relations than good progress in scientific research.

Question 28:

It is well established that modern humans evolved in Africa around 2 million years ago, and that the first humans were mainly hunter-gatherers, living off of hunted meat and plant foods collected from their environment. However, this poses an interesting question. Humans are relatively weak, small, feeble creatures, and around 2 million years ago most wildlife in Africa consisted of large, powerful creatures. Thus, it is unclear how humans were able to hunt successfully, and obtain meat for food. One theory is that humans are well-built for long-distance running, largely thanks to our ability to control our temperature via sweating.

This theory reasons that humans were able to pursue animals such as antelope, which run when challenged, and were able to keep on running until the antelope collapsed through heat exhaustion. Meanwhile, the humans were kept cool via sweating, and were able to then go in and butcher the defenceless antelope.

Recent evidence has emerged supporting this theory, showing that human feet are well-developed for long-distance running, with fleshy areas in the correct orientation to absorb the impact without causing joint damage, and a heart well evolved to keep pumping at a moderately fast pace for long periods. With the emergence of this powerful new evidence, we should accept this theory, known as "the persistence running theory" as true.

Which of the following identifies a flaw in this argument?

A. The emergence of evidence in support of the persistence running theory does not mean that this theory is true.

B. There is little evidence that the human body is well setup for long-distance running.

C. It has neglected to consider other theories for how humans obtained meat during their early evolution.

D. There are numerous issues with the theory of persistence running, but many of these have been resolved thanks to the new evidence that has emerged.

E. It has not considered evidence that humans evolved in Europe, where there are smaller animals which humans may have more easily been able to tackle.

Question 29:

Recent research suggests that people are becoming less inclined to follow medical advice to prevent ill-health. It is frequently argued that there is too much advice, and it is often contradictory. In spite of this the general population is living longer and is generally healthier. This suggests that individuals are more aware of what is good for their own health and wellbeing than the medical profession is.

Which one of the following, if true, weakens this argument?

A. Advances in medicine have meant that doctors give advice on a wider range of issues.

B. People now have easy access to websites giving information on health.

C. People believe that they know better than doctors how to improve their own health and wellbeing.

D. The health improvements are in areas that exactly match the medical advice given by doctors.

E. Doctors prefer to give advice rather than medication.

Question 30:

Over the last twenty years the number of people, including children, classed as overweight, and therefore at risk of serious health problems, has risen alarmingly. This trend could be caused by an increase in the amount people eat or by a decrease in the amount of exercise they take. Most of us exercise less than people did twenty years ago, and the average number of calories consumed per person is now less than it was twenty years ago. So, the increase in the number of overweight people is clearly caused by lack of exercise. The government therefore does not need to worry about trying to change people's diets.

Which one of the following identifies the flaw in this argument?

A. Some people may exercise more than the average.

B. Some individuals may have increased their calorie intake.

C. The government may need to worry about costs to the health service.

D. Children may use up more calories through exercise than adults.

E. Some individuals may have health problems which cause an increase in weight.

Question 31:

The general public cannot understand laws and legal documents unless they are written in clear and simple language. Therefore, the traditional style in which laws and legal documents are written must change. Citizens in a democracy must be able to understand what their legal rights and duties are.

Which one of the following best expresses the conclusion of this argument?

A. There must be a change in the style in which laws and legal documents are written.

B. It is necessary in a democracy for citizens to know their legal rights and duties.

C. Many laws and legal documents are written in old-fashioned and complicated language.

D. The general public can fully understand only those laws and documents written in simple language.

E. If citizens can understand laws and legal documents, they will be able to play their proper role in a democracy.

Question 32:

Many countries spent billions on vaccines in response to advice that a virus had the potential to kill millions. These countries are now trying to sell the stockpiles of vaccines which they do not need. There is concern that advice given by officials may have been influenced by pharmaceutical companies. Clearly such companies would have an interest in making sure that governments spend large quantities of money on vaccines. It is essential that an investigation into this matter takes place as soon as possible so that those responsible can be held to account.

Which one of the following is an assumption on which this argument depends?

A. The pharmaceutical companies influenced the advice given by officials.

B. The advice given by officials was not appropriate.

C. It will not be possible for the stockpiles of vaccines to be sold.

D. The pharmaceutical companies misjudged the dangers of the virus.

E. Groups with financial interests do not advise officials in other areas of decision making.

END OF SECTION

SECTION 2

Question 1:

Which of the following statements about the digestion of lipids in the small intestine is / are correct?

1. Bile acts as an emulsifier to make larger lipid droplets, each of which has a larger surface area.
2. Bile is neutral, to increase the pH of the material from the stomach.
3. Lipases break down lipids into their component fatty acid and glycerol molecules.

A. 1 only
B. 2 only
C. 3 only
D. 2 and 3 only
E. None of the above

Question 2:

The primary ions responsible for an action potential on a muscle cell membrane are sodium and potassium. Sodium concentration is higher than that of potassium outside the cell. Potassium concentration is higher than that of sodium inside the cell. Depolarisation occurs when the membrane potential increases (becomes more positive).

Which of the following **must** be true when a muscle cell membrane depolarises?

A. More potassium moves into the muscle cell than sodium.
B. More sodium moves into the muscle cell than potassium.
C. There is no net flow of sodium or potassium ions.
D. The membrane potential becomes more negative.
E. None of the above.

Question 3:

Calculate the radius of a sphere which has a surface area three times as great as its volume.

A. 0.5   C. 1.5   E. 2.5
B. 1   D. 2   F. More information is needed

Question 4:

A mechanical winch lifts up a bag of grain in a mill from the floor into a hopper.

Assuming that the machine is 100% efficient and lifts the bag vertically only, which of the following statements is / are **TRUE**?

1. This increases gravitational potential energy
2. The gravitational potential energy is independent of the mass of the grain
3. The work done is the difference between the gravitational potential energy at the hopper and when the grain is on the floor
4. The work done is the difference between the kinetic energy of the grain in the hopper and on the floor

A. I only
B. I and 3
C. I and 4
D. I, 2 and 3
E. I, 2 and 4
F. None of the above

Question 5:

A barometer records atmospheric pressure as 10^5 Pa. Recalling that the diameter of the Earth is 1.2×10^7 m, **estimate** the mass of the atmosphere. [Assume g = 10 ms^{-2}, the earth is spherical and that $\pi = 3$]

A. $3.6 \times 108 \, kg$
B. $4.32 \times 1010 \, kg$
C. $4.32 \times 1012 \, kg$
D. $3.6 \times 1013 \, kg$
E. $4.32 \times 1018 \, kg$
F. More information is required

Question 6:

Which of the following substances is **not** a polymer?

A. Polythene
B. Glycogen
C. Collagen
D. Starch
E. DNA
F. Triglyceride

Question 7:

SIADH is a metabolic disorder caused by an excess of Anti-Diuretic Hormone (ADH) released by the posterior pituitary gland.

Which row best describes the urine produced by a patient with SIADH?

	Volume	Salt Concentration	Glucose
A)	High	Low	Low
B)	High	High	Low
C)	High	High	High
D)	Low	Low	Low
E)	Low	High	Low
F)	Low	High	High

Question 8:
A 6kg missile is fired and decelerates at 6ms^{-2}.

What is the difference in resistive force compared to a 2kg missile fired and decelerating at 8ms^{-2}?

A. 8N B. 12N C. 16N D. 20N E. 24N

Question 9:
Place the following substances in order from most to least reactive:

1	Na
2	K
3	Zn
4	Cu
5	Au
6	Ca

A. 1 » 2 » 6 » 3 » 4 » 5   C. 2 » 1 » 6 » 3 » 4 » 5   E. 2 » 6 » 1 » 3 » 4 » 5
B. 1 » 2 » 6 » 3 » 5 » 4   D. 2 » 1 » 6 » 3 » 5 » 4

Question 10:
The normal cardiac cycle has two phases, systole and diastole.

During diastole, which of the following is **false**?
A. The aortic valve is closed.   D. The pressure in the aorta increases.
B. The ventricles are relaxing.   E. There is blood in the ventricles.
C. There is blood in the ventricles.

Question 11:

The figure below shows a schematic of a wiring system. All of the bulbs have equal resistance. The power supply is 24V.

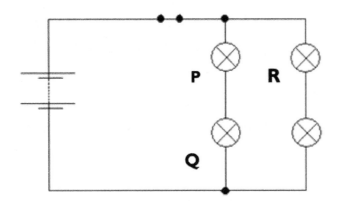

If headlight Q is replaced by a new one with twice the resistance, with the switch closed, which of these combinations of voltage drops across the four bulbs is possible?

	P	Q	R	S
A)	8V	16V	12V	12V
B)	8V	16V	16V	8V
C)	8V	16V	8V	16V
D)	12V	24V	24V	24V
E)	12V	12V	12V	12V
F)	16V	8V	12V	12V
G)	16V	8V	8V	16V
H)	24V	24V	24V	24V
I)	4V	8V	6V	6V
J)	8V	4V	6V	6V

Question 12:

A cup has 144ml of pure deionised water. How many electrons are in the cup due to the water? [Avogadro Constant = 6×10^{23}]

A. 8.64×10^{24}

B. 8.64×10^{25}

C. 1.2×10^{24}

D. 2.4×10^{24}

E. 4.8×10^{25}

Question 13:
Below is a graph showing the concentration of product over time as substrate concentration is increased, which can be visualised as Line 1. Some enzyme inhibitors are introduced.

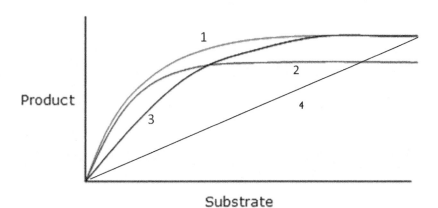

Which, if any, line represents the effect of competitive inhibition?

A. Line 2   B. Line 3   C. Line 4   D. None of these lines

Question 14:
Which of the following is **not** present in the plasma membrane?

A. Extrinsic proteins   C. Phospholipids   E. Nucleic Acids
B. Intrinsic proteins   D. Glycoproteins   F. They are all present

Question 15:
There are 1000 international airports in the world. If 4 flights take off every hour from each airport, estimate the annual number of commercial flights worldwide, to the nearest 1 million.

A. 20 million   C. 37 million   E. 42 million
B. 35 million   D. 40 million   F. 44 million

Question 16:
Steve's sports car requires 2.28kg of octane to travel to Pete's house 10 miles away. Calculate the mass of CO_2 produced during the journey. (Ar C = 12, Ar O = 16)

A. 0.88 kg   C. 2.64 kg   E. 5.28 kg
B. 1.66 kg   D. 3.52 kg   F. 7.04 kg

Question 17:
Given:

F + G + H = 1   F + G − H = 2   F − G − H = 3

Calculate the value of FGH.

A) -2   B) -0.5   C) 0   D) 0.5   E) 2

Question 18:
A pulmonary embolism occurs when a main artery supplying the lungs becomes blocked by a clot that has travelled from somewhere else in the body.

Which option best describes the path of a blood clot that originated in the leg and has caused a pulmonary embolism?

A) Inferior Vena cava   F) Left ventricle
B) Superior Vena cava   G) Pulmonary artery
C) Right atrium   H) Pulmonary vein
D) Right ventricle   I) Aorta
E) Left atrium   J) Coronary artery

A. C, D, H, G   C. I, E, F, G   E. A, C, D, J, G
B. B, C, D, H, G   D. A, C, D, G   F. A, C, D, J, E, F, G

Question 19:
The concentration of chloride in the blood is 100mM. The concentration of thyroxine is 1×10^{-10}kM. Calculate the ratio of thyroxine to chloride ions in the blood.

A) Chloride is 100,000,000 times more concentrated than thyroxine.
B) Chloride is 1,000,000 times more concentrated than thyroxine.
C) Chloride is 1000 times more concentrated than thyroxine.
D) Concentrations of chloride and thyroxine are equal.
E) Thyroxine is 1000 times more concentrated than chloride.
F) Thyroxine is 1,000,000 times more concentrated than chloride.

Question 20:

Put the following types of electromagnetic waves in ascending order of wavelength:

	Shortest -------------------------- Longest			
A)	Visible Light	Ultraviolet	Infrared	X-ray
B)	Visible Light	Infrared	Ultraviolet	X-ray
C)	Infrared	Visible Light	Ultraviolet	X-ray
D)	Infrared	Visible Light	X-ray	Ultraviolet
E)	X-ray	Ultraviolet	Visible Light	Infrared
F)	X-ray	Ultraviolet	Infrared	Visible Light
G)	Ultraviolet	X-ray	Visible Light	Infrared

Question 21:

How many seconds are there in 66 weeks? [n! = 1 x 2 x 3 x... x n].

A. 7!　　　B. 8!　　　C. 9!　　　D. 10!　　　E. 11!　　　F. 12!

Question 22:

Which of the following is **not** a hormone?

A. Insulin　　　C. Noradrenaline　　　E. Thyroxine　　　G. None of the
B. Glycogen　　　D. Cortisol　　　F. Progesterone　　　　above

Question 23:

In a lights display, a 100W water fountain shoots 1L of water vertically upward every second.
What is the maximum height attained by the jet of water, as measured from where it first leaves the fountain? (Assume that there is no air resistance, that the fountain is 100% efficient and $g=10$ ms^{-2}.)

A) 2m
B) 5m
C) 10m
D) 20m
E) The initial speed of the jet is required to calculate the maximum height

Question 24:
Which of the following statements regarding neural reflexes is **false**?

A) Reflexes are usually faster than voluntary decisions.
B) Reflex actions are faster than endocrine responses.
C) The heat-withdrawal reflex is an example of a spinal reflex.
D) Reflexes are completely unaffected by the brain.
E) Reflexes are present in simple animals.
F) Reflexes have both a sensory and motor component.

Question 25:
Study the diagram, comprising regular pentagons.
What is the product of **a** and **b**?

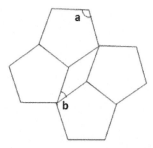

A. 580°   C. 3,888°   E. 9,255°
B. 1,111°   D. 7,420°   F. 15552°

Question 26:
The table below shows the results of a study investigating antibiotic resistance in staphylococcus populations.

Antibiotic	Number of Bacteria tested	Number of Resistant Bacteria
Benzyl-penicillin	10^{11}	98
Chloramphenicol	10^9	1200
Metronidazole	10^8	256
Erythtomycin	10^5	2

A single staphylococcus bacterium is chosen at random from a similar population. Resistance to any one antibiotic is independent of resistance to others.

Calculate the probability that the bacterium selected will be resistant to all four drugs.

A. 1 in 10^{12} C. 1 in 10^{20} E. 1 in 10^{30}
B. 1 in 10^6 D. 1 in 10^{25} F. 1 in 10^{35}

Question 27:
Which of the following units is **not** a measure of power?

A. W   C. Nms^{-1} E. $V^2\Omega^{-1}$
B. Js^{-1} D. VA   F. None of the above

END OF SECTION

SECTION 3

1) *'The concept of medical euthanasia is dangerous and should never be permitted within the UK'*

Explain the reasoning behind this statement. Suggest an argument against this statement. To what extent should legislation regarding the prohibition of medical euthanasia in the UK be changed?

2) *'The obstruction of stem-cell research is directly responsible for death arising from stem-cell treatable diseases.'*

Explain what this argument means. Argue the contrary. To what extent do you agree with the statement?

3) *'Imagination is more important than knowledge'*

Albert Einstein

Explain how this statement could be interpreted in a medical setting. Argue to the contrary that knowledge is more important than imagination in medicine. To what extent do you agree with the statement?

4) *"The most important quality of a good doctor is a thorough understanding of science"*

Explain what this statement means. Argue in favour of this statement. To what extent do you agree with it?

END OF PAPER

MOCK PAPER E

SECTION 1

Question 1:
In regions of a comparatively low altitude many birds, as is well known, fly to the far North to find the proper climatic conditions in which to rear their broods and spend their summer vacation. Some of them go to the subarctic provinces and others go beyond. How different among the sublime heights of the Rockies! Here they are required to make a journey of only a few miles, say from five to one hundred or slightly more, according to the locality selected, up the defiles and canons or over the ridges, to find the conditions as to temperature, food, nesting sites, etc., that are precisely to their taste.

Which of the following statement can be reliably concluded from the above passage?
A. A journey of 100 miles is too far for these birds to travel for food.
B. Rearing their young is the most important part of these birds' lives.
C. These birds fly north as all their needs are in a localised area.
D. Nesting in the Rockies keeps the birds away from predators.
E. The birds struggle to survive in the harsh cold temperatures further north.

Question 2:
Three ladies X, Y and Z marry three men A, B and C. X is married to A, Y is not married to an engineer, Z is not married to a doctor, C is not a doctor and A is a lawyer.

Which of the following statements is correct?
A. Y is married to C, who is an engineer.
B. Z is married to C, who is a doctor.
C. X is married to a doctor.
D. None of the statements is correct.
E. All three statements are correct.

Question 3:
The medical scientific establishment has a long-established system for naming body parts and medical phenomena. This system is based upon ease of understanding, such that a body part, or a process of the body, is named based on its clinical relevance. This means that features are named in a way which will help doctors understand and explain to patients what the body part is, or what is wrong with it in the case of a disease. However, this poses significant problems for scientific medical research. Often, the most important features of a body part from a scientific point of view are not the most clinically important features, leading to confusion within the scientific literature, as medical researchers misunderstand the purpose of a discussion, due to confusing nomenclature. Whilst it is important for doctors to be able to explain things clearly to patients, it is relatively easy for this to happen in spite of confusing nomenclature, whereas confusing names cause serious problems in the scientific world. Thus, the naming system for medical features should be edited, to reflect the scientifically important features of body parts, rather than the clinically important ones.

Which of the following best illustrates the main conclusion of this passage?

A. The naming system based on clinically important features causes problems in scientific literature.

B. Changing the naming system would allow faster progress to be made in scientific medical research.

C. The naming system should be changed to reflect the features of body parts which are most important scientifically.

D. The current naming system is sufficient and should not be changed to help lazy scientists who cannot be bothered to do fact-checking.

E. It is more important to have good doctor-patient relations than good progress in scientific research.

Question 4:

It is well established that modern humans evolved in Africa, around 2 million years ago, and that the first humans were mainly hunter-gatherers, living off hunted meat and plant foods collected from their environment. However, this poses an interesting question. Humans are relatively weak, small, feeble creatures, and around 2 million years ago most wildlife in Africa consisted of large, powerful creatures. Thus, it is unclear how humans were able to hunt successfully, and obtain meat for food. One theory is that humans are well-built for long-distance running, largely thanks to our ability to control our temperature via sweating.

This theory reasons that humans were able to pursue animals such as antelope, which run when challenged, and were able to keep on running until the antelope collapsed through heat exhaustion. Meanwhile, the humans were kept cool via sweating, and were able to then go in and butcher the defenceless antelope.

Recent evidence has emerged supporting this theory, showing that human feet are well-developed for long-distance running, with fleshy areas in the correct orientation to absorb the impact without causing joint damage, and a heart well evolved to keep pumping at a moderately fast pace for long periods. With the emergence of this powerful new evidence, we should accept this theory, known as "the persistence running theory" as true.

Which of the following identifies a flaw in this argument?

A. The emergence of evidence in support of the persistence running theory does not mean that this theory is true.

B. There is little evidence that the human body is well setup for long-distance running.

C. It has neglected to consider other theories for how humans obtained meat during their early evolution.

D. There are numerous issues with the theory of persistence running, but many of these have been resolved thanks to the new evidence that has emerged.

E. It has not considered evidence that humans evolved in Europe, where there are smaller animals which humans may have more easily been able to tackle.

Question 5:

Sam needs to measure out exactly 4 litres of water into a tank. He has two pieces of equipment – a bucket that holds 5 litres and one that holds 3 litres, with no intermediate markings.

Is it possible to measure out 4 litres? If so, how much water is needed in total in order to measure the 4 litres?

A. 4 litres

B. 7 litres

C. 8 litres

D. 10 litres

E. Not possible with this equipment

Question 6:

"A librarian is sorting books into their correct locations. All history books belong to the right of all science books. Science books are divided into five locations: engineering, biology, chemistry, physics and mathematics (in an uninterrupted order from right to left). Art books are located to the right of mathematics between engineering and sport, and sport books between art and history. Literature books are to the right of art books."

What can be certainly said about the location of literature books?

A. They are located between art and history books.

B. They are located to the left of history books.

C. They are located between mathematics and art books.

D. They are located to the right of engineering books.

E. They are not located to the left of sport books.

Question 7:

"Many people choose not to buy brand new cars, as buying brand new has significant disadvantages. Most importantly, a car's value drops substantially the moment it is first driven on the road. Even though a car is virtually unchanged by these first few miles, the potential resale value is significantly reduced. Therefore, it is better to buy second-hand cars, as their value does not drop so much immediately after purchase."

Which of the following best represents the main conclusion of this passage?

A. There are many equal reasons to avoid buying brand new cars.

B. Cars that have driven lots of miles should be avoided.

C. The rapid loss of value of new cars makes buying second-hand a wise choice.

D. Second-hand cars are at least as good as new ones.

E. New cars should not be driven to ensure they keep their resale value.

Question 8:

James is a wine dealer specialising in French wine. From his original stock of 2,000 bottles in one cellar, he sells 10% to one customer and 20% of the remaining wine to another customer. He makes £11,200 profit from the two transactions combined. What is the average profit per bottle?

A) £18   B) £20   C) £22   D) £24   E) £26

Question 9:

"Many good quality pieces of old furniture are considered 'timeless' – they are used and enjoyed by many people today, and this is expected to continue for many generations to come. However, most of this furniture dates back to previous eras, and modern furniture does not fall under the 'timeless' category of being enjoyed for many years to come."

Which of the following is the main flaw in the argument?

A. There may be many factors which make furniture good.
B. There used to be more furniture makers than today.
C. No evidence is given to tell us old furniture is better than new.
D. Old furniture is desirable for reasons other than its quality.
E. We cannot yet tell whether new furniture will become 'timeless'.

Question 10:

"Red wine is thought to be much healthier than beer because it contains many antioxidants, which have been shown to be beneficial to health. Many red wines are produced in Southern France and Italy, therefore it is no surprise that residents there have a greater life expectancy than in the UK and Germany, which are predominantly beer producing and drinking countries."

Which of the following is an assumption of the above argument?

A. Italian people drink red wine.
B. Antioxidants are beneficial for health.
C. British people prefer beer to red wine.
D. Beer is not produced in Italy.
E. Italian life expectancy is greater than in the UK.

Question 11:

Hannah, Jane and Tom are travelling to London to see a musical. Hannah catches the train at 1430. Jane leaves at the same time as Hannah, but catches a bus which takes 40% longer than Hannah's train. Tom also takes a train, and the journey time is 10 minutes less than Hannah's journey, but he leaves 45 minutes after Jane leaves. He arrives in London at 1620.

At what time will Jane arrive in London?

A. 1545   B. 1600   C. 1615   D. 1700   E. 1715

Question 12:

At a show, there are two different ticket prices for different seats. The cost is £10 for a standard seat, and £16 for a premium view seat. The total revenue from a show is £6,600, and the total attendance was 600 people.

How many premium view seats were purchased?

A. 60   B. 100   C. 140   D. 180   E. 240

Question 13:

The moon orbits the Earth once every 28 days. Between 20th January and 23rd May inclusive, how many degrees has the Moon turned through? (Assume this is not a leap year).

A) 720°   B) 880°   C) 1620°   D) 1790°   E) 1860°

Question 14:

Drama academies are special schools which students can go to in order to learn performing arts. These schools are only available to the most skilled young performers and aim to give students the best training in the arts, whilst still covering mainstream academic subjects. However, many parents are reluctant for their children to attend such academies, as they feel the academic teaching will be worse than at a standard school.

Which of the following, if true, would most weaken the above argument?
A) Most top actors attended a drama academy as children.
B) There is as much time dedicated to academic work in drama academies as there is in normal schools.
C) The academic work comprises a greater proportion of the study time than drama related activities.
D) Most children are keen to attend a drama academy if given the opportunity.
E) 80% of students at drama academies attain higher than average GCSE scores.

Question 15:

Anil and Suresh both leave point A at the same time. Anil travels 5km East then 10km North. Anil then travels a further 1km North before heading 3km West. Suresh travels East for 2km less than Anil's total journey distance. He then heads 13km North, before pausing and travelling back 2km South. What is the distance between the 2 men now?

A. 11km   B. 12km   C. 13km   D. 15km   E. 17km

Question 16:

Building foundations are covered by 14cm of concrete. A builder thinks this is too thick and grinds down the concrete by an amount three times the thickness of the concrete which he eventually leaves covering the foundations.

What is the thickness of the remaining concrete?

A. 1.5cm   B. 2.0cm   C. 2.5cm   D. 3.0cm   E. 3.5cm

Question 17:

Chris leaves his house to visit Laura, who lives 3 miles away. He leaves at 1730 and walks at 4mph towards Laura's house, stopping for 5 minutes to talk to a friend. Meanwhile Sarah also wants to visit Laura. She sets off from her house 6 miles away at 1810, driving in her car and averaging a speed of 24mph.

Who reaches the house first and with how long do they wait for the other person?

A. Chris, and waits 5 mins for Sarah

B. Chris, and waits 10 mins for Sarah

C. Sarah, and waits 5 mins for Chris

D. Sarah, and waits 10 mins for Chris

E. They both arrive at the same time

Question 18:

"Illegal film and music downloads have increased greatly in recent years. This causes significant harm to the associated industries. Many people justify this by telling themselves they are only diverting money away from wealthy and successful singers and actors, who do not need any more money anyway. But in reality, illegal downloads are deeply harming the music industry, making many studio workers redundant and making it difficult for less famous performers to make a living."

Which of the following best summarises the conclusion of this argument?

A. Unemployment is a problem in the music industry

B. Taking profits away from successful musicians does more harm than good

C. Studio workers are most affected by illegal downloads

D. Illegal downloads cause more harm than people often think

E. Buying music legally helps keep the music industry productive

Question 19:
"40,000 litres of water will extinguish two typical house fires. 70,000 litres of water will extinguish two house fires and three garden fires. There is no surplus water"

Which statement is **not** true?

A. A garden fire can be extinguished with 12,000 litres, with water to spare.
B. 20,000 litres is sufficient to extinguish a normal house fire.
C. A garden fire requires only half as much water to extinguish as a house fire.
D. Two house and four garden fires will need 80,000 litres to extinguish.
E. Three house and ten garden fires will need 140,000 litres to extinguish.

Question 20:
A car travels at $20 ms^{-1}$ for 30 seconds. It then accelerates at a constant rate of $2 ms^{-2}$ for 5 seconds, then proceeds at the new speed for 20 seconds before braking with constant deceleration of $3 ms^{-2}$ to a stop. What distance is covered in total?

A. 1325m   B. 1350m   C. 1375m   D. 1425m   E. 1475m

Question 21:
"Plans are in place to install antennae underground, so that users of underground trains will be able to pick up mobile reception. There are, as usual, winners and losers from this policy. Supporters of the policy argue that it will lead to an increase in workforce productivity and increased convenience in day-to-day life. Critics respond by saying that it will lead to an annoying environment whilst travelling, it will facilitate the ease of conducting a terrorist threat and it will decrease levels of sociability. The latter camp seems to have the greatest support and so a reconsideration of the policy is urged."

Which of the following **best** summarises the conclusion of this passage?
A. The disadvantages of installing underground antennae outweigh the benefits.
B. The cost of the scheme is likely to be prohibitive.
C. The policy must be dropped, since a majority does not want it.
D. More people don't want this scheme than do want it.
E. A detailed consultation process should take place.

Question 22:

"Ecosystems in the oceans are changing. Recently, restrictions on fishing have been imposed to tackle the decline in fish populations. As a result, farm fishing and the price of fish have increased, whilst the seas recover. It is hoped that these changes will lead to a brighter future for all."

Which of the following statements is an assumption of this argument?

A. People will still buy farmed fish at a higher price.

B. The population of wild fish can recover.

C. Fishermen will benefit from working on this scheme.

D. Ecosystems have been altered as a result of climate change.

E. None of the above

Question 23:

Brian is tossing a coin. He tosses the coin 5 times. What is the probability of tossing exactly 2 heads?

A. $^1/_{16}$　　　　B. $^5/_{32}$　　　　C. $^4/_{16}$　　　　D. $^5/_{16}$　　　　E. $^7/_{16}$

Question 24:

The amount of a cleaning powder to be added to a bucket of water is determined by the volume of water, such that exactly 40g is added to each litre. A bucket contains 5 litres of water, and is required to have cleaning powder added. However, the markings on the bucket are only accurate to the nearest 2%. Calculate the difference between the maximum and minimum amounts of cleaning powder which might be required to make up the solution correctly.

A. 4g　　　　B. 6g　　　　C. 8g　　　　D. 12g　　　　E. 20g

Question 25:

International telephone calls are charged at a rate per minute. For a call between two European countries, the rate is 22p per minute off-peak and 32p per minute at peak hours, rounded up to the nearest whole minute. In addition, there is a connection fee of 18p for every call.

What is the cost of an off-peak call from France to Germany, lasting 1.4 hours?

A. £18.48　　　B. £18.66　　　C. £26.88　　　D. £27.06　　　E. £30.98

Question 26:

"UV radiation is harmful to the skin, and can lead to the development of skin cancers. Despite this, many people sunbathe and use tanning salons, exposing themselves to dangerous radiation. If people took more sensible decisions about their health, many serious diseases, such as skin cancers, could be avoided."

What is the main conclusion of this passage?

A. UV radiation is harmful to the skin.

B. Many people like to get tanned, despite the risks.

C. People do not always consider the health risks of choices they make.

D. Skin cancer is a serious disease.

E. Sunbathing is risky, and people should avoid it.

Question 27:

Jim washes windows for pocket money. Washing a window takes two minutes. Between one house and the next, it takes Jim 15 minutes to pack up, walk to the next house and get ready to start washing again. Each resident pays Jim £3 per house, regardless of how many windows the house has. In one day, Jim washes 8 houses, with an average of 10 windows per house.

What is his equivalent hourly pay rate?

A. £4.38   B. £4.86   C. £5.33   D. £5.78   E. £6.67

Question 28:

"Bottled water is becomingly increasingly popular, but it is hard to see why. Bottled water costs many hundreds of times more than a virtually identical product from the tap, and bears a significant environmental cost of transportation. Those who argue in favour of bottled water may point out that the flavour is slightly better – but would you pay 300 times the price for a car with just a few added features?"

Which of the following, if true, would most weaken the above argument?

A. Bottled water has many health benefits in addition to tasting nicer.

B. Bottled water does not taste any different to tap water.

C. The cost of transportation is only a fraction of the costs associated with bottling and selling water.

D. Some people do buy very expensive cars.

E. Buying bottled water supports a big industry, providing many jobs to people.

Question 29:

"There are no marathon runners that aren't lean, nor no cyclists that aren't marathon runners."
Which of the following **must** be true?

A. Cyclists do not run marathons.
B. Cyclists are all lean.
C. Any lean person is also a cyclist.
D. Marathon runners must all be cyclists.
E. None of the above.

Question 30:

"Langham is east of Hadleigh but west of Frampton. Oakton is midway between Langham and Stour.
Frampton is west of Stour. Manley is not east of Langham."

Which of the following cannot be concluded?

A. Oakton is east of Langham and Hadleigh.
B. Frampton is west of Stour and east of Manley.
C. Stour is east of Hadleigh and Langham.
D. Oakton is east of Langham and west of Frampton
E. Manley is west of Oakton and west of Frampton.

Question 31:

A pot of paint is enough to cover 12m² of wall area. The inner surface of a planetarium must be painted.
The planetarium consists of a hemispheric dome of internal diameter 14 metres. How many pots of
paint are required to give the dome two full coats of paint? [Assume π=3]

A. 25   B. 36   C. 49   D. 64   E. 98

Question 32:

A planetarium has just been painted as in **31**, above. Assuming each pot of paint is 2 litres, and that the
solid component of the paint is 40%, calculate the percentage decrease in the volume of the planetarium,
due to the painting.

A. 0.0029%   B. 0.0057%   C. 0.029%   D. 0.057%   E. 2.86%

END OF SECTION

SECTION 2

Question 1:
The buoyancy force of an object is the product of its volume, density and the gravitational constant, g. A boat weighing 600 kg with a density of $1000 kgm^{-3}$ and hull volume of 950 litres is placed in a lake. What is the minimum mass that, if added to the boat, will cause it to sink? Use $g = 10 ms^{-1}$.

A. 3.55kg

B. 35 kg

C. 350 kg

D. 355 kg

E. 3,550 kg

F. None, the boat has already sunk

Question 2:
Which of the following below is **not** an example of an oxidation reaction?

A. $Li^+ + H_2O \rightarrow Li^+ + OH^- + \frac{1}{2}H_2$

B. $N_2 \rightarrow 2N^+ + 2e^-$

C. $2CH_4 + 2O_2 \rightarrow 2CH_2O + 2H_2O$

D. $2N_2 + O_2 \rightarrow 2N_2O$

E. $I_2 + 2e^- \rightarrow 2I^-$

F. All of the above are oxidation reactions

Question 3:
Regarding normal human digestion, which of the following statements (if any) is **false**?

A. Amylase is an enzyme which breaks down starch.

B. Amylase is produced by the pancreas.

C. Bile is stored in the gallbladder.

D. The small intestine is the longest part of the gut.

E. Insulin is released in response to feeding.

F. None of the above.

Question 4:
Mr Khan fires a bullet at a speed of $310 ms^{-1}$ from a height of 1.93m parallel to the floor. At the same time, Mr Weeks drops an identical bullet from the same height.

What is the time difference between the bullets first making contact with the floor? [Assume that there is negligible air resistance; $g = 10 ms^{-2}$]

A. 0 s

B. 0.2 s

C. 1.93 s

D. 2.1 s

E. More information is needed

Question 5:

A 1.4kg fish swims through water at a constant speed of 2ms^{-1}. Resistive forces against the fish are 2N. Assuming $g = 10$ms^{-2}, how much work does the fish do in one hour?

A. 7,200 J

B. 10,080 J

C. 14,400 J

D. 19,880 J

E. 22,500 J

F. More information is needed

Question 6:

Jane is one mile into a marathon. Which of the following statements (if any) is **not** true, relative to before she started?

A. Blood flow to the skin is increased.
B. Blood flow to the muscles is increased.
C. Blood flow to the gut is decreased.
D. Blood flow to the kidneys is decreased.
E. Cardiac Output Increases.
F. None of the above.

Question 7:

Balance the following chemical equation. What is the value of **x**?

w HIO$_3$ + 4FeI$_2$ + **x** HCl → **y** FeCl$_3$ + **z** ICl + 15H$_2$O

A. 4   B. 5   C. 9   D. 15   E. 22   F. 25

Question 8:

A newly discovered species of beetle is found to have 29.6% Adenine (A) bases in its genome. What is the percentage of Cytosine (C) bases in the beetle's DNA?

A. 20.4%

B. 29.6%

C. 40.8%

D. 59.2%

E. 70.6%

F. More information is required

Question 9:

Study the following diagram of the human heart. What is true about structure **A**?

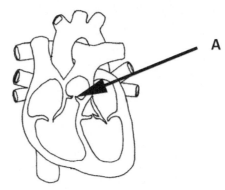

A. It is closed during systole.

B. It prevents blood flowing into the left ventricle during systole.

C. It prevents blood flowing into the right ventricle during systole.

D. It prevents blood flowing into the left ventricle during diastole.

E. It opens due to left ventricular pressure being greater than aortic pressure.

F. It is open when the right ventricle is emptying.

Question 10:

Carbon monoxide binds irreversibly to the oxygen binding site of haemoglobin. Which of the following statements is true regarding carbon monoxide poisoning?

A. Carbon monoxide poisoning has no serious consequences.

B. Haemoglobin is heavier, as both oxygen and carbon monoxide bind to it.

C. Affected individuals have a raised heart rate.

D. The CO_2 carrying capacity of the blood is decreased.

E. The O_2 carrying capacity of the blood is unchanged as it dissolves in the plasma instead.

Question 11:

A crane is 40 m tall. The lifting arm is 5m long and the counterbalance arm is 2m long. The beam joining the two weighs 350kg and is of uniform thickness. The lifting arm lifts a 2000 kg mass. What counterbalance mass is required to balance exactly around the centre point? Use $g = 10$ ms^{-2}.

A. 4,220 kg   B. 4,820 kg   C. 5,013 kg   D. 5,263 kg   E. 10,525 kg

Question 12:

For Christmas, Mr James decorates his house with 20 strings of lights containing 150 bulbs each. Each 150-bulb string of lights is rated at 50 Watts. Mr James turns the lights on at 8pm and off at 6am each night. The lights are used for 20 days in total.

If 100 kJ of energy costs 2p, what is the total cost Mr James has to pay for the running of these lights?

A. £2160.00   B. £144.00   C. £14.40   D. £0.72   E. £0.24

Question 13:

Calculate the perimeter of a regular polygon where each interior angle is 150° and each side is 15 cm.

A. 75 cm   C. 180 cm   E. 1,500 cm

B. 150 cm   D. 225 cm   F. More information is needed.

Question 14:

The diagram shown below depicts an electrical circuit with multiple resistors, each with equal resistance, Z. The total resistance between X and Y is 22 MΩ. Calculate the value of Z.

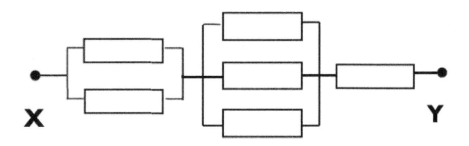

A. 3.33 MΩ   C. 7.33 MΩ   E. 12 MΩ

B. 4.33 MΩ   D. 11 MΩ

Question 15:

A cylindrical candle of diameter 4cm burns steadily at a rate of 1cm per hour. Assuming the candle is composed entirely of paraffin wax ($C_{24}H_{52}$) of density 900 kgm^{-3} and undergoes complete combustion, how much energy is transferred in 30 minutes? (You may assume the molar combustion energy is 11,000 kJmol^{-1}, and that $\pi=3$).

A. 140,000J   C. 185,000J   E. 215,000J

B. 175,000J   D. 200,500J   F. 348,000J

Question 16:

A different candle to that in question **15** is used to heat a bucket of water. The candle burns for 45 minutes, releasing 250KJ of energy as heat. It is used to heat a 2-litre bucket of water at 25°C.

Assuming the bucket is completely insulated, what is the water temperature after 45 minutes? (For reference: One calorie heats one cm^3 of water by one degree Celsius, 1kCal = 4,200J).

A. 35°C   B. 45°C   C. 55°C   D. 65°C   E. 75°C   F. 85°C

Question 17:

A person responds to the starting gun of a race and begins to run. Place the following order of events in the most likely chronological sequence. Which option shows a correct sequence of the events that occur after this gunshot?

1	Blood CO_2 increases	5	Impulses travel along relay neurones
2	The eardrum vibrates to the sound	6	Quadriceps muscles contract
3	Impulses travel along motor neurones	7	Glycogen is converted into glucose
4	Impulses travel along sensory neurones	8	Creatine phosphate rapidly re-phosphorylates ADP

A. $2 \to 5 \to 4 \to 3 \to 6 \to 7$ D. $2 \to 4 \to 3 \to 1 \to 6 \to 7$
B. $2 \to 4 \to 3 \to 8 \to 6 \to 1$ E. $2 \to 4 \to 3 \to 6 \to 8 \to 7$
C. $2 \to 3 \to 4 \to 6 \to 7 \to 1$

Question 18:

On analysis, an organic substance is found to contain 55% Carbon, 36% Oxygen and 9% Hydrogen by mass. Which of the following could be the chemical formula of this substance?

A. $C_3 H_6 O_3$ C. $C_4 H_8 O_2$ E. $C_6 H_{12} O_3$
B. $C_3 H_8 O_2$ D. $C_2 H_4 O$ F. More information needed

Question 19:

Simplify and solve: (e - a) (e + b) (e − c) (e + d)...(e - z)?

A. 0   C. e^{26} (a-b+c-d...+z)   E. e^{26} (abcd...z)
B. e^{26} D. e^{26} (a+b-c+d...-z)   F. None of the above.

Question 20:

Which of the following best describes the events that occur during expiration?

A. The ribs move up and in; the diaphragm moves down.
B. The ribs move down and in; the diaphragm moves up.
C. The ribs move up and in; the diaphragm moves up.
D. The ribs move down and out; the diaphragm moves down.
E. The ribs move up and out; the diaphragm moves down.
F. The ribs move up and out; the diaphragm moves up.

Question 21:

Simplify fully: $1 + \left(3\sqrt{2} - 1\right)^2 + \left(3 + \sqrt{2}\right)^2$

A. $30 + 6\sqrt{2} - 2\sqrt{18}$
B. $30 + 6\sqrt{2} + 2\sqrt{18}$
C. $3\left[2\left(\sqrt{2} - 1\right) + 2\right]$
D. 24
E. 29
F. 31

Question 22:

$200\ cm^3$ of a $1.8\ moldm^{-3}$ solution of sodium nitrate ($NaNO_3$) is used in a chemical reaction. How many moles of sodium nitrate is this?

A. 0.09 mol   B. 0.36 mol   C. 9.00 mol   D. 36.0 mol   E. 360 mol

Question 23:

A tourist at Victoria Falls accidentally drops her 400g camera. It falls 125 metres into the water below. Assuming resistive forces to be zero and $g = 10ms^{-1}$, what is the momentum of the camera the instant before it strikes the water? [Momentum = mass x velocity]

A. 4 kgms^{-1}
B. 13 kgms^{-1}
C. 16 kgms^{-1}
D. 20 kgms^{-1}
E. 50 kgms^{-1}
F. 20,000 kgms^{-1}

Question 24:

Antibiotics can have serious side effects such as liver failure and renal failure. Therefore, scientists are always trying to develop antibiotics to minimise these effects by targeting specific cellular components. Which of these cellular components would offer the best target for drugs to treat infection whilst minimising side effects?

A. Mitochondria
B. Cell membrane
C. Nucleic acids
D. Cytoskeleton
E. Flagellum

Question 25:

A is a group 3 element and B is a group 6 element. Which row best describes what happens to A when it reacts with B?

	Electrons are	Size of Atom
A)	Gained	Increases
B)	Gained	Decreases
C)	Gained	Unchanged
D)	Lost	Increases
E)	Lost	Decreases
F)	Lost	Unchanged

Question 26:

Each vertex of a square lies directly on the edge of a circle with a radius of 1cm. Calculate the area of the circle that is not occupied by the square. Use $\pi = 3$.

A. 0.25cm²   C. 0.75cm²   E. 1.25cm²

B. 0.5cm²   D. 1.0cm²   F. 1.5cm²

Question 27:

A funicular railway, like the one illustrated, lifts a full carriage weighing 3600kg up an incline. The distance travelled is 200m, and the vertical ascent, **v**, is 80m. Ten passengers weighing an average of 72kg disembark, then the carriage descends. As a result of efficient design, the energy from the descent is stored to drive the next ascent.

Assuming the same load of 10 passengers then enters the car, how powerful an engine is required to move the carriage at 4ms⁻¹?

A) 9.2 kW
B) 11.5 kW
C) 28.8 kW
D) 46.1 kW
E) 57.6 kW

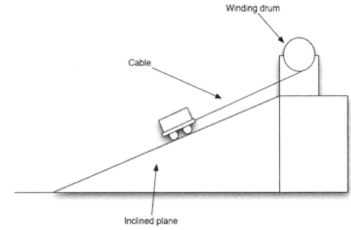

END OF SECTION

SECTION 3

1) *'Doctors know best and should decide which treatment a patient receives'*

Explain what this statement alludes to. Argue to the contrary that patients know best and should be able to choose their management plan. To what extent do you agree with this statement?

2) *'World peace will be achieved in the future'*

Explain what this statement means. Argue to the contrary, that world peace will never be achieved. To what extent, if any, do you agree with the statement?

3) *'Medicine is a science; not an art'*

Explain what this statement means. Argue to the contrary that medicine is in fact an art using examples to illustrate your answer. To what extent, if any, is medicine a science?

4) *'People should live healthier lives to reduce the financial burden of healthcare to the taxpayer.'*

Explain what this statement means. Argue to the contrary. To what extent do you agree with the statement?

END OF PAPER

MOCK PAPER F

SECTION 1

Question 1:

Every year, there are tens of thousands of motor crashes, causing a serious number of fatalities. Indeed, this represents the leading non-disease cause of death in the UK. In spite of this horrendous statistic, there are still thousands of uninsured drivers. The government is under a moral obligation to clamp down on uninsured drivers, to reduce the incidence of such crashes. That they have not acted is arguably the most outrageous failing of the present government.

Which of the following represents a flaw in this passage?

A. It has made unsupported claims that the government's failure to act is morally outrageous.

B. It has not provided any evidence to support its claims that motor crashes are the leading cause of death in the UK outside of diseases.

C. Even if motor crashes were prevented, it would not save lives of people who die from other causes.

D. It has implied that lack of insurance is related to the incidence of motor crashes.

E. It has fabricated an obligation on the government's part to intervene and reduce the number of uninsured drivers.

Question 2:

Several years ago, the Brazilian government held a referendum to decide whether they should enact a law banning the ownership of guns. The Brazilian people voted strongly against this proposal. When asked why this had happened, one commentator said he believed the reason was that 90% of criminals who use guns to commit crimes buy their weapons on the black market, illegally. Thus, if Brazil were to ban the legal sale of guns, this would remove the ability of law-abiding citizens to purchase protection, whilst doing little to remove weapons from the hands of criminals.

Some commentators have pointed to this statistic, and claimed that the UK should also legalise guns, to allow citizens to protect themselves. However, in the UK the black market for weapons is not as widespread as in Brazil. Most people in the UK have little reason to fear gun attacks and legalising the sale of guns would simply make it much easier for criminals to acquire weapons.

Which of the following best expresses the main conclusion of this passage?

A) The UK should not follow Brazil's lead on gun legislation.

B) Efforts to reduce gun ownership should focus on the black market.

C) Violent crime is a more pressing concern in Brazil than in the UK.

D) Legalising the sale of guns in the UK would result in widespread ownership.

E) Criminals will always find a way to obtain firearms.

Question 3:

Hannah is buying tiles for her new bathroom. She wants to use the same tiles on the floor and all 4 walls and for all the walls to be completely tiled excluding the door. The bathroom is 2.4 metres high, 2 metres wide and 2 metres long, and the door is 2 metres high, 80cm wide and at the end of one of the 4 identical walls. The tiles she wants to use are 40cm x 40cm.

How many of these tiles does she need to fully tile the bathroom?

A. 110   B. 120   C. 135   D. 145   E. 150

Question 4:

Jane and Trevor are both travelling south, from York to London. Jane is driving, whilst Trevor is travelling by train. The speed limit on the roads between York and London is 70mph, and the train travels at 90mph. Thus, we should expect that Trevor will arrive first.

Which of the following would most weaken this passage's conclusion?

A. The train takes a direct route, whilst the road from York to London goes through several major cities and zig-zags somewhat on its way down the country.
B. Trevor left before Jane.
C. Jane is a conscientious driver, who never exceeds the speed limit.
D. Trevor's train makes a lot of stops on the way and spends several minutes at each stop waiting for new passengers to board.
E. Meanwhile, Raheem is making the same journey by plane, and will arrive before either Trevor or Jane.

Question 5:

A recipe for 20 cupcakes needs 200g of butter, 200g of sugar, 200g of flour and 4 eggs. Jeremy has two 250g packs of butter, a bag of 600g of sugar, a kilogram bag of flour and a pack of 12 eggs.

How many cupcakes can he make and how many eggs does he have left over?

A. 50, 2   B. 50, 3   C. 60, 0   D. 60, 2   E. 60, 3

Question 6:

ABC taxis charges a rate of 15p per minute, plus £4. XYZ taxis charges a rate of £4 plus 30p per mile. I live 6 miles from the station.

What would the average speed of the taxi have to be on my journey home from the station for the two taxi firms to charge exactly the same fare?

A. 25   B. 30   C. 45   D. 55   E. 60

Question 7:

King Arthur has been issued a challenge by Mordac, his nephew who rules the adjacent Kingdom. Mordac has challenged King Arthur to select a knight to complete a series of challenging obstacles, battling a number of dark creatures along the way, in a test known as the Adzol. The King's squire reports that there are tales told by the elders of the court meaning that only a knight with tremendous courage will succeed in Adzol, and all others will fail. He therefore suggests that Arthur should select Lancelot, the most courageous of all Arthur's knights. The squire argues that due to what the Elders have said, Lancelot will succeed in the task, but all others will fail.

Which of the following is **not** an assumption in the squire's reasoning?
A. Lancelot has sufficient courage to succeed in the Adzol.
B. No other knights in Arthur's command also have tremendous courage, so will all fail Adzol.
C. Great courage is required to be successful in the Adzol.
D. The tales told by the elders of the court are correct.
E. None of the above – they are all assumptions.

Question 8:

An archaeologist is examining a recently excavated hall beneath a medieval castle. She finds that there are a series of arch-shaped gaps along one length of the wall, surrounded by a different pattern of bricks to that seen elsewhere in the walls. These are found to be where windows were once located, looking out onto one side of the castle. However, the site is now underground. Underground halls in castles never had windows, so the archaeologist reasons that this hall must once have been located above the ground. Therefore, the ground level must have changed since the castle was built.

Which of the following represents the main conclusion of this passage?

A. Windows are never found in underground halls.
B. Arch-shaped gaps always indicate that windows were once present.
C. It is unexpected for windows to be found in halls in castles.
D. The hall was once located above ground.
E. The ground level must have changed since this hall was built.

Question 9:

Adam's grandmother sent him to the shop to buy bread rolls. Usually, bread rolls are 30p for a pack of 6 and so his grandmother has given him the exact amount to buy a certain number of bread rolls. However, today there is a special offer whereby if you buy 3 or more packs of rolls, the price per roll is reduced by 1p. He can now buy 1 more pack than before and get no change.
How many bread rolls was he originally supposed to buy?

A. 4   B. 5   C. 6   D. 24   E. 30

Question 10:

The England men's cricket team have recently been knocked out of the world cup after a very poor performance that saw them eliminated at the group stage, managing only 1 win and losing against teams well below them in the rankings. The board of English cricket is sitting down to discuss why the team's performance was so poor, and what can be done to ensure that future world cups have a more positive outcome. The chairman of the board says that the current crop of players is not good enough, and that the team's performance should improve soon, as more able players come through the ranks in the county teams, so no action is needed.

However, the sporting director takes a different view, saying that England have not gone further than the group stage of any cricket world cup for the last 25 years, during which time numerous players have come and gone from the team. The sporting director argues that this long period of poor performance indicates that there is a problem with English cricket, meaning that not enough talented players are being produced in the country. He argues that therefore, steps should be taken to reform English cricket to actively foster the development of more talented players.

Which of the following, if true, would most strengthen the sporting director's argument?
A. The English cricket team is regarded as one of the best in the world, with some of the most talented players.
B. England have been steadily falling lower in the world cricket rankings for the last 25 years, due to poor performances across the board in various cricket competitions.
C. A skilled batsman, who was ranked as the 4th best player in the world, has recently retired from the England team. Now, there are no English cricket players in the top 10 of the world cricket player rankings, which is the first time this has happened in over 70 years.
D. Despite not performing well in world cups, England have performed well in other cricket competitions over the last 20 years.
E. Cricket was invented in England, so everybody expects that England should have a lot of good players in their team, regardless of the trainee players' status.

Question 11:

Karl is making cupcakes for a wedding. It takes him 25 minutes to prepare each batch of 12 cupcakes. Only 12 can go in the oven at a time, and each batch takes 20 minutes in the oven to cook. While a batch is in the oven, Karl can start preparing the next batch.

What is the latest time Karl can start if he needs to make 100 cupcakes by 4pm?
A. 11:55am   B. 12:20pm   C. 12:40pm   D. 13:20pm   E. 14:00pm

Question 12:

	Boys Absenteeism	Girls Absenteeism	Pupils on Roll	Average
Hazelwood Grammar	7%	Boys' School	300	7%
Heather Park Academy	5%	6%	1000	5.60%
Holland Wood Comprehensive	5%	6%	500	5.60%
Hurlington Academy	Girls' School		200	
Average		7%		

Some of the information is missing from the table above. What is the rate of girls' absenteeism at Hurlington Academy?

A. 6.5%   B. 7%   C. 9%   D. 11.5%   E. 13%

Question 13:

Up until the 20th century, all watches were made by hand, by watchmakers. Watchmaking is considered to be one of the most difficult and delicate of manufacturing skills, requiring immense patience, meticulous attention to detail and an extremely steady hand. However, due to the advent of more accurate technology, most watches are now produced by machines, and only a minority are made by hand, for specialist collectors. Thus, some watchmakers now work for the watch industry, and only perform *repairs* by hand on watches that are initially produced by machines.

Which of the following cannot be reliably concluded from this passage?

A. Most watches are now produced by machines, not by hand.

B. Watchmaking is considered one of the most difficult of manufacturing skills.

C. Most watchmakers now work for the watch industry, repairing watches by hand rather than making new ones.

D. The advent of more accurate technology caused the situation today, where most watches are made by machines.

E. Some watches are now made by hand for specialist collectors.

Question 14:

Many vegetarians claim that they do not eat meat, poultry or fish because it is unethical to kill a sentient being. Most agree that this argument is logical. However, some pescatarians have also used this argument, that they do not eat meat because they do not believe in killing sentient beings, but they are happy to eat fish. This argument is clearly illogical. There is powerful evidence that fish fulfil just as much of the criteria for being sentient as do most commonly eaten animals, such as chicken or pigs, but that all these animals lack certain criteria for being "sentient" that humans possess. Thus, pescatarians should either accept the killing of beings less sentient than humans, and thus be happy to eat meat and poultry, or they should not accept the killing of any partially sentient beings, and thus not be happy to eat fish.

Which of the following best illustrates the main conclusion of this passage?

A. The argument that it is unethical to eat meat due to not wishing to kill sentient beings but eating fish is acceptable is illogical.

B. Pescatarians cannot use logic.

C. Fish are just as sentient as chickens and pigs, and all these beings are less sentient than humans.

D. It is not unethical to eat meat, poultry or fish.

E. It is unethical to eat all forms of meat, including fish and poultry.

Question 15:

Recent research into cultural attitudes in Britain has revealed a striking hypocrisy. When asked whether foreign people travelling to Britain on holiday should learn some English, 60% of respondents answered yes. However, when asked if they would attempt to learn some of the local language before travelling to a non-English speaking country, only 15% of the respondents answered yes. This is a shocking double-standard on the part of the British public and is symptomatic of a deeper underlying issue that British people feel themselves superior to other cultures.

Which of the following can be reliably concluded from this passage?

A. 60% of people in Britain think that foreign people travelling to Britain for a holiday should learn English but would not learn the language themselves when going on holiday to a country which did not speak English.

B. The British public do not feel that it is important to learn some of the language before travelling to a non-English speaking country.

C. There are numerous issues of racism amongst the British public, stemming from the fact they feel superior to other cultures.

D. Less than 10% of the British public would attempt to learn some of the language before travelling to a non-English speaking country

E. Some people in Britain think that foreign people travelling to Britain for a holiday should learn English but would not learn the language themselves when going on holiday to a country which did not speak English.

Question 16:

Harriet is a headmistress, and she is making 400 information packs for the sixth form open evening. Each information pack needs to have 2 double sided sheets of A4 of general information about the school. She also needs to produce 50 A5 single sided sheets about each of the 30 A Level courses on offer. Single sided A5 costs £0.01 per sheet. Double sided costs twice as much as single sided. A4 printing costs 1.5 times as much as A5.

How much does she spend altogether on the printing?

A. £27   B. £31   C. £35   D. £39   E. £43

Question 17:

Kirkleatham Town football club are currently top of the league. One week they play a crucial match against Redcar Rovers, who are in second place. The points tally of the teams in the table means that if Kirkleatham Town win this game, they will win the league. Before the game, the manager of Kirkleatham Town says that Redcar Rovers are a tough opponent, and that if his team do not play with desire and commitment, they will not win the game. After the game, the manager is asked for comment on the game, and says he was pleased that his team played with so much desire and showed high levels of commitment. Therefore, Kirkleatham will win the league.

Which of the following best illustrates a flaw in this passage?
A. It has assumed that Kirkleatham will not win the game if they do not play with desire and commitment.
B. It has assumed that if Kirkleatham play with desire and commitment, they will win the game.
C. It has assumed that Kirkleatham played with desire and commitment.
D. It has assumed that Redcar Rovers are a tough opponent, and that Kirkleatham will not be able to easily win the game against this team.
E. It has assumed that if Kirkleatham win the match against Redcar Rovers, they will win the league.

Question 18:

Two councillors are considering planning proposals for a new housing estate, to be built on the edge of Bluedown Village. Councillor Johnson argues for a proposal for houses to be built upon brownfield land, land which has previously been built on, rather than greenbelt land, which has not previously been built on. He argues that this will both lower the cost of building the estate, as the land would already have some underlying infrastructure and would not need as much preparation and will also ensure a minimal impact on wildlife around the area.

Which of the following would most weaken the councillor's argument?

A. Brownfield land is often not as appealing as greenbelt land visually, and it is likely that houses built on brownfield land will not sell for as high a price as houses built on greenbelt land.

B. An area of brownfield land on the edge of the village, originally built as an outdoor leisure complex, has since become run down, and ironically is now a haven for various types of rare newts, lizards and birds.

C. Much of the brownfield land around the edge of the village has undergone substantial underground development, with a good system of electricity cables, gas pipes and plumbing in place.

D. The village is surrounded by several greenbelt areas designated as areas of outstanding natural beauty, supporting an abundance of wildlife.

E. The village mayor, who has ultimate control over the planning proposal, agrees with councillor Johnson's argument. Thus, it is likely his recommendations will be followed.

Question 19:

	Pool A	Pool B	Pool C	Pool D
1st	France	Argentina	England	South Africa
2nd	Holland	Mexico	Nigeria	Brazil
3rd	United States	Denmark	Germany	Japan
4th	India	Korea	Ghana	Algeria
5th	Australia	Switzerland	Portugal	Serbia
6th	Greece	New Zealand	Honduras	Uruguay
7th	Chile	Slovakia	Cameroon	Paraguay

The table above shows the final standings in the pool stages of a football competition. The top two teams from each pool progress into the quarterfinals. The fixtures for the quarterfinals are determined as follows:

QF1: Winners Pool A vs. Runner up Pool B
QF2: Winners Pool B vs. Runner up Pool C
QF3: Winners Pool C vs. Runner up Pool D
QF4: Winners Pool D vs. Runner up Pool A

The winners of QF1 then play the winners of QF3 in one semifinal, and the winners of QF2 and winners of QF4 play each other in the other semifinal. The winners of the semi-finals progress to the final. Which of these teams could England play in the final?

A. Nigeria   B. France   C. Mexico   D. Denmark   E. Brazil

Question 20:

The pie chart shows the voting intentions of some constituents interviewed by a polling group, prior to an upcoming election. Y = yellow, B = blue, R = red, G = green.

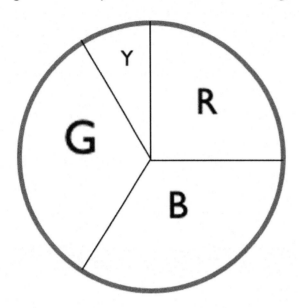

How many times more people said their intention was to vote for the red party than the yellow party?

A. 2   B. 3   C. 4   D. 5   E. 6

Question 21:

A pizza takeaway is having a sale. If you spend £30 or more at full price, you can get 40% off.
Prices are as follows:

- Basic cheese and tomato pizza: £8 small, £10 large
- All other toppings are £1 each
- Sides include: Garlic bread £3, Potato wedges £2.50, Chips £1.50 and Dips £1 each

Ellie and Mike want to order a large pizza with mushrooms and ham, garlic bread, 2 portions of chips and a dip.

Which of these additional items can they order to minimise the amount they have to pay?

A. Small pizza with pineapple and onion
B. Large pizza with mushroom
C. Barbecue dip
D. 4 portions of potato wedges
E. Garlic bread

Question 22:

	Goals Scored	Goals Conceded
City	10	4
United	8	5
Rovers	1	10

The table above shows the goal scoring record of teams in a football tournament. Each team plays against the other teams twice, once at home and once away. Here are the results of the first 4 matches:

- United 2 – 2 City
- Rovers 0 – 3 City
- City 2 – 1 Rovers
- Rovers 0 – 3 United

What were the results of the final two fixtures?

A. United 2 – 0 Rovers, City 0 – 0 United
B. United 1 – 0 Rovers, City 1 – 1 United
C. United 0 – 0 Rovers, City 2 – 1 United
D. United 1 – 0 Rovers, City 2 – 2 United
E. United 2 – 0 Rovers, City 3 – 1 United

Question 23:

The M1 Abrams tank is widely regarded as the most fearsome tank in the world. Highly advanced depleted uranium composite armour makes it difficult to damage from range, whilst a good top speed in excess of 50kmph and a large fuel capacity make it difficult to catch and contain the tank in an operational context. Whilst the tank does have weak spots that can be exploited at close range, a formidable 122m smoothbore gun as the main armament makes this an incredibly dangerous tactic for opposing tanks.

Country X is developing a new main battle tank to boost the prowess of their armoured formations, and have released a statement describing how they will implement next-generation armour into this new tank, to boost its defensive capacity. The government of country X believe this will allow their new tank to compete with the best tanks in the world. However, this view is mistaken. The M1 Abrams clearly demonstrates that a *combination* of different factors, including protection, manoeuvrability and firepower, is responsible for its status as the world's most formidable tank. Simply increasing the defensive capabilities of a tank is not sufficient to compete with the M1 Abrams. Thus, Country X's government is clearly incorrect in this matter.

Which of the following best illustrates the main conclusion of this passage?

A. Increasing the defensive capacity of a tank is not sufficient to make it equal to the best tanks in the world.

B. Multiple factors are required to make a tank equal to the best tanks in the world.

C. The new tank will not be as good as the MI Abrams, as its defensive capacity will not be as good.

D. The view of Country X's government, that increasing the defensive capacity of a tank will make it equal to the best in the world, is clearly incorrect.

E. No tank is able to compete with the MI Abrams, which will always be the world's most formidable tank.

Question 24:

The table below shows the balances of my bank accounts in pounds. Interest is paid at the end of the calendar year. My salary, which is the same every month, is paid into my current account on the 2nd of each month. All the money I have is in one or other of my bank accounts.

	Current Account	Savings	ISA
1st March	1300	5203	2941
1st April	3249	2948	2941
1st May	4398	9384	0
1st June	3948	8292	0

In which month did I spend the most money?

A. February

B. March

C. April

D. May

E. 2 or more months are the same

Question 25:

On Monday, my son developed a disease; no one else in the house has the disease. The doctor gave me some medicine and told me that everyone in the house who does not have the disease should also take half the dose. We need to take the medicine for 10 days, and the dosage is based on weight.

Weight	Dosage
Under 30kg	0.1ml per kg, 3 times a day
30kg – 60kg	0.2ml per kg, 4 times a day
60kg +	0.1ml per kg, 6 times a day

My son is 40 kg. I also have a daughter who is 20 kg. I am 75 kg and my husband is 80 kg. How many 200 ml bottles of medicine will we need for the whole 10 days?

A. 4   B. 5   C. 6   D. 7   E. 8

Question 26:

At Tina's nursery school, they have red, yellow or blue plastic cutlery. They have just enough forks and just enough knives for the 21 children there. There are the same number of forks as knives of each colour. Twice as many pieces of cutlery are yellow as are blue. Half as many pieces of cutlery are red as are blue. Tina takes a fork and a knife at random. What is the probability that she will get her favourite combination, a red fork and a yellow knife?

A. 4/49   B. 1/9   C. 36/49   D. 3/9   E. 3/49

Question 27:

The UK's taxation and public spending programme is horrendously flawed, with various immoral features. One example of such a flaw is the subsidy of public transport with money raised via taxation. According to recent research, public transport is only used by 65% of the population, and since there is no economic benefit stemming from a good public transport system, the other 35% of the population gets no benefit from public transport, but are still required to pay towards it via taxation. The system is in urgent need of reform, such that taxation is only used to support services and systems which are of benefit to everyone.

Which of the following is the best application of the principle used in this passage?

A. Only 48% of the population have ever visited an art gallery, so public funds should not be used to subsidise art galleries, as not all the population use it.

B. Primary and secondary education provides an economic benefit to the whole country, so public funds should be used to support schools.

C. Although many people never use a hospital, we should still use public funds to provide them, because many people cannot afford private healthcare, and thus we need a publically available health service for those people.

D. There is no evidence that the fire service provides any benefit to the majority of the public, who will never experience a house fire in their lifetime. Thus, the fire service should not be publically funded via taxation.

E. The police service is a vital service for the country so should be publically funded regardless of how few people benefit from its presence.

Question 28:

SpicNSpan Inc is a cleaning company offering a range of cleaning services across the UK. The board has recently acquired a new chairman, who has called a meeting of the board to assess how the company can move forwards, expanding its services and increasing its market share. One of the things the new chairman is looking at is the type of services the company provides. He argues that their "all inclusive" service, where customers pay a fixed amount to clean a house throughout as a one-off event, is more popular than their "hourly" services, where customers pay for a cleaner to carry out a certain number of hours each week. The new chairman argues that they should therefore focus on the "all inclusive services", rather than the "hourly" services, in order to increase profits.

Which of the following best illustrates a flaw in the chairman's argument?

A. The company offers other services which may bring in even more profit than all inclusive services.

B. The fact that all inclusive services are more popular than hourly services does not mean that they are more profitable. Hourly services may be more profitable.

C. He has assumed that hourly services are more popular than all inclusive services.

D. He has assumed that all inclusive services are more popular than hourly services.

E. The rest of the board may have other strategies to increase profits, which are better than the new chairman's.

Question 29:

The effects of fossil fuels such as oil, coal and natural gas on the environment are plain and clear for everybody to see. The long-term use of such non-renewable fuels to produce power has led to devastating climate change, and will continue to cause damage as long as it continues. With this in mind, the European Commission has devised a set of targets to promote energy production by different types of fuels. However, there is a glaring problem with these targets. Shockingly, the Commission has targeted a "150% increase in the amount of energy produced by nuclear power by 2025". This is an outrageous misjudgement, because nuclear power is a non-renewable fuel, just like oil, coal and natural gas. If we wish to protect the environment and halt climate change, we need to switch to renewable fuels, which are proven not to cause damage to the environment, not non-renewables such as nuclear power.

Which of the following best illustrates a flaw in this passage?

A. It has assumed that all non-renewable power sources cause environmental damage.

B. It has assumed that renewable energy sources do not cause environmental damage.

C. It has assumed that the targets will be met, when in fact there is no guarantee that this will happen.

D. It has neglected to consider other problems with the targets set by the Commission.

E. It has assumed that the climate change caused by burning of oil, coal and natural gas cannot be offset or prevented by other strategies.

Question 30:

Despite the overwhelming evidence to certify that vaccines are a miracle of modern medicine, and are responsible for saving a great number of lives, there remains a stubborn section of society that refuses to take vaccinations against important diseases and insist that they are unsafe and ineffective. This group maintain this view in spite of extremely strong evidence that vaccines are safe, and against advice given by doctors. This group is particularly strong in the USA, where they pose a very real concern. Over the last 5 years, the proportion of the population that is unvaccinated has been rising by 1% each year, such that today a staggering 6% of Americans have not received any vaccinations against infectious diseases.

Experts have advised that due to the way diseases are spread, if less than 90% of the population at any given time is unvaccinated, then it is almost certain that we will see an outbreak of measles, a highly contagious and damaging disease. Thus, we expect that there will likely be an outbreak of measles in the next 5 years in the USA, and we should take steps to prepare for this.

Which of the following, if true, would most strengthen this argument?

A. New and powerful evidence of the safety of vaccinations is due to be released to the public next year.

B. Measles is a highly damaging disease, which frequently causes death or severe permanent injury in those affected.

C. Throughout the last half-century, the number of people who are not vaccinated has risen and fallen continuously. Usually, the increases in non-vaccinated individuals occur over a 6-year period, after which time vaccination becomes more popular, and this number falls.

D. The number of doctors advising against vaccination has been rising for the last 10 years, and shows no signs of decreasing.

E. The rise in unvaccinated individuals has been increasing steadily for 5 years. The only time such a rate of increase has occurred in history was during the 1950s/1960s. In this case, a similar rate of increase in non-vaccinated individuals was maintained for a staggering 13 years.

Question 31:

PREDICTED

		A	B	C	D	E	U
ACTUAL	**A**	7	4	2	1	0	0
	B	3	8	2	2	1	0
	C	2	4	5	7	3	1
	D	2	2	2	6	5	0
	E	1	2	2	1	7	2
	U	1	1	0	3	5	6

The table above shows the actual and predicted AS grades for 100 AS mathematics students at Greentown Sixth Form. Each student is only predicted one grade. What percentage of students had their grades correctly predicted?

A. 14%   B. 16%   C. 39%   D. 61%   E. 78%

Question 32:

In one year, Mike lowers his workers' wages by x%. The next year, he lowers their wages by x%. The year after this, he raises the wages by x%. In the final year, he raises their wages by x%. In all these stages, x is a constant positive number.

Compared to the workers' original wages before any raising or lowering, what are their new wages?
A. The same as the original wages.
B. Lower than the original wages.
C. Higher than the original wages.
D. Can't tell from the provided information even if we know what x was.
E. Can't tell from the provided information but would be able to tell if we knew what x was.

END OF SECTION

SECTION 2

Question 1:

Which of the following options are reasons for cells to undergo mitosis?

1. Asexual reproduction
2. Sexual reproduction
3. Growth of the human embryo
4. Replacement of dead cells

A. 1 only

B. 2 only

C. 3 only

D. 4 only

E. 2 and 3

F. 1, 2, and 3

G. 1, 3, and 4

H. 2, 3, and 4

Question 2:

A ball of radius 2m and density 3 kg/m³ is released from the top of a frictionless ramp of height 20m and rolls down. What is its speed at the bottom? Take $\pi = 3$ and $g = 10m^{-2}$.

A. 1 ms⁻¹

B. 4 ms⁻¹

C. 7 ms⁻¹

D. 9 ms⁻¹

E. 14 ms⁻¹

F. 20 ms⁻¹

Question 3:

In a healthy person, which one of the following has the highest blood pressure?

A. The vena cava

B. The systemic capillaries

C. The pulmonary artery

D. The pulmonary vein

E. The aorta

F. The coronary artery

Question 4:

Which of the following statements is true regarding waves?

A. Waves can transfer mass in the direction of propagation.

B. All waves have the same energy.

C. All light waves have the same energy.

D. Waves can interfere with each other.

E. None of the above.

The following information applies to questions 5 - 6:

Professor Huang accidentally touches a hot pan and her hand moves away in a reflex action. The diagram below shows a schematic of the reflex arc involved.

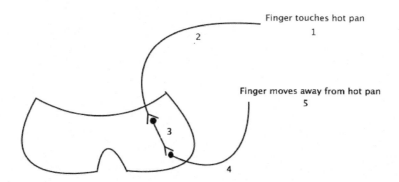

Question 5:

Which option correctly identifies the labels in the pathway?

	Muscle	Sensory Neurone	Receptor	Motor Neurone
A)	1	2	3	4
B)	2	3	1	5
C)	5	2	1	4
D)	1	4	5	2
E)	3	4	5	2
F)	4	2	1	3

Question 6:

Which one of the following statements is correct?

1. Information passes between 1 and 2 chemically.
2. Information passes between 2 and 3 electrically.
3. Information passes between 3 and 4 chemically.

A. 1 only

B. 2 only

C. 3 only

D. 1 and 2

E. 2 and 3

F. 1 and 3

G. All of the above

H. None of the above

Question 7:

Which of the following correctly describes the product of the reaction between hydrochloric acid and but-2-ene?

A. CH_3-CH_2-$C(Cl)H$-CH_3
B. CH_3-$C(Cl)$-CH_2-CH_3
C. $C(Cl)H_2$-CH_2-CH_2-CH_3
D. CH_3-CH_2-CH_2-$C(Cl)H_2$
E. None of the above.

Question 8:

Rearrange $\frac{(7x+10)}{(9x+5)} = 3z^2 + 2$, to make x the subject.

A. $x = \dfrac{15\,z^2}{7 - 9(3z^2+2)}$

B. $x = \dfrac{15\,z^2}{7 + 9(3z^2+2)}$

C. $x = -\dfrac{15\,z^2}{7 - 9(3z^2+2)}$

D. $x = -\dfrac{15\,z^2}{7 + 9(3z^2+2)}$

E. $x = -\dfrac{15\,z^2}{7 + 3(3z^2+2)}$

F. $x = \dfrac{15\,z^2}{7 + 3(3z^2+2)}$

Question 9:

The electrolysis of brine can be represented by the following equation: $2\,NaCl + 2\,X = 2\,Y + Z + Cl_2$

What are the correct formulae for X, Y and Z?

	X	Y	Z
A)	H_2O	H_2	O_2
B)	H_2O	NaOH	O_2
C)	H_2O	NaOH	H_2
D)	H_2	H_2O	O_2
E)	H_2	NaOH	O_2
F)	H_2	NaOH	H_2
G)	NaOH	H_2O	H_2
H)	NaOH	H_2O	O_2

Question 10:

Element $^{188}_{90}X$ decays into two equal daughter nuclei after a single round of alpha decay and the release of gamma radiation. What is the daughter element?

A. $^{91}_{45}D$ B. $^{92}_{44}D$ C. $^{184}_{88}D$ D. $^{186}_{90}D$ E. $^{186}_{45}D$

Question 11:

An unknown element has two isotopes: ^{76}X and ^{78}X. $A_r = 76.5$. Which of the statements below is / are true of X?

1. ^{76}X is three times as abundant as ^{78}X.
2. ^{78}X is three times as abundant as ^{76}X.
3. ^{76}X is more stable than ^{78}X.

A. 1 only	C. 3 only	E. 2 and 3
B. 2 only	D. 1 and 3	F. None of the above.

Question 12:

For the following reaction, which of the statements below is true?

$$6CO_{2\ (g)} + 6H_2O \rightarrow C_6H_{12}O_6 + 6O_{2\ (g)}$$

A. Increasing the concentration of the products will increase the reaction rate.
B. Whether this reaction will proceed at room temperature is independent of the entropy.
C. The reaction rate can be monitored by measuring the volume of gas released.
D. This reaction represents aerobic respiration.
E. This reaction represents anaerobic respiration.

Question 13:

Which of the following is/are true about the formation of polymers?

1. They are formed from saturated molecules.
2. Water is released when polymers form.
3. Polymers only form linear molecules.

A. Only 1
B. Only 2
C. Only 3
D. 1 and 2
E. 1 and 3
F. 2 and 3
G. All of the above.
H. None of the above.

Question 14:

The diagram below shows a series of identical sports fields:

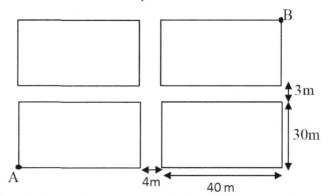

Calculate the shortest distance between points A and B.

A. 100 m

B. 105 m

C. 146 m

D. 148 m

E. 154 m

F. None of the above.

Question 15:

Calculate $\frac{1.25 \times 10^{10} + 1.25 \times 10^9}{2.5 \times 10^8}$

A. 0

B. 1

C. 55

D. 110

E. 1.25×10^8

F. $\times 10^7$

G. $\times 10^8$

The following information applies to questions 16 - 17:

Duchenne muscular dystrophy (DMD) is inherited in an X-linked recessive pattern. A man with DMD and a female carrier have 2 male offspring.

Question 16:

What is the probability that both boys have DMD?

A. 100%   B. 75%   C. 50%   D. 25%   E. 12.5%   F. 0%

Question 17:

If the same couple had two more children, what is the probability that they are both girls with DMD?

A. 100%   B. 75%   C. 50%   D. 25%   E. 12.5%   F. 0%

Question 18:

Which row of the table is correct regarding the cell shown below?

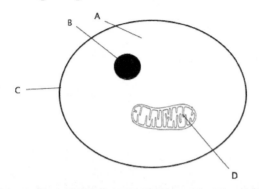

	Most Chemical Reactions occur here	Involved in Energy Release	Cell Type
A)	A	B	Animal
B)	A	B	Bacterial
C)	A	D	Animal
D)	B	D	Bacterial
E)	B	B	Animal
F)	B	A	Bacterial
G)	D	D	Animal
H)	D	B	Bacterial

Question 19:

Where do the following lines, when plotted, cross?

$$y = 2x - 1$$
$$y = x^2 - 1$$

A. (0, -1) and (2, 3)

B. (1, -1) and (2, 2)

C. (1, 4) and (3, 2)

D. (2, -3) and (4, 5)

E. (3, -1) and (3, 1)

F. (4, -2) and (-2, 4)

Question 20:

Tim stands at the waterfront and holds a 30cm ruler horizontally at eye level one metre in front of him. It lines up so it appears to be exactly the same length as a cruise ship 1km out to sea. How long is the cruise ship?

A. 299.7 m   B. 300.0 m   C. 333.3 m   D. 29,970 m   E. 30,000 m

Question 21:

Which of the following Energy-Temperature graphs best represents the melting of ice to water?

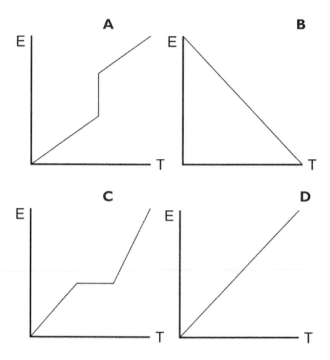

Question 22:

Which of the following statements (if any) about white blood cells is / are correct?

1. They act by engulfing pathogens such as bacteria.
2. They are able to kill pathogens.
3. They transport carbon dioxide away from dying cells.

A. Only 1
B. Only 2
C. Only 3
D. 1 and 2
E. 2 and 3
F. 1 and 3
G. All
H. None

Question 23:

Which of the following statements is true regarding the Doppler effect?

A) The Doppler effect applies only to sounds.
B) The Doppler effect makes ambulances appear to have a higher pitch when driving towards you.
C) The Doppler effect makes ambulances sound higher-pitched when driving away from you.
D) The Doppler effect means you never hear the real siren sound as an ambulance drives past.

Question 24:
A 1.2 V battery is rated at 2500 mA hours and is used to power a 30 W light. How many batteries will it take to power the light for 1 hour?

A. 1   B. 6   C. 10   D. 60   E. 100

Question 25:
When electricity flows through a metal, which of the following statements is / are true?

1. Ions move through the metal to create a current.
2. The lattice in the metal is broken.
3. Only electrons which were already free of their atoms will flow.

A. 1 only   C. 3 only   E. 1 and 3
B. 2 only   D. 1 and 2   F. 2 and 3

Question 26:
A man cycles along a road at the rate shown in the graph below.

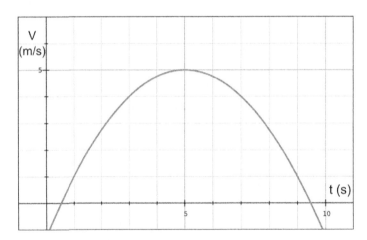

Calculate his displacement at t = 10 seconds.

A. 5 m   B. 10 m   C. 25 m   D. 30 m   E. 35 m   F. 40m

Question 27:
Bob is twice as old as Kerry, and Kerry is three times as old as Bob's son. Their ages, when added together, make 50 years. How old was Bob when his son was born?

A. 15   B. 20   C. 25   D. 30   E. 35

END OF SECTION

SECTION 3

1) *'A doctor should never disclose medical information about his patients'*
What does this statement mean? Argue to the contrary using examples to strengthen your response. To what extent do you agree with this statement?

2) *'Science is nothing more than just a thought process'*
Explain what this statement means. Argue to the contrary, that science is much more than just a thought process. To what extent, if any, do you agree with the statement?

3) *'With an ageing population, it's necessary to increase the individual's contribution to the healthcare system in order to maintain standards.'*
Explain what this statement means. Argue to the contrary. To what extent do you agree with the statement?

4) *'Assisted suicide allows those suffering from incurable diseases to die with dignity and without unnecessary pain.'*
Explain what this statement means. Argue the contrary. To what extent do you agree with the statement?

END OF PAPER

MOCK PAPER G

SECTION 1

Question 1:

Irish Folk Band, The Willow, have recently signed a contract with a new manager, and are organising a new musical tour. They and their manager are discussing which country would be best to organise their tour in. The lead singer of The Willow would like to organise a tour in Germany, which has a rich history of folk music. However, the new manager finds that ticket sales for folk music concerts in Germany have been steadily declining for several years, whilst France has recently seen a significant increase in ticket sales for folk music concerts. The manager says that this means the group's ticket sales would be higher if they organise a tour in France rather than Germany.

Which of the following is an assumption that the manager has made?

A. The band should prioritise profits and organise a tour in the most profitable country possible.

B. The band should not embark upon a new tour and should instead focus on record sales.

C. The decrease of ticket sales in Germany and the increase in France means that the band will sell fewer tickets in Germany than in France.

D. There will not be other countries which are even more profitable than France to organise the tour in.

E. Folk music is popular in France.

Question 2:

Wendy is sending 50 invitations to her housewarming party by first class post. Every envelope contains an invitation weighing 70g, and some who are going to family and friends who live further away also contain a sheet of directions, which weighs 25g. The table below gives the prices of sending letters of certain weights by first or second class post.

If the total cost of sending the invitations is £33, how many of the invitations contain the extra information?

	First Class	**Second Class**
Less than 50g	£0.50	£0.30
Less than 75g	£0.60	£0.40
Less than 100g	£0.70	£0.50
Less than 125g	£0.80	£0.60
Less than 150g	£0.90	£0.70

A. 15   B. 20   C. 25   D. 30   E. 35

Question 3:

Grace and Rose have both been attending an afterschool gymnastics class, which finishes at 5pm. After the class has finished, Grace and Rose cool down and change out of their gym clothes before heading home. Both girls depart at 5:15pm. Grace and Rose both live a 1.5 mile walk away from the local gymnasium. Therefore, they will definitely arrive home at the same time.

Which of the following is not an assumption made in this argument?

A. Both girls will walk at the same speed.

B. Both girls departed at the same time.

C. The gymnastics class is being held at the local gymnasium.

D. Grace will not get lost on the way home.

E. Both girls are walking home.

Question 4:

John is a train enthusiast, who has been studying the directions in which trains travel after departing from various London Stations. He finds that trains departing from King's Cross station in London head north on the East Coast Mainline, and travel to Edinburgh. Trains departing from Waterloo Station head west on the Southwest Mainline and travel to Plymouth. Trains departing from Victoria Station head south and travel to Kent. John surmises that presently, in order to travel on a train from London to Edinburgh, he must get on at King's Cross Station.

Which of the following is an assumption that John has made?

A. The East Coast mainline has the fastest trains.

B. It would not be quicker to take a train from Waterloo to Southampton Airport, then travel to Edinburgh on an aeroplane.

C. Rail lines will not be built that will allow trains to travel from Waterloo Station or Victoria Station to Edinburgh.

D. King's Cross trains do not have any other destinations other than Edinburgh.

E. There are no other train stations in London from which trains may travel to Edinburgh.

Question 5:

Crystal and Nancy are playing a game of "noughts and crosses". Each player is assigned either "noughts" (O) or "crosses" (X) and they take it in turns to choose an empty box of the 3x3 grid to put their symbol in. The winner is the first person to get a line of 3 of their symbol, in any direction in the grid (vertically, horizontally or diagonally). Crystal starts the game. The current position is shown below:

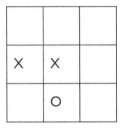

Assuming Nancy now plays her symbol in the square which will stop Crystal being able to win the game straight away, Crystal should play in either of which 2 boxes to ensure she is able to win the game on the next turn no matter what Nancy does?

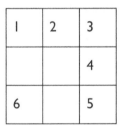

A. 1 and 3   B. 1 and 5   C. 1 and 6   D. 2 and 4   E. 3 and 5

Question 6:

Tanks and armoured vehicles were a hugely influential factor in all battles in World War Two. German tanks were highly superior to the tanks used by France, and this was an essential reason why Germany was able to defeat France in 1940. However, Germany was later defeated in World War Two by the Soviet Union. Germany lost a number of key battles such as the Battle of Stalingrad and the Battle of Kursk. These victories were essential for the eventual victory of the Soviet Union over Germany. Therefore, the Soviet Union's tanks in the battles of Stalingrad and Kursk must have been superior to those of Germany.

Which of the following is an assumption made in this argument?

A. Tanks were hugely influential in the Battle of Stalingrad.

B. The Battles of Stalingrad and Kursk were essential for the Soviet Union's victory over Germany.

C. The reasons why the Soviet Union defeated Germany in battle were the same as the reasons why Germany defeated France in battle.

D. German tanks being superior to those used by France was an essential reason why Germany was able to defeat France.

E. If the Soviet Union's tanks were superior to Germany's tanks, the Soviet Union's armoured vehicles must also have been superior to Germany's armoured vehicles.

Question 7:

In the Battle of Waterloo, in 1815, French Emperor Napoleon Bonaparte's army was defeated by a British army commanded by British General Arthur Wellesley, Duke of Wellington. Essential to the British Army's victory was the arrival of a group of Prussian reinforcements led by Field Marshal Von Blucher, which joined up with the British Army and allowed them to overwhelm Bonaparte's left flank. Bonaparte had been aware of the threat posed by Von Blucher's Prussians and had detached a force of French soldiers several days earlier under the command of Field Marshal Grouchy, with orders to engage the Prussians led by Von Blucher, and prevent them joining up with the British Army.

However, whilst dining at a local inn, Grouchy mistook the sounds of gunfire for thunder, and believed that the battle had been cancelled. He therefore disobeyed his orders and did not engage the Prussians commanded by Von Blucher. Therefore, if Field Marshal Grouchy had not made this mistake and had engaged the Prussian force as commanded, The British would not have won the Battle of Waterloo.

Which is the best statement of a flaw in this argument?

A. It implies Field Marshal Grouchy was an incompetent commander, when in fact he was a highly respected general of the day.

B. It assumes that had Grouchy engaged the Prussian force, he would have been able to successfully prevent them joining up with the British Army.

C. It assumes that the British Army would not have been victorious without the arrival of the Prussian reinforcements.

D. It ignores the other mistakes made by Napoleon which contributed to the British Army being victorious in the Battle of Waterloo.

E. It implies that thunder and gunshot sounds are frequently mistaken by generals.

Question 8:

A cruise ship is sailing from Southampton to Barcelona, making several stops along the way at Calais and Bordeaux, in France, Bilbao in Spain, and Porto in Portugal. At each stop, the ship must wait in a queue to be assigned a dock at which it can pull in, refuel and resupply. The busier the port, the longer the ship will have to queue to be assigned a dock. The captain of the ship is planning the journey and knows he must work out which ports will have the longest queues.

The captain made the same journey last year and found out that Bilbao was the busiest port in Europe during the course of the journey. He also knows that Bordeaux is the busiest port in France, and that Porto is the busiest port in Portugal. Whilst he is planning the journey, he discovers that Calais is busier than Porto. The captain concludes that he must plan for Bilbao to have the longest queue in the journey, Bordeaux to have the second longest queue, Calais to have the third longest queue, and Porto to have the fourth longest queue.

Which of the following best illustrates a flaw in the captain's Reasoning?

A. Porto is less busy than Calais but may be busier than Bordeaux.

B. The rankings may have changed, and Bilbao may no longer be the busiest port in Europe.

C. Just because a port is busier does not necessarily mean it will have the longest queues.

D. The ship may not have time to make all the stops.

E. The captain has forgotten to consider how many passengers will embark and disembark at each stop.

Question 9:

A packaging company wishes to make cardboard boxes by taking a flat 1.2 m by 1.2 m square piece of cardboard, cutting square sections out of each corner, as shown by the picture below, and folding up the sections remaining on each side to make a box. The company experiments with different size boxes by cutting differently sized squares from the corners each time. It makes a box with 10 cm by 10 cm squares cut out of each corner, a box with 20 cm by 20 cm squares cut out of each corner and so on up to one with 50 cm by 50 cm squares cut out of each corner.

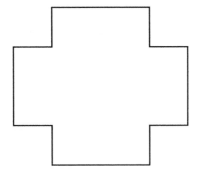

Which side length cut out would result in a box with the largest volume?

A. 10 cm   B. 20 cm   C. 30 cm   D. 40 cm   E. 50 cm

Question 10:

The aeroplane was a marvel of modern engineering when it was first developed in the early 20th Century and was a testament to human ingenuity. Throughout the 20th Century, the aeroplane allowed humans to travel more freely and widely than ever before and allowed people to see and appreciate the stunning natural beauty the world has to offer. However, aeroplanes also produce lots of pollution, such as carbon dioxide and sulphur oxide. High levels of carbon dioxide in the atmosphere are currently causing global warming, which is destroying or damaging many natural environments throughout the world.

Therefore, it is clear that the aeroplane, which once offered such opportunity to appreciate the world's natural beauty, has been largely responsible for damage to various natural environments throughout the world. We must now seek to curb air traffic in order to save the world's remaining natural environments.

Which of the following is the best statement of a flaw in this argument?

A. It assumes that aeroplanes are a major reason for the high levels of carbon dioxide in the atmosphere which are currently causing global warming.

B. It assumes that aeroplanes offer greater opportunity to appreciate the world's natural environments.

C. It assumes that high levels of carbon dioxide are responsible for global warming.

D. It does not consider the effects of sulphur dioxide pollution released by aeroplanes.

E. It implies that we should take action to prevent damage to the world's natural environments.

Question 11:

Professors from the department of pathology at Oxford University are conducting research into possible new treatments for malaria, which is caused by a microbe known as plasmodium. Research from Sierra Leone, a third world country with a high rate of malaria, has found that liver cells in malaria patients are reactive to the antibody tarpulin. Plasmodium is known to infect liver cells, and thus liver cells would react to tarpulin if plasmodium itself was reactive to tarpulin. Thus, the professors at Oxford begin to research how tarpulin can be used to target plasmodium and treat malaria.

However, this research will not be successful, because liver cells would also react to tarpulin if the wrong solution is used whilst conducting the experiments. Since malaria is not prevalent in Oxford, the professors must rely on the data from Sierra Leone. If the experiments in Sierra Leone used the wrong solutions, then the liver cells would react to tarpulin even if plasmodium does not react to tarpulin.

Which of the following best illustrates a flaw in this argument?

A. From the fact that plasmodium infects liver cells, it cannot be inferred that infected liver cells would react to tarpulin if plasmodium does.

B. From the fact that the research was carried out in Sierra Leone, it cannot be inferred that the wrong solutions were used.

C. From the fact that the wrong solutions are used, it cannot be inferred that the liver cells would react to tarpulin.

D. From the fact that plasmodium is reactive to tarpulin it cannot be assumed that tarpulin can be used to combat plasmodium.

E. From the fact that liver cells react to tarpulin, it cannot be inferred that plasmodium is reactive to tarpulin.

Question 12:

Ancient Egypt was one of the world's most powerful nations for several thousand years, and wondrous structures such as the Sphinxes and the Great Pyramids serve as a permanent reminder of its stature. Many other powerful nations throughout the ages have also built magnificent structures, such as the Colosseum built by the Romans, the Hanging Gardens of Babylon built by the Persians and the Great Wall of China built by the Chinese. As well as building magnificent structures, Rome, Persia and China had one other thing in common, namely a very strong military. Thus, history clearly shows us that in ancient times, for a nation to be a powerful nation, it must have had a very strong military. In addition to building great structures such as the pyramids, Ancient Egypt must have also possessed a very strong military.

Which of the following best illustrates the main conclusion of this argument?

A. In order to be a powerful nation, a nation must build magnificent structures.

B. In ancient times a very strong military was required to be a powerful nation.

C. Ancient Egypt built magnificent structures; therefore, it must have been a powerful nation.

D. Rome, Persia and China were all powerful nations.

E. Ancient Egypt was a powerful nation; therefore, it must have had a very strong military.

Question 13:

Global warming is widely presented in modern society as a cause for significant concern. One particular area often thought to be at risk is the ice caps of the North and South Poles, which are often presented to be at risk of melting due to increased temperature. Environmentalist groups often campaign for energy consumption to be reduced, thus reducing CO_2 emissions, which is the leading cause of global warming. However, recent research shows that the North and South Poles are actually becoming cooler, not warmer, thanks to mysterious and unexplained weather patterns. Clearly, high energy consumption is not contributing to damage to the polar ice caps.

Which of the following statements can be reliably inferred from this argument?

A. There is no point in reducing energy consumption for environmental reasons.

B. Reducing energy consumption will not reduce CO_2 emissions.

C. We should trust the recent research stating that the North and South poles are becoming cooler.

D. Reducing energy consumption will not contribute to saving the polar ice caps.

E. We should not be concerned about damage to the polar ice caps.

Question 14:

In 1957, the drug thalidomide was released and used to relieve nausea and morning sickness during pregnancy. The pharmaceutical company that released thalidomide had carried out extensive testing of the drug and had carried out more tests than was required for new drugs in the 1950s. No adverse effects were reported, and the drug was thought to be safe and effective. However, after it was released, thalidomide was found to be responsible for severe deformities in thousands of babies whose mothers had taken the drug whilst pregnant with them. When further research was carried out, it was found that the molecules in thalidomide could adopt 2 molecular structures, known as isomers. One of these isomers was perfectly safe, but the other caused significant biological problems in pregnant women and had been responsible for the deformities in the babies. The company producing thalidomide had not been aware of this 2nd isomer when developing the drug.

Which of the following is a conclusion that can be drawn from this passage?

A. The company that produced thalidomide had acted irresponsibly by not carrying out the required level of testing for the drug.

B. No isomers of thalidomide are safe.

C. The drug testing requirements in 1950s were not sufficient to identify all possible isomers of a given drug.

D. Thalidomide was not effective at relieving nausea and morning sickness.

E. The dangerous isomer of thalidomide was not effective at relieving nausea and morning sickness.

Question 15:

A teacher is trying to arrange the 5 students in her class into a seating plan. Her classroom contains 2 tables, arranged one behind the other, which each sit 3 people. Ashley must sit on the front row on the left-hand side nearest the board because she has poor eyesight. Bella and Caitlin must not be sat in the same row as each other because they talk and disrupt the class. Danielle needs to be sat next to an empty seat as she sometimes has help from a teaching assistant. Emily should be sat on the end of a row because she has poor mobility, and it is hard for her to get into a middle seat.

Who is sitting in the front right seat?

A. Empty   B. Bella   C. Caitlin   D. Danielle   E. Emily

Question 16:

The release of CO_2 from the consumption of fossil fuels is the main reason behind global warming, which is causing significant damage to many natural environments throughout the world. One significant source of CO_2 emissions is cars, which release CO_2 as they use up petrol. In order to tackle this problem, many car companies have begun to design cars with engines that do not use as much petrol. However, engines which use less petrol are not as powerful, and less powerful cars are not attractive to the public. If a car company produces cars which are not attractive to the public, they will not be profitable.

Which of the following best illustrates the main conclusion of this argument?

A. Car companies which produce cars that use less petrol will not be profitable.

B. The public prefer more powerful cars.

C. Car companies should prioritise profits over helping the environment.

D. Car companies should seek to produce engines that use less petrol but are still just as powerful.

E. The public are not interested in helping the environment.

Question 17:

Penicillin is one of the major success stories of modern medicine. Since its discovery in 1928, it has grown to become a crucial foundation of medicine, saving countless lives and introducing the age of antibiotics. Alexander Fleming is today given most of the credit for introducing and developing antibiotics, but in fact Fleming played a relatively minor role. Fleming initially discovered penicillin, but was unable to demonstrate its clinical effectiveness, or discern ways of reliably and consistently producing it. Two other scientists called Howard Florey and Ernst Chain were actually responsible for developing penicillin to the point where it could be reliably produced and used in medicine, to treat infections in patients. Clearly, the credit for the wonders worked by penicillin should not go to Fleming, but to Florey and Chain.

Which of the following best illustrates the main conclusion of this argument?

A. Fleming was unable to develop penicillin to the point of being a viable medical treatment.

B. The credit for penicillin should go to Ernst Chain and Howard Florey, not to Alexander Fleming.

C. Without Chain and Florey, penicillin would not have been developed into a viable treatment.

D. Alexander Fleming only played a small role in the process of penicillin becoming a feature of modern medicine.

E. Alexander Fleming is not given enough credit for his role in the development of penicillin.

Question 18:

I write my 4-digit pin number down in a coded format, by multiplying the first and second number together, dividing by the third number, then subtracting the fourth number. If my code is 3, which of these could my pin number be?

A. 3461   B. 9864   C. 5423   D. 7848   E. 6849

Question 19:

Worcestershire Aquatic Centre is a business seeking to recruit a new dolphin trainer. They interview several candidates and find that there are 2 candidates who are clearly more suitable than the others. They give both of these candidates a 2nd interview, with further questions about their experience and qualifications. They discern that Candidate 1 has a proven capability to perform well to crowds, which is likely to bring in more profit to the Aquatic Centre as more people will come and watch a more entertaining dolphin show. However, unlike Candidate 1, Candidate 2 has experience at handling dolphins, and a proven ability to maximise their welfare standards. The manager of the aquatic centre tells the recruiting officer to prioritise profits, and therefore to hire Candidate 1.

Which of the following statements, if true, would most *weaken* the manager's argument?

A. Market research conducted by an external organisation showed that 60% of members of the public would be more likely to attend a dolphin show presented by a charismatic host.

B. Candidate 1's performance experience was not in the aquatic industry.

C. Other aquatic centres with poor welfare standards have been subject to negative media attention and boycotts.

D. A local charity-run aquatic centre has decided to prioritise donkey welfare and their manager recommends such a strategy.

E. A well-respected business analyst predicts that profit will rise under Candidate 2.

Question 20:

Rental yield for buy to let properties is calculated by dividing the potential rent per year paid for a house by the amount it cost to buy the house and get it in a rentable condition. Tina is considering five houses as possible buy-to-let investments. House A is in good condition and could be rented as it is for £700 a month, and costs £168,000 to buy. House B is also in good condition but it is a student house so Tina would need to buy furniture for it. The house would cost £190,000 to buy and £10,000 to furnish but could be rented for 40 weeks of the year to 4 students at a rent of £125 a week each. House C needs a lot of work doing to it. It costs £100,000 but would need £44,000 of renovations and would rent for £600 a month. House D costs £200,000 and would need £40,000 of renovations and would rent out for £2000 a month. House E costs £80,000 and would need £20,000 of renovations and could be rented out for £200 a week.

Which house has the highest rental yield?

A. A   B. B   C. C   D. D   E. E

Question 21:

There has recently been an election in the UK, and the new government is pondering what policy to adopt on the railway system in the UK. The Chancellor argues that the best policy is to have an entirely privatised railway system, which will encourage different train companies to be competitive, and try and attract customers by providing the best service at the lowest price, thus driving down costs and increasing quality for customers. However, the Transport Minister argues that this is a short-sighted policy. She argues that privatised companies will only run services on the most profitable lines, where there are lots of passengers. Under this system, train companies may not choose to run many services to rural areas. This will lead to rural communities being cut off from the trainline, with a consequent lack of opportunities for people in these communities. She argues that public funding should be put towards rail services in order to ensure that people in rural communities are adequately served by rail services.

Which of the following, if true, would most strengthen the Transport Minister's argument?

A. The Transport Minister has ultimate power over railway policy, and can overrule the Chancellor if she sees fit.

B. Many train services to rural communities currently have low passenger numbers, so are unlikely to be profitable.

C. French rail services receive a high level of public funding, and users of these services enjoy good quality services and low prices.

D. American railway services are privatised with no public funding, and yet rural communities in America are well served by railway services.

E. The Prime Minister agrees with the Transport Minister's line of argument. He sympathises with rural communities and does not believe in a privatised rail system.

Question 22:

Niall can choose whether to join the gym as a member or pay per session. Gym membership costs £30 per month but attending classes or gym sessions is free. Pay as you go gym sessions cost £4 and attending classes are £2 each. Niall works out that it will cost him £2 more if he pays per session than it will to buy membership. Which of these is a possible combination of Niall's gym sessions and classes for one month?

A. 5 gym sessions, 4 classes

B. 4 gym sessions, 4 classes

C. 5 gym sessions, 6 classes

D. 4 gym sessions, 6 classes

E. 5 gym sessions, 8 classes

Question 23:

If the mean of 5 numbers is 8, the median is 6 and the mode is 4, what must the two largest numbers in the set of numbers add up to?

A. 13   B. 16   C. 22   D. 26   E. 28

Question 24:

The North York Moors is one of several National Parks in England. The management team has been awarded a grant from the National Lottery looking for a way to attract more visitors to the Moors. Sam suggests that they invest in enhancing the natural landscapes present in the Moors, thus creating more beauty, and making people more inspired to visit. However, Lucy disagrees, and feels that they should invest in more visitor centres and information points. Lucy's argument that whilst this will be more costly in terms of staffing these centres, the increase in visitor numbers will bring in more income for the Moors, and will counteract this extra cost.

Which of the following, if true, would most weaken Lucy's argument?

A. Information Centres in other National Parks do not generally generate as much revenue as they cost to staff.

B. National Lottery grants have a history of being badly spent by National Parks such as the North York Moors.

C. There are large numbers of people who are interested in volunteering to help the North York Moors and would be happy to staff visitor centres.

D. Another National Park, the Yorkshire Dales, has recently opened up 5 new visitor centres and seen their profits increase significantly.

E. The North York Moors is currently struggling to attract visitors.

Question 25:

How many different squares (of either 1, 2, 3 or 4 grid squares inside length) can be made using the grid below?

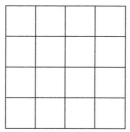

A. 16   B. 20   C. 25   D. 26   E. 30

Question 26:

When I board my train at York at 15:30, the announcer tells me it is 120 minutes to London Kings Cross. Assuming the announcement is accurate to the nearest 10 minutes and that the train is on time, what is the earliest time I might arrive at my destination which is 10 minutes walk from Kings Cross?

A. 17:20   B. 17:25   C. 17:30   D. 17:35   E. 17:40

Question 27:

My sister does 2 loads of washing per week plus an extra one for everyone who is living in the house that week. When her son is away at university, she buys a new carton of washing powder every 6 weeks, but when her son is home, she has to buy a new one every 5 weeks. The amount of washing powder used directly correlates to the number of washing loads done. How many people are living in the house when her son is home?

A. 2   B. 3   C. 4   D. 5   E. 6

Question 28:

A train shuttle service runs between the city centre and the airport between 5:30am and 11:30pm on weekdays and 6:00am and midnight on weekends. There are two trains used to operate the service and each journey from the airport to the city centre or vice versa takes 24 minutes. The train stops for 4 minutes at the airport to unload and reload, and it stops again for 2 minutes at the city centre, to unload and reload.

What is the maximum number of single journeys that can be made by the shuttle service in one day?

A. 36   B. 48   C. 60   D. 72   E. 96

Question 29:

Northern Line trains arrive at Waterloo station every 6 minutes, Jubilee Line trains every 2.5 minutes and Bakerloo Line trains every 4 minutes.

If trains from all 3 lines arrived at the station 4 minutes ago, how long will it be before they do so again?

A. 20 minutes   C. 30 minutes   E. 60 minutes
B. 26 minutes   D. 56 minutes

Question 30:

Sam is deciding whether to make her wedding invitations herself or get them professionally made at a cost of £1 each. She decides to work out how much it will cost to make them herself. Each invitation uses 1 sheet of cream card, 4 sheets of red paper and 1 metre of gold ribbon. She will also use a gold sticker on each invitation and stamp them with a stamper she will buy. The stamper needs a pad of ink which will last for 70 invitations. The table of stationery costs is shown below:

Product	Price
Red paper (pack of 100)	£2
15m roll of gold ribbon	£3
Pack of 30 gold stickers	£1
Stamper	£8
Ink pad	£4
Cream card (pack of 20)	£2

She wants to send 90 invitations and wants to have enough supplies for 4 spares only. How much will she save by making the invitations herself?

A. £15　　　　　B. £19　　　　　C. £29　　　　　D. £31　　　　　E. £33

Question 31:

Half of the boys in Mrs Nelson's class have brown eyes and two thirds of the class have brown hair. At least as many boys in the class as girls have brown hair. There are at least as many boys as girls in the class. There are 36 children in the class in total.

What's the minimum number of boys that have both brown hair and brown eyes?

A. 2　　　　　B. 3　　　　　C. 4　　　　　D. 5　　　　　E. 6

Question 32:

Mandy is making orange squash for her daughter's birthday party. She wants to have a 300ml glass of squash for each of the 8 children attending and a 400ml glass of squash each for her and for 2 parents who are helping out. She has 600ml of the concentrated squash. What ratio of water to concentrated squash should she use in the dilution to ensure she has the right amount to go around?

A. 7:1　　　　　B. 6:1　　　　　C. 5:1　　　　　D. 4:1　　　　　E. 3:1

END OF SECTION

SECTION 2

Question 1:

Which of the following statements is / are true?

1. Natural selection always favours organisms that are faster or stronger.
2. Genetic variation leads to different adaptations to the environment.
3. Variation is purely due to genetics.

A. Only 1   C. Only 3   E. 2 and 3   G. All of the above.

B. Only 2   D. 1 and 2   F. 1 and 3   H. None of the above.

Question 2:

Which of the following statements regarding the electrolysis of brine is / are true?

1. It describes the reduction of 2 chloride ions to Cl_2.
2. The amount of NaOH produced increases in proportion with the amount of NaCl present in solution, provided there is enough H_2 present to dissolve the NaCl.
3. The redox reaction of the electrolysis of brine results in the production of dissolved NaOH, which is a strong acid.

A. Only 1   C. Only 3   E. 1 and 3   G. All of the above.

B. Only 2   D. 1 and 2   F. 2 and 3   H. None of the above.

The following information applies to questions 3 – 4:

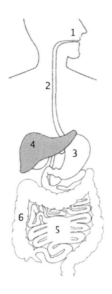

Question 3:

Which of the following number(s) indicate where amylase functions?

A. 1 only

B. 2 only

C. 1 and 3

D. 1 and 5

E. 2 and 4

F. 3 and 4

G. 5 and 6

Question 4:

In which of the following does the majority of chemical digestion occur?

A. 1

B. 2

C. 3

D. 4

E. 5

F. 6

G. None of the above.

Question 5:

Which of the following correctly describes the product of the reaction between propene and hydrofluoric acid (HF)?

A. $C(F)H_3-CH_2-CH_3$

B. $CH_3-C(F)H-CH_3$

C. $CH_3-C(F)H_2-CH_2$

D. $CH_3-C(F)H_2-CH_3$

E. None of the above.

Question 6:

Which of the following statements is false?

A. A nuclear power plant may have an accident if free neutrons in a fuel rod aren't captured.

B. Humans cannot currently harness the energy from nuclear fusion.

C. Uncontrolled nuclear fission leads to a large explosion.

D. Mass is conserved during nuclear explosions caused by nuclear bombs.

E. Nuclear fusion produces much more energy than nuclear fission.

Question 7:

Which of the following statements about the reaction between alkenes and hydrogen halides is / are true?

1. The product formed is fully saturated.
2. The hydrogen halide binds at the alkene's saturated double bond.
3. The hydrogen halide forms ionic bonds with the alkene.

A. Only 1

B. Only 2

C. Only 3

D. 1 and 2

E. 2 and 3

F. 1 and 3

G. All of the above.

H. None of the above.

Question 8:
Rearrange the following to make m the subject.

$$T = 4\pi \sqrt{\frac{(M + 3m)l}{3(M + 2m)g}}$$

A. $m = \dfrac{16\pi^2 M - 3gMT^2}{48\pi^2 l - 6gT^2}$

B. $m = \dfrac{16\pi^2 lM - 3gMT^2}{6gT^2 - 48\pi^2 l}$

C. $m = \dfrac{3gMT^2 - 16\pi^2 lM}{6gT^2 - 48\pi^2 l}$

D. $m = \dfrac{4\pi^2 lM - 3gMT^2}{6gT^2 - 16\pi^2 l}$

E. $m = \left(\dfrac{16\pi^2 lM - 3gMT^2}{6gT^2 - 48\pi^2 l}\right)^2$

Question 9:
Which of the following correctly describes the product of the polymerisation of chloroethene molecules?

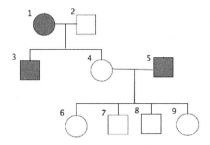

The following information applies to questions 10 – 11:
The diagram below shows the genetic inheritance of colour-blindness, which is inherited in an X-linked recessive manner. X^B is the normal allele and X^b is the colour-blind allele.

Question 10:
What is the genotype of the individual marked 4?

A. $X^B X^b$

B. $X^B X^B$

C. $X^b X^b$

D. $X^B Y$

E. $X^b Y$

Question 11:

If 8 were to reproduce with a heterozygote female, what is the probability of producing a colour-blind boy?

A. 100%   B. 75%   C. 50%   D. 25%   E. 12.5%   F. 0%

Question 12:

The mean of a set of 11 numbers is 6. Two numbers are removed, and the mean is now 5. Which of the following could be a possible combination of removed numbers?

A. 1 and 19   B. 15 and 6   C. 11 and 12   D. 15 and 8   E. 18 and 2

Question 13:

For the following reaction, which of the statements below, if any, is / are true?

$$N_{2(g)} + 3H_{2(g)} \leftrightarrow 2NH_{3(g)}$$

1. Increasing pressure will cause the equilibrium to shift to the right.
2. Increasing pressure will form more ammonia gas.
3. Increasing the concentration of N_2 will create more ammonia.

A. 1 only   C. 3 only   E. 2 and 3   F. All of the above.
B. 2 only   D. 1 and 2   G. None of the above.

Question 14:

Find the values of angles b and c.

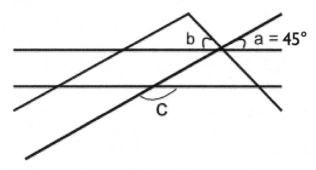

A. 45° and 135°   C. 50° and 135°   E. More information needed.
B. 45° and 130°   D. 55° and 130°

Question 15

When sodium and chlorine react to form salt, which of the following best represents the bonding and electron configurations of the products and reactants?

	Sodium (s)		Chlorine (g)		Salt (s)	
	Intra-element bond	Element electron configuration	Intra-element bond	Element electron configuration	Compound bond	Compound electron configuration
A)	Ionic	2, 8, 1	Covalent	2, 8, 8, 1	Ionic	2, 8, 1 : 2, 8, 8, 1
B)	Metallic	2, 7	Covalent	2, 8, 1	Ionic	2, 8 : 2, 8
C)	Covalent	2, 8, 2	Ionic	2, 8, 8	Covalent	2, 8 : 2, 8, 8
D)	Ionic	2, 7	Ionic	2, 8, 8, 7	Covalent	2, 7 : 2, 8, 8, 7
E)	Metallic	2, 8, 1	Covalent	2, 8, 7	Ionic	2, 8 : 2, 8, 8

Question 16:

Evaluate: $\dfrac{3.4 \times 10^{11} + 3.4 \times 10^{10}}{6.8 \times 10^{12}}$

A. 5.5×10^{-12} C. 5.5×10^{1} E. 5.5×10^{10}

B. 5.5×10^{-2} D. 5.5×10^{2} F. 5.5×10^{12}

The following information applies to questions 17 – 18:

In pea plants, colour and stem length are inherited in an autosomal manner. The allele for yellow colour, Y, is dominant to the allele for green colour, y. Furthermore, the allele for tall stem length, T, is dominant to short stem length, t.

When a pea plant of unknown genotype is crossed with a green short-stemmed pea plant, the progeny are 25% yellow + tall-stemmed plants, 25% yellow + short-stemmed plants, 25% green + tall-stemmed plants and 25% green + short-stemmed plants.

Question 17

What is the genotype of the unknown pea plant?

A. Yytt   C. YyTT   E. yyTT

B. YyTt   D. yyTt   F. yytt

Question 18:

Taking both colour and height into account, how many different combinations of genotypes and phenotypes are possible?

A. 6 genotypes and 3 phenotypes

B. 8 genotypes and 3 phenotypes

C. 8 genotypes and 4 phenotypes

D. 9 genotypes and 4 phenotypes

E. 9 genotypes and 3 phenotypes

F. 10 genotypes and 3 phenotypes

Question 19:

Which of the following statements is true regarding electrolysis?

A. Using an AC-current is most effective.

B. Using a DC-current is most effective.

C. An AC-current causes cations to gather at the cathode.

D. A DC-current would plate the anode in copper from a copper sulphate solution.

E. No current is used in electrolysis.

Question 20:

Evaluate the following expression:

$$\left(\left(\tfrac{6}{8}\times\tfrac{7}{3}\right)\div\left(\tfrac{7}{5}\times\tfrac{2}{6}\right)\right)\ \times\ 0.40\ \times\ 15\%\ \times\ 5\%\ \times\ \pi\ \times\ \left(\sqrt{e^2}\right)\ \times\ 0.20\ \times\ (e\pi)^{-1}$$

A. $\dfrac{4}{55}$

B. $\dfrac{8}{770}$

C. $\dfrac{9}{4,000}$

D. $\dfrac{8}{54,321}$

E. $\dfrac{9}{67,800}$

Question 21:

Which will have a greater current, a circuit with two identical resistors in series or one with the same two resistors in parallel?

A. Series will have greater current than parallel.

B. Parallel will have greater current than series.

C. Same current in both.

D. It depends on the battery.

Question 22:

A 2000 kg car is driving down the road at 36 km per hour. A deer runs out into the road 105 m in front of the car. It takes the driver 0.5 seconds to react to the deer and start to engage the brakes. The car stops just in time. What is the average braking force exerted?

A. 20 N   B. 100 N   C. 200 N   D. 1,000 N   E. 2,000 N

Question 23:

What is the most important reason for each cell in the human body to have an adequate blood supply?

A. To allow protein synthesis.
B. To receive essential minerals and vitamins for life.
C. To kill invading bacteria.
D. To allow aerobic respiration to take place.
E. To maintain an optimum cellular temperature.
F. To maintain an optimum cellular pH.

Question 24:

A man drives along a road as shown in the figure below.

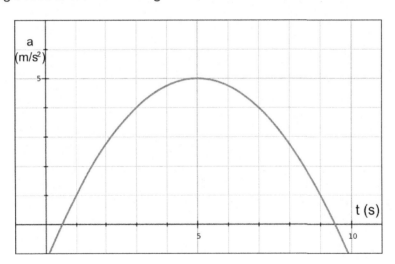

Which of the following statements is true?

A. He drives a total of 30 m.
B. He has an average velocity of 30 m/s.
C. He has a final velocity of 30 m/s.
D. He has an average acceleration of 30 m/s^2.
E. His velocity decreases between 5 and 9 seconds.

Question 25:

A circle has a radius of 3 metres. A line passes through the circle's centre and intersects with a tangent 4 metres from its tangent point. How far is this point of intersection from the centre of the circle?

A. I metre   B. 3 metres   C. 5 metres   D. 7 metres   E. 9 metres

Question 26:

Which of the following statements is false?

A. Energy cannot be created nor destroyed.
B. Energy can be turned into matter.
C. Efficiency is the ratio of useful energy to wasted energy.
D. Energy can be dispersed through a vacuum.
E. There are always losses when energy is transformed from one type to another.

Question 27

Which of the following statements is false?

A. A beam of light exits a pane of glass at a different angle than it entered.
B. A beam of light reflects at an angle dependent on the angle of incidence.
C. Light travels a shorter distance to reach the bottom of a pool filled with water than a pool without water.
D. Any neutrally charged atom has the potential to emit light.
E. Photons are particles without a mass.

END OF SECTION

SECTION 3

1) *'Doctors will eventually become obsolete as a result of advancing medical technologies.'*

Explain what this statement means. Argue to the contrary. To what extent do you agree with the statement?

2) *"Science is a procedure for testing and rejecting hypotheses, not a compendium of certain knowledge."*

Stephen Jay Gould

What do you understand from the statement above? Explain why it might be argued that science does rely on a compendium of certain knowledge? To what extent is science defined by the challenging of preconceived hypotheses?

3) *'Animal euthanasia should be made illegal'*

Explain what this statement means. Argue to the contrary that animal euthanasia should remain legal. To what extent do you agree with the statement?

4) *'The primary duty of a doctor is to prolong life'*

What does this statement mean? Argue to the contrary, that the primary duty of a doctor is not to prolong life. To what extent do you agree with this statement?

END OF PAPER

MOCK PAPER H

SECTION 1

Question 1:

A chemical change may add something to a substance, or subtract something from it, or it may both subtract and add, making a new substance with entirely different properties. Sulphur and carbon are two stable solids. The chemical union of the two forms a volatile liquid. A substance may be at one time a solid, at another a liquid, at another a gas, and yet not undergo any chemical change, because in each case the chemical composition is identical.

Which of the following statements cannot be reliably concluded from the above passage?

A. The chemical composition of a compound may influence its physical nature.

B. Substances can exist as solid, liquid or gas, without their chemical composition changing.

C. Chemicals can be combined to create a new substance with similar or very different properties.

D. Combining two substances in one state can lead to a compound in a completely different state.

E. The transition from solid to liquid is not a chemical one.

Question 2:

In the sequence B Y F U I R K P ...   ... Which two letters come next?

A. U N   B. M N   C. L O   D. H O   E. N M

Question 3:

An insect differs from a horse, for example, as much as a modern printing press differs from the printing press Franklin used. Both machines are made of iron, steel, wood, etc., and both print; but the plan of their structure differs throughout, and some parts are wanting in the simpler press, which are present and absolutely essential in the other. So, with the two sorts of animals; they are built up originally out of protoplasm, or the original jelly-like germinal matter, which fills the cells composing their tissues, and nearly the same chemical elements occur in both, but the mode in which these are combined, the arrangement of their products: the muscular, nervous and skin tissues, differ in the two animals.

Which of the following statements can be reliably concluded from the above passage?

A. The printing press has adapted from the press Franklin used, due to the designers observing differences in nature.

B. Horses and insects differ as they are made up of completely different chemical elements.

C. The muscular, nervous and skin tissues are what define an organism.

D. Chemical elements make up protoplasm, which is the building block for all major organisms.

E. It is the manner in which chemicals are arranged that determine an organism as a final product.

Question 4:

What day comes two days after the day, which comes four days after the day, which comes immediately after the day, which comes two days before Monday?

A. Monday   B. Tuesday   C. Thursday   D. Saturday   E. Sunday

Question 5:

Cellulose is distinguished by its inherent constructive functions, and these functions take effect in the plastic or colloidal condition of the substance. These properties are equally conspicuous in the synthetical derivatives of the compound.

Which of the following statements, if true, would weaken the above passage?
A. Cellulose has a constructive role in nature.
B. Synthetic cellulose is made from natural cellulose.
C. Synthetic and natural cellulose are structurally very similar.
D. Synthetic cellulose only actually shares some of its properties with natural cellulose.
E. Synthetics cellulose is more useful in industry than natural cellulose.

Question 6:

If John gives Michael £20, the ratio of their money is 2:1. If Michael gives John £5, the ratio of John's money to Michael's is 5:1. How much money do they have combined?

A. £180   B. £120   C. £90   D. £210   E. £150

Question 7:

From the primitive pine-torch to the paraffin candle, how wide an interval! Between them how vast a contrast! The means adopted by man to illuminate his home at night, stamp at once his position in the scale of civilisation. The fluid bitumen of the far East, blazing in rude vessels of baked earth; the Etruscan lamp, exquisite in form, yet ill adapted to its office; the whale, seal, or bear fat, filling the hut of the Esquimaux or Lap with odour rather than light; the huge wax candle on the glittering altar, the range of gas lamps in our streets, all have their stories to tell.

Which of the following statements best summarises the above passage?
A. Burning animal fat was the original way to produce fire.
B. The use of fire has spread to all corners of the Earth.
C. Using fire for light is what defines us as being human.
D. Each light source over the globe is able to tell its own tale.
E. The development and evolution of the use of fire helps to define mankind as a civilisation.

Question 8:

972 patients ordered food for lunch. They could choose roast chicken, mac and cheese, vegetable chilli or cottage pie. Half chose the roast chicken, 1/3 chose the mac and cheese and 1/12 chose the cottage pie.

How many opted for a vegetarian option?

A. 81   B. 324   C. 405   D. 486   E. 567

Question 9:

It was a little late to search for the philosophers' stone in 1669, yet it was in such a search that phosphorus was discovered. Wilhelm Homberg (1652-1715) described it in the following manner: "a man little known, of low birth, with a bizarre and mysterious nature in all he did, found this luminous matter while searching for something else."

What can be reliably concluded about the above passage?

A. Phosphorous was easy to identify as a result of its luminous nature.
B. Phosphorous was found as a result of this man's low social status.
C. Phosphorous was identified by accident, in the search for the philosophers' stone.
D. Wilhelm Homberg discovered phosphorous.
E. Phosphorous was discovered in the 18th century.

Question 10:

How many minutes past noon is it, if 3 times this many minutes before 3pm is 28 minutes later than this many minutes past noon?

A. 54   B. 32   C. 45   D. 38   E. 18

Question 11:

Everyone is familiar with the main facts of such a life-story as that of a moth or butterfly. The form of the adult insect is dominated by the wings—two pairs of scaly wings, carried respectively on the middle and hindmost of the three segments that make up the thorax or central region of the insect's body. Each of these three segments carries a pair of legs.

Which of the following statements can be concluded from the above statement?

A. The wings of the insects alternate patterns when the insect flies.
B. The wings that attach to the segments of the insect's body are the most prominent feature of the butterfly or moth.
C. Wings attach to each of the three segments of the thorax.
D. Moths and butterflies are very similar in that each segment of their thorax carries a pair of legs.
E. Scaly wings protect these creatures from predators.

Question 12:

John and Mary are selling cakes at a cake sale. John has 8 cupcakes and 56 brownies, where as Mary has 12 cupcakes and 24 brownies. What is the difference between the percentages of brownies in the two stalls?

A. 6%   B. 115/3%   C. 19.25%   D. 125/6%   E. 22.2%

Question 13:

In 2007 AD, Halley's Comet and Comet Encke were observed in the same calendar year. Halley's Comet is observed on average once every 73 years; Comet Encke is observed on average once every 104 years. Based on this, estimate the calendar year in which both Halley's Comet and Comet Encke will next be observed in the same year.

A. 9559 AD   B. 2114 AD   C. 5643 AD   D. 3562 AD   E. 1757 AD

Question 14:

The supreme court of Judicature at Athens punished a boy for putting out the eyes of a poor bird; and parents and masters should never overlook an instance of cruelty to anything that has life, however minute, and seemingly contemptible the object may be.

Which of the following statements best summarises the above passage?
A. The boy was prosecuted because the bird is a large enough organism.
B. Putting out the eyes of an organism is the most unacceptable form of animal cruelty.
C. The more important to mankind the animal, the worse the animal cruelty crime is.
D. Any cruelty to any creature is an action that should not be tolerated.
E. It is only acceptable to harm an animal so long as it benefits a human.

Question 15:

In a school of only boys and girls, there are 40 more girls than there are boys. The boys make up a percentage of 40% of the school. What is the number of students in the school?

A. 150   B. 200   C. 300   D. 500   E. 720

Question 16:

5 cars are travelling down a road in a line. The red car is following the blue car; the yellow car is in front of the green car. The purple car is between the green car and the blue car. What colour is the car second in line?

A. Red   B. Blue   C. Yellow   D. Green   E. Purple

Question 17:

To get to school, Joanne takes the school bus every morning. If she misses this, then she can take the public bus to school. The school bus arrives at 08:15, which if she misses will come again at 08:37. The public bus comes every 17 minutes, starting at 06:56. The school bus takes 24 minutes to get to her school; the public bus takes 18 minutes. If she arrives at the bus stop at 08:25, which bus must she catch to get to school first?

A. The 08:37 school bus
B. The 08:26 public bus
C. The 08:38 public bus
D. The 08:31 public bus
E. More information needed

Question 18:

Puddle ducks are typically birds of fresh, shallow marshes and rivers rather than of large lakes and bays. They are good divers, but usually feed by dabbling or tipping rather than submerging. The speculum, or coloured wing patch, is generally iridescent and bright, and often a tell-tale field mark. Any duck feeding in croplands will likely be a puddle duck, for most of this group are sure-footed and can walk and run well on land. Their diet is mostly vegetable, and grain-fed mallards or pintails or acorn-fattened wood ducks are highly regarded as food.

Which of the following statements summarises the above passage best?
A. Other ducks are often eaten by puddle ducks in both large lakes and shallower waters.
B. Puddle ducks feed mainly without diving to gain vegetarian food sources.
C. Puddle ducks are the most common duck seen in croplands because they are vegetarian.
D. Other ducks are prone to predate on puddle ducks.
E. Puddle ducks live in large lakes as they can access vegetable food sources easily.

Question 19:

When the earth had to be prepared for the habitation of man, a veil, as it were, of intermediate being was spread between him and its darkness, in which were joined, in a subdued measure, the stability and the insensibility of the Earth, and the passion and perishing of mankind.

Which of the following statements best summarises the above statement?
A. The veil discussed is what links the good and evil of the human race.
B. Without this veil, mankind would not exist.
C. The Earth has more good than evil.
D. The veil keeps the human race alive.
E. Mankind would be better off without such a veil.

Question 20:

What is the value of x in the following sequence?

3 1 6 8

8 4 5 0

4 2 7 8

9 2 3 x

A. 5   B. 4   C. 8   D. 2   E. 7

Question 21:

Metformin has been thought to inhibit the process of fat cell growth. This is because *in vitro* metformin causes fat cells to stop growing. However, when a metformin inhibitor is used alongside metformin, the fat cells still don't grow. Thus, we can conclude that metformin does not inhibit fat cell growth.

Which of the following statements highlights the flaw in the argument?

A. Metformin doesn't inhibit fat cell growth.

B. The mechanism by which metformin inhibits fat cell growth is poorly understood.

C. We are not aware of how this inhibitor acts to inhibit the actions of metformin.

D. Fat cell growth has not been quantified here.

E. Metformin does not inhibit fat cell growth *in vivo*.

Question 22:

All dancers are strong. Some dancers are pretty. Alexandra is strong, and Katie is pretty.

Using the information above, which of the following statements is correct?

A. Alexandra is a dancer

B. Katie is not a dancer

C. A dancer can be strong and pretty

D. A dancer can be strong and ugly

Question 23:

During the last fifteen years the subject of bacteriology has developed with a marvellous rapidity. At the beginning of the ninth decade of the century bacteria were scarcely heard of outside of scientific circles, and very little was known about them even among scientists. Today they are almost household words, and everyone who reads is beginning to recognise that they have important relations to everyday life.

Which of the following statements, if true, would best support the above passage?

A. Bacteriology has improved due to the advancements in our ability to see and study such organisms.

B. Bacteria are too small to see in everyday life.

C. The development of antibiotics has helped us to understand bacteria better.

D. Every household understands the problems with bacterial infections.

E. Bacteria were much scarcer in the ninth decade than they are today.

Question 24:

Read the following statements. Which of the options is correct regarding whether the conclusions drawn from the statements are true or not?

Statements:

- No man is a lion.
- Joseph is a man.

Conclusions:

- **I:** Joseph is not a lion.
- **II:** All men are not Joseph.

A. Conclusion I is true, and conclusion II is true.

B. Conclusion I is true, and conclusion II is false.

C. Conclusion I is true, and we cannot tell if conclusion II is true or false.

D. Conclusion I is false, and we cannot tell if conclusion II is true or false.

E. Conclusion I is false, and conclusion II is true.

Question 25:

We may define food as any substance, which will repair the functional waste of the body, increase its growth, or maintain the heat, muscular, and nervous energy. In its most comprehensive sense, the oxygen of the air is a food; although it is admitted by the lungs, it passes into the blood, and there, it reacts with the other food, which has passed through the stomach. It is usual, however, to restrict the term 'food' to such nutrients as enters the body by the intestinal canal. Water is often spoken of as being distinct from food, but for this, there is no sufficient reason.

Which of the following statements highlights the weakness in the above argument?

A. Oxygen also is absorbed in the digestive tract.

B. Water is only made up of two elements, which is why it is not classified as food.

C. Water is needed for bodily functions and therefore must be food.

D. It is not explained why water is not classified as a food.

E. Any substance that is involved in a physiological process must have originated from food.

Question 26:

Billy is James's father. 4 years ago, Billy's age was 4 times that of James. 6 years from now, the ages of the Billy and James are in the ratio of 5:2. How old is Billy?

A. 12   B. 13   C. 14   D. 15   E. 16

Question 27:

The word 'soap' appears to have been originally applied to the product obtained by treating tallow with ashes. In its strictly chemical sense, it refers to combinations of fatty acids with metallic bases, a definition which includes not only sodium stearate, oleate and palmitate, which form the bulk of the soaps of commerce, but also the linoleates of lead, manganese, etc., used as driers, and various pharmaceutical preparations, e.g., mercury oleate, zinc oleate and lead plaster, together with a number of other metallic salts of fatty acids.

What can be reliably concluded from the above passage?
A. All metallic salts of fatty acids are classified as soaps.
B. Soaps are only used in industry for commercial use, driers and as pharmaceutical preparations.
C. Treating tallow with acids forms a soap as it results in fatty acids combining with a metallic acid.
D. All soaps are fatty acids combined with metallic bases.
E. All metals form soaps used in industry.

Question 28:

The average marks scored by 12 students is 73. If the scores of Bea, Bay and Boe are included, the average becomes 73.6. If Bea scored 68 marks and Boe scored 6 more than Bay, what was Bay's score?

A. 75   B. 76   C. 77   D. 78   E. 79

Question 29:

A building company employs 90 men to work for 8 hours per day to complete some building work. The company wants to finish work in 200 days but after 120 days, the work is only a third complete. If the men start working 12-hour days, how many more men are required to complete the work on time?

A. 170   B. 180   C. 190   D. 200   E. 210

Question 30:

The organs that form the digestive tract are the mouth, pharynx, oesophagus, stomach, intestines and the annexed glands, viz.: the salivary, liver, and pancreas. The development of these organs differs in the different species of animals. For example, solipeds possess a small, simple stomach and capacious, complicated intestines. Just the opposite is true of ruminants. The different species of ruminants possess a large, complicated stomach, and comparatively simple intestines. In swine we meet with a more highly developed stomach than that of solipeds and a more simple intestinal tract. Of all domestic animals the most simple digestive tract occurs in the dog.

What can be reliably concluded from the above passage?

A. Dogs have the simplest digestive tracts of domesticated animals due to their size.

B. Solipeds and ruminants differ only in their digestive tracts.

C. The more complex the digestive tract, the more complex the organism.

D. Mammals have varying digestive tracts that are adapted for their environments.

E. The mammalian digestive tract is vital for the survival of the animal.

Question 31:

If every alternative letter starting from A of the English alphabet is written in lower case, and the rest are all written in upper case, how would the day "Wednesday" be written?

A. wEdNEsdAy

B. weDNesDay

C. WEdnESdAY

D. weDneSDaY

E. WedNesdAY

Question 32:

A man covers a distance in 1hr 24min by covering 2/3 of the distance at 4 km/h and the rest at 5km/h. What distance does he cover?

A. 5km

B. 6km

C. 7km

D. 8km

END OF SECTION

SECTION 2

Question 1:

Hydrogen Bicarbonate (HCO_3^-) acts as a buffer in the blood i.e. to keep the PH close to 7.

Which statement is true regarding bicarbonate?

A. It acts as a base.

B. It is an acidic molecule.

C. If the pH of the blood drops below 7, bicarbonate will release the H^+ ion to stabilise the pH.

D. It is only released when the pH drops below 7.

E. It is bound to protein in the blood.

Question 2:

Which of the statements regarding this series circuit is true?

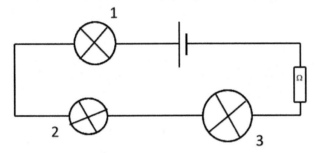

A. Current is different at different points in the circuit.

B. Potential difference is shared between the three lightbulbs.

C. Resistance is constant throughout the circuit.

D. The current is higher in bulb 1 than in bulbs 2 and 3.

E. None of the above.

Question 3:

Which of the following statements regarding breathing is / are correct?

1. The diaphragm plays no part in breathing.
2. The intercostal muscles relax during exhalation to allow the ribcage to move inwards and downwards.
3. The total pressure inside the chest decreases relative to the pressure outside the body during inhaling to draw air inside the lungs.

A. 1 only   C. 3 only   E. 1 and 3

B. 2 only   D. 2 and 3   F. None

Question 4:

Bill wants to lay down laminate flooring in his living room, which has an in-built circular fish tank that he will have to lay the flooring around. He has decided to buy planks that he can cut to fit the dimensions of his room. He must, however, buy whole planks and cut them down himself. The room's dimensions are given below, as are those of one plank.

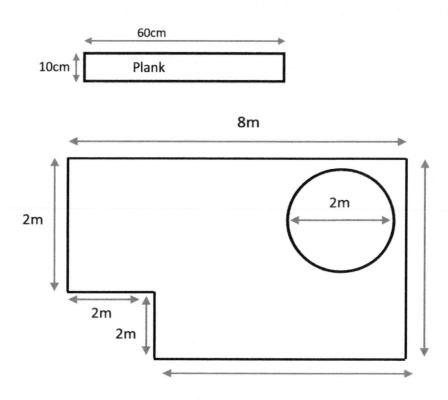

Calculate the number of planks needed to cover the whole floor. Use π = 3.

A. 30   B. 417   C. 600   D. 589   E. 43

Question 5:

In pregnancy the foetus is supplied with blood from the mother via the umbilical cord. This cord is comprised of one vein and two arteries. The table below shows which vessel carries which type of blood in which direction.

	Vessel	Direction	Blood
1.	Vein	Mother to foetus	Oxygenated
2.	Artery	Foetus to Mother	Deoxygenated
3.	Artery	Foetus to Mother	Oxygenated
4.	Vein	Mother to Foetus	Deoxygenated

Which options are correct?

A. 1 only   C. 3 only   E. 1 and 2   G. 4 and 1

B. 2 only   D. 4 only   F. 2 and 3   H. 3 and 1

Question 6:

Solve $y = x^2 - 3x + 4$ and $y - x = 1$ for x and y.

A. (-1, 2) and (3,4)   C. (7,-2) and (6,5)   E. (1,-1) and (-7,-1)

B. (1,2) and (3,4)   D. (2,-3) and (4,-1)

Question 7:

A ball of mass 5kg is at rest at the top of a 5m slope. Calculate the velocity of the ball as it travels down the slope. Take $g = 10kgm^{-1}$ and assume there is no resistance.

A. $10ms^{-1}$ B. $45ms^{-1}$ C. $100ms^{-1}$ D. $5ms^{-1}$ E. $6ms^{-1}$

Question 8:

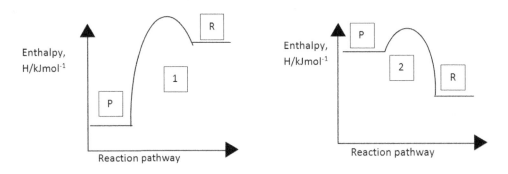

The two graphs shown above are enthalpy profile diagrams. Which best describes an endothermic reaction?

	Graph	ΔH	Heat energy	Stability of reactants
A)	1	Negative	Absorbed from surroundings	P is more stable than R
B)	2	Negative	Released to surroundings	R is more stable than P
C)	1	Positive	Absorbed from surroundings	P is more stable than R
D)	2	Positive	Absorbed from surroundings	R is more stable than P

Question 9:

Which of the following statements describe(s) a function of the kidney?

1. Ultrafiltration
2. Kill bacteria in the blood
3. Reabsorption
4. Release of waste
5. Store water
6. Produce hormones
7. Blood glucose regulation

A. 1 only   C. 3 only   E. 5 only   G. 3 and 5   I. 4, 5 and 6

B. 2 only   D. 4 only   F. 6 and 7   H. 1, 3 and 4

Question 10:

Mike and Vanessa are two healthy adults. They have two children. Their first child, Rory, was born with Haemophilia B, an X linked recessive disorder that causes problems with blood clotting. They have just had another baby, a girl, and want to get her tested for the condition. What is the likelihood of the baby girl having the condition, providing no spontaneous mutation occurs?

A. 0%   B. 25%   C. 50%   D. 75%   E. 34%

Question 11:

Pyrite, also known as Fool's Gold, is an ore of iron containing sulphur in the form of iron (II) disulphide, FeS_2.
The ore contains 75% FeS_2 by mass.

Calculate the maximum mass of iron that can be extracted from 480kg of ore.
[A_r: Fe = 55; S = 32]

A. 167.7kg   B. 200kg   C. 360.5kg   D. 118kg   E. 120.2kg

Question 12:

$X_{(s)} + FeSO_{4(aq)} \rightarrow XSO_{4(aq)} + Fe_{(S)}$

Which metal can be correctly be substituted in X's place?

A. Tin (Sn)

B. Zinc (Zn)

C. Lead (Pb)

D. Silver (Ag)

E. Copper (Cu)

Question 13:

Which of the following statements regarding electromagnetic radiation is / are true?

1. Visible light has a smaller wavelength than microwaves.
2. The speed of visible light in a vacuum is lower than the speed of radiowaves.
3. The wavelength of red light is shorter than the wavelength of yellow light.

A. 1 only

B. 2 only

C. 3 only

D. 1 and 2

E. None of the above

Question 14:

Which of the following statements regarding X-rays is true?

A. X-rays do not pass through denser materials like bone and that's why they show up as white on the X-ray film.

B. X-rays pass through bone but not skin and soft tissue, and that's why bones show up white on the X-ray film.

C. X-rays don't ionise cells and thus are safe.

D. Gamma rays are safer than X-rays.

E. None of the above.

Question 15:

Rearrange $\frac{(16x+11)}{(4x+5)} = 4y^2 + 2$ to make x the subject

A. $x = \frac{20y^2-1}{[16-4\,(4y^2+2)]}$

B. $= \frac{20y^2-8}{[16-6\,(4y^2+2)]}$

C. $= \frac{6y^2-1}{[16-4\,(4y^2+2)]}$

D. $= \frac{21y^2-1}{[16-4\,(2y^2+2)]}$

E. $= \frac{7y^2-1}{[6-14\,(6+7)]}$

~ 578 ~

Question 16:

The element shown below is Germanium, which forms ion with a charge of +4. How many electrons does one atom of Germanium have?

```
73
Ge
32
```

A. 32   B. 73   C. 36   D. 41   E. 4

Question 17:

Bacteria invade the body and produce toxins that kill cells.

Which of the following is / are first line defences used by the body to prevent entry of bacteria?

1. Mucus lining the airways
2. Heat produced by the body
3. Skin
4. Antibodies produced by the immune system
5. Toxins produced by the body
6. Hydrochloric acid in the stomach

A. 1 only   C. 3 only   E. 4, 5 and 6   G. 2 and 4
B. 2 only   D. 1, 3, 4 and 6   F. 1, 3 and 6

Question 18:

If $(3p + 5)^2 = 24p + 49$, calculate p.

A. -5 or -9   B. -3 or -6   C. -4 or 6   D. -6 or 4   E. 4 or -2

Question 19:

Reaction rates are explained by the Collision Theory. This theory states that particles have to collide (hard enough) in order to react.

Which of the following statements is / are true?

1. An increase in the temperature of the system can cause more collisions with greater force, therefore causing more reactions.
2. Smaller particles are don't collide or react as well.
3. Increasing the pressure of the system does not cause more collisions.
4. The collision theory only applies to gasses.
5. Increasing the surface area of the reactant increases the chances of collision.

A. 1 only	C. 3 only	E. 2 and 3	G. 1 and 5
B. 2 only	D. 4 only	F. 3, 4 and 5	

Question 20:

When someone is lost in the mountains, the rescue team often wraps an aluminium covered plastic sheet around them in order to keep them warm.

Which of the following statements is / are true regarding the effects that heat loss may have on their body?

1. There is less heat loss through conduction.
2. Air is trapped closer to the body, and this means that there is less heat loss due to convection.
3. Aluminium absorbs more sunlight and thus this keeps the person warm as more heat is absorbed.

A. 1 only	C. 3 only	E. 1, 2, 3	G. None
B. 2 only	D. 1 and 2	F. 2 and 3	

Question 21:

Which of the following is true with regards to osmosis?

A. It does not require a concentration gradient.
B. It can apply to any substance, not just water.
C. It is the movement of water across a partially permeable membrane.
D. It is an active process.
E. Transporters move water molecules across the membrane of cells.

Question 22

For the following reaction, which of the statements is true?

$$CH_{4(g)} + 2O_{2(g)} \rightarrow 2H_2O_{(aq)} + CO_{2(g)}$$

A. This is an example of complete combustion.

B. By increasing the concentration of CO_2 you can increase the rate of combustion.

C. The reaction is anaerobic.

D. Combustion of a gas always produces a liquid like water.

E. If you remove some of the oxygen you get more product.

Question 23:

Which of the following is a unit of resistance?

A. $V.A^{-1}$ B. $C.A$ C. $C.\Omega$ D. $V.\Omega^{-1}$ E. $W.V^{-1}$ F. J

Question 24:

To screw a piece of wood into a plank of wood, Bob uses a 20cm spanner. The moment of the force used to twist the screw into the plank is 40Nm.

How much force does Bob need to exert on the screw?

A. 2   B. 0.2   C. 80   D. 200   E. 0.5   F. 820

Question 25:

The carbon cycle is the cycle regarding the intake and release of carbon by organisms. Which of these statements are true?

A. Plants intake carbon via photosynthesis and taking nutrients from the soil, which have come from decayed organisms.

B. Animals give off carbon via respiration, waste, eating and death.

C. The CO_2 in the air comes from the burning of plant or animal products, and from respiration of living organisms only.

D. Trees do not store any carbon as they give it all off as carbon dioxide.

E. None of the above.

Question 26:

Refraction occurs when a wave passes from a material of low density to a material of high density or vice versa.

Which of these statements regarding refraction is true?

A. If a wave hits a different medium at an angle, the wave does not change direction.

B. If a wave hits a boundary face on, it slows down but carries on in the same direction. Thus, it has a shorter wavelength but the same frequency.

C. Waves can be refracted even if they hit the boundary head on.

D. Light is the only type of wave that can be refracted.

E. Glass to air slows down the wave.

Question 27:

There are 2 theories concerning how enzymes function – the 'lock and key' theory, or the 'induced fit' theory. The 'lock and key' theory states that the active site of an enzyme is already perfectly shaped for the substrate, whereas the 'induced fit' theory states that the enzyme's active site moulds itself around the substrate's shape. Which of these statements is true?

A. Enzymes are substrate specific.

B. The 'induced fit' theory allows multiple, different types of substrates to be acted on by one enzyme.

C. The 'induced fit' theory allows multiple, different types of enzymes to work on the same substrate.

D. The 'lock and key' theory does not allow space for catatonic reactions (breaking the substrate up).

END OF SECTION

SECTION 3

1) *"Time and time again, throughout the history of medical practice, what was once considered as "scientific" eventually becomes regarded as "bad practice"."* - David Stewart

What does this statement mean? Give some examples of times when scientific practice has become bad practice and describe how this has had an impact on medicine.

2) *"Formerly, when religion was strong and science was weak, men mistook magic for medicine; now, when science is strong and religion is weak, men mistake medicine for magic."* – Thomas Szasz

What does this statement mean? Do you think it is correct in assuming all men mistake medicine for magic?

3) *'Approximately 26.9% of the adult population in the UK is obese. We should be offering bariatric surgery to every obese person that walks through the doors.'*

Explain what this statement means. Argue to the contrary. To what extent do you agree with the statement?

4) *'Placebos may solve the problem of patients demanding medication they do not need.'*

Explain what this statement means. Argue the contrary. To what extent do you agree with the statement?

END OF PAPER

PRACTICE PAPER

ANSWERS

ANSWER KEY

Paper A				Paper B				Paper C			
Section 1		Section 2		Section 1		Section 2		Section 1		Section 2	
1	C	1	B	1	A	1	D	1	D	1	E
2	D	2	D	2	C	2	E	2	D	2	C
3	B	3	A	3	C	3	D	3	B	3	C
4	D	4	E	4	D	4	A	4	B	4	C
5	D	5	D	5	D	5	D	5	B	5	C
6	B	6	B	6	B	6	A	6	C	6	E
7	E	7	D	7	D	7	A	7	D	7	B
8	D	8	E	8	C	8	D	8	D	8	B
9	C	9	D	9	B	9	A	9	A	9	A
10	E	10	E	10	A	10	D	10	C	10	C
11	A	11	B	11	E	11	A	11	D	11	D
12	D	12	E	12	D	12	B	12	D	12	D
13	B	13	C	13	E	13	C	13	E	13	B
14	C	14	A	14	C	14	A	14	A	14	C
15	D	15	B	15	B	15	D	15	E	15	E
16	D	16	A	16	E	16	D	16	C	16	C
17	D	17	C	17	E	17	C	17	E	17	B
18	B	18	B	18	D	18	C	18	E	18	C
19	B	19	C	19	A	19	D	19	B	19	E
20	D	20	A	20	B	20	E	20	A	20	E
21	E	21	C	21	B	21	C	21	A	21	A
22	B	22	B	22	B	22	E	22	B	22	E
23	C	23	C	23	D	23	E	23	C	23	A
24	C	24	B	24	C	24	D	24	B	24	C
25	C	25	A	25	B	25	D	25	C	25	C
26	B	26	C	26	A	26	B	26	A	26	C
27	C	27	C	27	E	27	D	27	E	27	E
28	C			28	C			28	C		
29	B			29	A			29	C		
30	E			30	D			30	B		
31	D			31	D			31	B		
32	A			32	C			32	D		

Paper D				Paper E				Paper F			
Section 1		Section 2		Section 1		Section 2		Section 1		Section 2	
1	B	1	C	1	C	1	D	1	D	1	G
2	D	2	B	2	D	2	E	2	A	2	F
3	E	3	B	3	C	3	F	3	C	3	E
4	C	4	B	4	A	4	A	4	D	4	D
5	C	5	E	5	B	5	C	5	A	5	C
6	C	6	F	6	D	6	D	6	B	6	C
7	E	7	E	7	C	7	F	7	E	7	A
8	E	8	D	8	B	8	A	8	E	8	A
9	D	9	C	9	E	9	F	9	D	9	C
10	D	10	D	10	A	10	C	10	B	10	B
11	B	11	A	11	C	11	D	11	A	11	A
12	C	12	E	12	B	12	B	12	D	12	C
13	E	13	A	13	C	13	C	13	C	13	H
14	C	14	E	14	E	14	E	14	A	14	B
15	C	15	B	15	D	15	B	15	E	15	C
16	B	16	F	16	E	16	C	16	D	16	D
17	D	17	D	17	A	17	E	17	B	17	E
18	D	18	D	18	D	18	F	18	B	18	C
19	C	19	B	19	E	19	A	19	A	19	A
20	B	20	E	20	E	20	B	20	B	20	B
21	C	21	E	21	D	21	F	21	B	21	A
22	D	22	B	22	B	22	B	22	E	22	D
23	E	23	C	23	D	23	D	23	D	23	B
24	D	24	D	24	C	24	E	24	D	24	C
25	C	25	C	25	B	25	E	25	B	25	C
26	E	26	D	26	C	26	D	26	A	26	D
27	C	27	F	27	C	27	B	27	D	27	C
28	A			28	A			28	B		
29	D			29	A			29	A		
30	B			30	D			30	E		
31	A			31	A			31	A		
32	B			32	A			32	B		

Paper G				Paper H			
Section 1		Section 2		Section 1		Section 2	
1	C	1	B	1	A	1	A
2	D	2	H	2	C	2	B
3	B	3	D	3	E	3	D
4	E	4	E	4	D	4	B
5	C	5	B	5	D	5	E
6	C	6	D	6	E	6	B
7	B	7	A	7	E	7	A
8	B	8	B	8	C	8	C
9	B	9	A	9	C	9	H
10	A	10	A	10	D	10	A
11	B	11	D	11	B	11	A
12	E	12	B	12	D	12	B
13	D	13	F	13	A	13	A
14	C	14	E	14	D	14	A
15	E	15	E	15	B	15	A
16	A	16	B	16	D	16	A
17	B	17	B	17	C	17	F
18	E	18	D	18	B	18	D
19	C	19	B	19	A	19	G
20	E	20	C	20	E	20	E
21	B	21	B	21	C	21	C
22	C	22	D	22	C	22	A
23	D	23	D	23	A	23	A
24	A	24	C	24	C	24	D
25	E	25	C	25	D	25	B
26	D	26	E	26	C	26	B
27	C	27	A	27	D	27	A
28	D			28	C		
29	D			29	B		
30	D			30	D		
31	B			31	B		
32	C			32	B		

RAW TO SCALED SCORES

Section 1								Section 2					
1	1	11	2.8	21	5.4	31	8.3	1	1	11	3.5	21	6.6
2	1	12	3.0	22	5.7	32	9.0	2	1	12	3.7	22	6.9
3	1	13	3.2	23	6.0			3	1.3	13	4	23	7.3
4	1	14	3.5	24	6.3			4	1.6	14	4.3	24	7.6
5	1.2	15	3.7	25	6.6			5	1.9	15	4.6	25	8
6	1.5	16	4.0	26	6.9			6	2.2	16	5	26	8.5
7	1.8	17	4.2	27	7.1			7	2.5	17	5.3	27	9
8	2.0	18	4.5	28	7.4			8	2.8	18	5.6		
9	2.3	19	4.8	29	7.7			9	3.0	19	5.9		
10	2.5	20	5.1	30	8.0			10	3.2	20	6.2		

BMAT MOCK PAPER WORKED SOLUTIONS

We didn't have space to include all of the worked solutions for the mock papers in this book - and we wanted to prioritise giving you as many practice papers to work through as possible!

You can get detailed worked solutions to *every question* from *every mock paper* absolutely free, by going to uniadmissions.co.uk/bmat-book

FINAL ADVICE

Arrive well rested, well fed and well hydrated

The BMAT is an intense test, so make sure you're ready for it. Unlike the UCAT, you'll have to sit this at a fixed time (normally at 9AM). Thus, ensure you get a good night's sleep before the exam (there is little point cramming) and don't miss breakfast. If you're taking water into the exam then make sure you've been to the toilet before so you don't have to leave during the exam. Make sure you're well rested and fed in order to be at your best!

Move on

If you're struggling, move on. Every question has equal weighting and there is no negative marking. In the time it takes to answer one hard question, you could gain three times the marks by answering several easier ones. Be smart to score maximum points- especially in section two where some questions are far easier than others.

Make Notes on your Essay

Some universities may ask you questions on your BMAT essay at the interview. Sometimes you may have the interview as late as March which means that you **MUST** make short notes on the essay title and your main arguments after the essay. This is especially important if you're applying to UCL and Cambridge where the essay is discussed more frequently.

Afterword

Remember that the route to a high score is your approach and practice. Don't fall into the trap that "*you can't prepare for the BMAT*"– this could not be further from the truth. With knowledge of the test, some useful time-saving techniques and plenty of practice you can dramatically boost your score.

Work hard, never give up and do yourself justice.
Good Luck!

What's Next?

100+ successful medical school personal statements, so you can see what admissions tutors at your chosen medical schools like (and don't like)

✓ Complete guide to writing your own statement- from opening sentence to finishing touches

✓ Advice from 25+ Specialist Medical Doctors & Admissions Interviewers

✓ Learn how to carefully structure your statement so that you can accurately predict the questions in your interview

✓ 150 real interview questions with fully explained model answers.

✓ Written by more than 30 Medical Doctors and Admissions Tutors.

✓ Includes core syllabus for interview preparation e.g. NHS Structure, Medical Ethics, Audit Cycle, current Affairs, Junior Doctor Strikes, COVID

✓ Includes dedicated section on MMI style interviews.

ACKNOWLEDGEMENTS

I would like to thank Rohan and the UniAdmissions Tutors for all their hard work and advice in compiling this book, and both my parents and Meg for their continued unwavering support.

Matthew

ABOUT US

UniAdmissions currently publishes over 85 titles across a range of subject areas – covering specialised admissions tests, examination techniques, personal statement guides, plus everything else you need to improve your chances of getting on to competitive university courses such as medicine and law, as well as into universities such as Oxford and Cambridge.

This company was founded in 2013 by Dr Rohan Agarwal and Dr David Salt, both Cambridge medical graduates with several years of tutoring experience. Since then, every year, hundreds of applicants and schools work with us on our programmes. Through the programmes we offer, we deliver expert tuition, exclusive course places, online courses, best-selling textbooks and much more.

With a team of over 1,000 Oxbridge tutors and a proven track record, UniAdmissions have quickly become the UK's number one admissions company.

Visit and engage with us at:
Website (UniAdmissions): www.uniadmissions.co.uk
Facebook: www.facebook.com/uniadmissionsuk

YOUR FREE BOOK

Thanks for purchasing this Ultimate Book. Readers like you have the power to make or break a book – hopefully you found this one useful and informative. *UniAdmissions* would love to hear about your experiences with this book. As thanks for your time we'll send you another ebook from our Ultimate Guide series absolutely <u>FREE</u>!

How to Redeem Your Free Ebook

1) Find the book you have on your Amazon purchase history or your email receipt to help find the book on Amazon.

2) On the product page at the Customer Reviews area, click 'Write a customer review'. Write your review and post it! Copy the review page or take a screen shot of the review you have left.

3) Head over to www.uniadmissions.co.uk/free-book and select your chosen free ebook!

Your ebook will then be emailed to you – it's as simple as that!
Alternatively, you can buy all the titles at

<u>www.uniadmissions.co.uk</u>

MEDICINE PROGRAMME
Oxbridge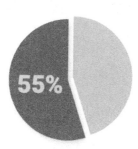

UNIADMISSIONS

55%

| UNIADMISSIONS 2019 Oxbridge
| Medicine Programme Success Rate

13%

| The Average Oxford & Cambridge
| Medicine Success Rate

300+
**Students successfully placed
at Oxbridge in the last 3 years**

50
**Places available on our Oxbridge
Medicine Programme in 2020**

WHY DO OUR STUDENTS SEE SUCH HIGH SUCCESS RATES?

1 **30 HOURS OF EXPERT TUITION.**
UniAdmissions will guide you through a comprehensive, tried & tested syllabus that covers all aspects of the application - you are never alone.

2 **UNPARALLED RESOURCES.**
UniAdmissions' resources are the best available for your Admissions Test. You will get access to all of our resources, including the Online Academy, books and ongoing tutor support.

3 **WEEKLY ENRICHMENT SEMINARS.**
You'll get access to weekly enrichment seminars which will help you think like and become the ideal candidate that admissions tutors are looking for.

4 **INTENSIVE COURSE PLACES.**
By enrolling onto our Oxbridge Programme you will get reserved places for all of the Intensive Courses relevant to your application, such as the Oxbridge Interview Intensive Course.

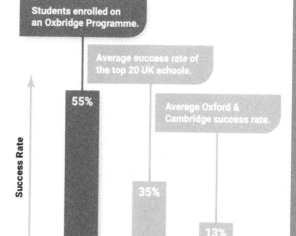

Students enrolled on an Oxbridge Programme.

Average success rate of the top 20 UK schools.

Average Oxford & Cambridge success rate.

55%

35%

13%

Success Rate

| UNIADMISSIONS Oxbridge Medicine
| Programme Average Success Rate

Printed in Great Britain
by Amazon

27245974R00328